Understanding International Broadcasting
A Reference for Shortwave Radio

Contents

Chapter 1

Introduction

1.1 International broadcasting

International broadcasting is broadcasting that is deliberately aimed at a foreign, rather than a domestic, audience. It usually is broadcast by means of longwave, mediumwave, or (more usually) shortwave radio, but in recent years has also used direct satellite broadcasting and the internet as means of reaching audiences.

Although radio and television programs do travel outside national borders, in many cases reception by foreigners is accidental. However, for purposes of propaganda, transmitting religious beliefs, keeping in touch with colonies or expatriates, education, improving trade, increasing national prestige, or promoting tourism and goodwill, broadcasting services have operated **external services** since the 1920s.

Guglielmo Marconi carried out the first short wave transmissions over a long distance.

1.1.1 History

International broadcasting, in a limited extent, began during World War I, when German and British stations broadcast press communiqués using Morse code. With the severing of Germany's undersea cables, the wireless telegraph station in Nauen, Germany was the sole means of long-distance communication.

The US Navy Radio Service radio station in New Brunswick, Canada, transmitted the 'Fourteen Points' by wireless to Nauen in 1917.[1] In turn, Nauen station broadcast the news of the abdication of Kaiser Wilhelm II on November 10, 1918.[2]

Origins

Guglielmo Marconi pioneered the use of short wave radio for long distance transmissions in the early 1920s. Using a system of parabolic reflector antennae, Marconi's assistant, Charles Samuel Franklin, rigged up a large antenna at Poldhu Wireless Station, Cornwall, running on 25 kW of power. In June and July 1923, wireless transmissions were completed during nights on 97 meters from Poldhu to Marconi's yacht *Elettra* in the Cape Verde Islands.[3] High speed shortwave telegraphy circuits were then installed from London to Australia, India, South Africa and Canada as the main element of the Imperial Wireless Chain from 1926.[3]

The Dutch began conducting experiments in the shortwave frequencies in 1925 from Eindhoven. The radio station PCJJ began the first international broadcasting on March 11, 1927 with programmes in Dutch for colonies in the Dutch West Indies and Dutch East Indies and in German, Spanish and English for the rest of the world. The popular Happy Station show was inaugurated in 1928.[4]

In 1927, Marconi also turned his attention toward long distance broadcasting on shortwave. His first such broadcasts took place to commemorate Armistice Day in the same year. He continued running a regular international broadcast that was picked up around the world, with programming from the 2LO station, then run by the BBC. The success of this operation caught the BBC's attention, and rented out a shortwave transmitting station in Chelmsford, with the callsign G5SW, to Marconi.[5] The BBC Empire Service was finally inaugurated on December 19, 1932, with transmissions aimed towards Australia and New Zealand.[6]

Expansion

Other notable early international broadcasters included Vatican Radio (February 12, 1931), Radio Moscow, the official service of the Soviet Union (this has since been renamed the Voice of Russia, following the collapse of the Soviet Union). Clarence W. Jones started transmitting on Christmas Day, 1931 from Christian missionary radio station HCJB in Quito, Ecuador. Broadcasting in South Asia was launched in 1925 in Ceylon - Radio Ceylon, now the Sri Lanka Broadcasting Corporation is the oldest in the region.

Joseph Goebbels **headed** *Nazi Germany's Ministry of Public Enlightenment and Propaganda. International broadcasting was an important element in Nazi propaganda.*

Shortwave broadcasting from Nauen in Germany to the USA, Central and South America, and the Far East began in 1926. A second station, Zeesen, was added in 1931.[7] In January 1932, the German Reichspost assumed control of the Nauen station and added to its shortwave and longwave capacity.[8] Once Adolf Hitler assumed power in 1933, shortwave, under the *Auslandsrundfunk* (Foreign Radio Section), was regarded as a vital element of Nazi propaganda.

German shortwave hours were increased from two hours a day to 18 per day, and eventually twelve languages were broadcast on a 24-hour basis, including English. A 100 kilowatt transmitter and antenna complex was built at Zeesen, near Berlin. Specialty target programming to the United States began in 1933, to South Africa, South America, and East Asia in 1934, and South Asia and Central America in 1938. German propaganda was organized under Joseph Goebbels, and played a key role in the German annexation of Austria and the Munich Crisis of 1938.

In 1936, the International Radio Union recognized Vatican Radio as a "special case" and authorized its broadcasting without any geographical limits. On December 25, 1937, a Telefunken 25-kW transmitter and two directional antennas were added. Vatican Radio broadcast over 10 frequencies.[9]

During the Spanish Civil War, the Nationalist forces received a powerful Telefunken transmitter as a gift of Nazi Germany to aid their propaganda efforts, and until 1943 Radio Nacional de España collaborated with the Axis powers to retransmit in Spanish news from the official radio stations of Germany and Italy.

World War II

During the Second World War, Russian, German, British, and Italian international broadcasting services expanded. In 1942, the United States initiated its international broadcasting service, the Voice of America. In the Pacific theater, General Douglas MacArthur used shortwave radio to keep in touch with the citizens of the Japanese-occupied Philippine Islands.

Several announcers who became well known in their countries included British Union of Fascists member William Joyce, who was one of the two "Lord Haw-Haw"s; Frenchmen Paul Ferdonnet and André Olbrecht, called "the traitors of [Radio] Stuttgart"; and Americans Frederick William Kaltenbach, "Lord Hee-Haw", and Mildred Gillars, one of the two announcers called "Axis Sally". Listeners to German programs often tuned in for curiosity's sake—at one time, German radio had half a million listeners in the U.S.--but most of them soon lost interest. Japan had "Tokyo Rose", who broadcast Japanese propaganda in English, along with American music to help ensure listeners.

During World War II, Vatican Radio's news broadcasts were banned in Germany. During the war, the radio service operated in four languages.[9]

The British launched Radio SEAC from Colombo, Ceylon (Sri Lanka) during World War II. The station broadcast radio programs to the allied armed forces across the region from their headquarters in Ceylon.

Following the war and German partition, each Germany developed its own international broadcasting station: Deutsche Welle, using studios in Cologne, West Germany, and Radio Berlin International (RBI) in East Germany. RBI's broadcasts ceased shortly before the reunification of Germany on October 3, 1990, and Deutsche Welle took over its transmitters and frequencies.

Cold War Era

The Cold War led to increased international broadcasting (and jamming), as Communist and non-Communist states attempted to influence each other's domestic population. Some of the most prominent Western broadcasters were the Voice of America, the BBC World Service, and the (covertly) CIA-backed Radio Free Europe/Radio Liberty. The Soviet Union's most prominent service was Radio Moscow (now the Voice of Russia) and China used Radio Peking (then Radio Beijing, now China Radio International). In addition to the U.S.-Soviet cold war, the Chinese-Russian border dispute led to an increase of the numbers of transmitters aimed at the two nations, and the development of new techniques such as playing tapes backwards for reel-to-reel recorders.

West Germany resumed regular shortwave broadcasts using Deutsche Welle on May 3, 1953. Its Julich transmitter site began operation in 1956, with eleven 100-kW Telefunken transmitters. The Wertachtal site was authorized in 1972 and began with four 500-kW transmitters. By 1989, there were 15 transmitters, four of which relayed the Voice of America.[10] Meanwhile in East Germany, the Nauen site began transmitting Radio DDR, later Radio Berlin International, on October 15, 1959.[11]

In addition to these states, international broadcast services grew in Europe and the Middle East. Under the presidency of Gamal Nasser, Egyptian transmitters covered the Arab world; Israel's service, Kol Yisrael, served both to present the Israeli point of view to the world and to serve the Jewish diaspora, particularly behind the Iron Curtain.

Radio RSA, as part of the South African Broadcasting Corporation, was established in 1966 to promote the image of South Africa internationally and reduce criticism of apartheid.[12] It continued in 1992, when the post-apartheid government renamed it Channel Africa.

Ironically, the isolationist Albania under Enver Hoxha, virtually a hermit kingdom, became one of the most prolific international broadcasters during the latter decades of the Cold War, with Radio Tirana one of the top five broadcasters in terms of hours of programming produced.

Today

At the end of the Cold War, many international broadcasters cut back on hours and foreign languages broadcast, or reemphasized other language services. For example, in 1984, Radio Canada International broadcast in English, French, German, Spanish, Czech/Slovak, Hungarian, Polish, Russian, and Ukrainian. In 2005, Canada broadcast in English, Chinese, Arabic, Russian, and Spanish There is a trend towards more TV (e.g. BBC World, NHK World, CCTV-9), and news websites. Some services, such as Swiss Radio International, left shortwave altogether and exist in Internet form. In addition, new standards, such as Digital Radio Mondiale, are being introduced, as well as sending programs over the Web to be played back later, as "podcasts".

International broadcasting using the traditional audio only method will not cease any time soon due to its cost efficiencies. However, international broadcasting via television is considered more strategically important at least since the early 2000s.

The BBC World Service was the first broadcaster to consider setting up a satellite television news and information channel as far back as 1976, but ceded being the first to CNN (that had primary access to Canada soon after launch). The defunct BBC World Service Antigua Relay Station was built in 1976, but its setup costs were not known to have been part of the BBCWS decision processes at the time.

In the early 1990s, many international (as well as domestic) 24-hour news and information channels launched as part of the post-Cold War prosperity bubble. There was another burst of global news channels launching in the late 2000s as part the developing world trying to catch up with the developed world in this area.

1.1.2 Reasons for international broadcasting

Broadcasters in one country have several reasons to reach out to an audience in other countries. The examples given below are not meant to be exhaustive, but are illustrative.

One clear reason is for ideological, or propaganda reasons. Many government-owned stations portray their nation in a positive, non-threatening way. This could be to encourage business investment and/or tourism to the nation. Another reason is to combat a negative image produced by other nations or internal dissidents, or insurgents. Radio RSA, the broadcasting arm of the apartheid South African government, is an example of this. A third reason is to promote the ideology of the broadcaster. For example, a program on Radio Moscow from the 1960s to the 1980s was *What is Communism?*

A second reason is to advance a nation's foreign policy interests and agenda by disseminating its views on international affairs or on the events in particular parts of the world. During the Cold War the American Radio Free Europe and Radio Liberty were founded to broadcast news from "behind the Iron Curtain" that was otherwise being censored and promote dissent and occasionally, to disseminate disinformation. Currently the US operates similar services aimed at Cuba and the People's Republic of China.

The BBC World Service, the Voice of America and other western broadcasters have emphasized news broadcasts, particularly to countries that are experiencing repression or civil unrest and whose populations are unable to obtain news from non-government sources. In the case of emergencies, a nation may broadcast special programs overseas to inform listeners what is occurring. During Iraqi missile strikes on Israel during the 1991 Gulf War, Kol Israel relayed its domestic service on its shortwave service.

Besides ideological reasons, many stations are run by religious broadcasters and are used to provide religious education, religious music, or worship service programs. For example, Vatican Radio, established in 1931, broadcasts such programs. Another station, such as HCJB or Trans World Radio will carry brokered programming from evangelists. In the case of the Broadcasting Services of the Kingdom of Saudi Arabia, both governmental and religious programming is provided.

Stations also broadcast to international audiences for cultural reasons. Often a station has an official mandate to keep expatriates in touch with the home country. Many broadcasters often relay their national domestic service on shortwave for that reason. Other reasons include teaching a foreign language, such as Radio Exterior de España's Spanish class, *Un idioma sin fronteras*, or the Voice of America's broadcasts in Special English. In the case of major broadcasters such as the BBC World Service or Radio Australia, there is also an educational outreach.

An additional reason for international broadcasting is to maintain contact with a country's citizens travelling abroad or expatriates who have emigrated and share news from home as well as cultural programming. This role of external shortwave broadcasting has declined as advances in communications have allowed expatriates to read news from home and listen and watch to domestic broadcasts in their own language via the internet and satellite. A number of international services such as the original BBC Empire Service, Radio Netherlands, France's Poste Colonial (now Radio France International) and others were founded in part with the goal of helping draw overseas empires closer to the mother country and provide closer cultural and communication connections between the home country and its colonies,

a role that became largely obsolete due to decolonization.

Notable networks

- CNN International (English)

- BBC World News (Arabic, Azeri, Bengali, Burmese, Cantonese, English, French for Africa, Hausa, Hindi, Indonesian, Kinyarwanda, Kirundi, Kyrgyz, Nepali, Pashto, Perian, Portuguese for Brazil, Russian, Sinhala, Somali, Spanish for Latin America, Swahili, Tamil, Turkish, Ukrainian, Urdu, Uzbek, Vietnamese)

- Sky News (English, Arabic)

- France24 (French, English, Arabic)

- TV Globo (Portuguese, English)

- Rede Record (Portuguese, English, Spanish)

- Al Jazeera (English, Arabic)

- teleSUR (Spanish, Portuguese)

- Deutsche Welle (German, English, Arabic, and 27 other languages)

- TRT (Turkish, English, Arabic and 39 other languages)

- Press TV (English)

- TV5MONDE (French)

- RT (English, Russian, Arabic, Spanish)

- Voice of Indonesia (English, Spanish, German, French, Indonesian, Japanese, Arabic, Chinese)

- Australia Network (English)

- i24news (English, French, Arabic)

- NHK World (English, Japanese)

- CCTV News (English)

- Arirang TV (English, Korean)

1.1.3   Means to reach an audience

Because of this many broadcasters are discovering they can reach a wider audience through other methods (particularly the internet and satellite television) and are cutting back on (or even entirely dropping) shortwave.

An international broadcaster has several options for reaching a foreign audience:

- If the foreign audience is near the broadcaster, high-power longwave and mediumwave stations can provide reliable coverage.

- If the foreign audience is more than 1,000 kilometers away from the broadcaster, shortwave radio is reliable, but subject to interruption by adverse solar/geomagnetic conditions.

- An international broadcaster may use a local mediumwave or FM radio or television relay station in the target country or countries.

- An international broadcaster may use a local shortwave broadcaster as a relay station.

- Neighboring states, such as Israel and Jordan, may broadcast television programs to each other's viewing public.

An international broadcaster such as the BBC, Radio France International or Germany's Deutsche Welle, may use all the above methods. Several international broadcasters, such as Swiss Radio International, have abandoned shortwave broadcasting altogether, relying on Internet transmissions only. Others, such as the BBC World Service, have abandoned shortwave transmissions to North America, relying on local relays, the Internet, and satellite transmissions

Mediumwave and longwave broadcasts

Most radio receivers in the world receive the mediumwave band (530 kHz to 1710 kHz), which at night is capable of reliable reception from 150 to 2,500 km distance from a transmitter. Mediumwave is used heavily all over the world for international broadcasting on a formal and informal basis.

In addition, many receivers used in Europe and Russia can receive the longwave broadcast band (150 to 280 kHz), which provides reliable long-distance communications over continental distances.

Shortwave broadcast

Yet other receivers are capable of receiving shortwave transmissions (2,000 to 30,000 kHz or 2 to 30 MHz). Depending on time of day, season of year, solar weather and Earth's geomagnetic field, a signal might reach around the world.

In previous decades shortwave (and sometimes high-powered mediumwave) transmission was regarded as the main (and often the *only*) way in which broadcasters could

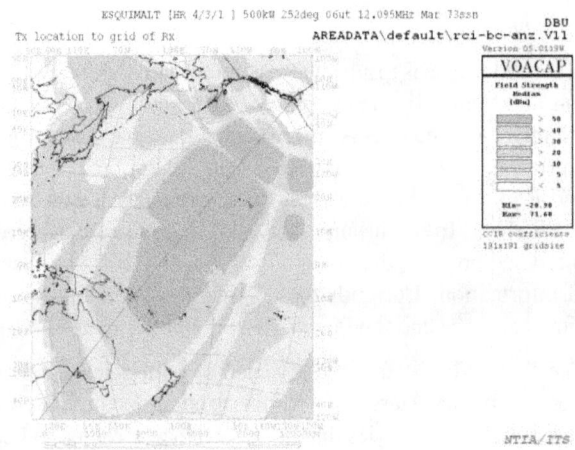

This sort of map is used by radio engineers to determine the best frequencies to reach international audiences on shortwave bands. In this case, a transmitter is sited in the Southern Vancouver Island, using a frequency of 12095 kHz and transmitting at the 500 kW power level. The picture shows a good signal over the Southern Pacific. The signal fades out as it approaches the East Coast of Australia.

reach an international audience. In recent years the proliferation of technologies such as satellite broadcasting, the Internet, and rebroadcasts of programming on AM and FM within target nations has meant that this is no longer necessarily the case.

Transmitter output power has increased since 1920. Higher transmitter powers do guarantee better reception in the target area. Higher transmitter power in most cases counteracts the lesser effects of jamming.

- 1950s : 100 kW

- 1960s : 200 kW, early 1960s (2 x 100 kW 'twinned')

- 1970s : 300 kW, but many 250 kW transmitters sold

- 1980s : 500 kW sometimes transmitters were "doubled up" to produce 1000 kW output

- 1980s-present: 600 kW single, 1200 kW from twinned transmitters.

International stations generally use special directional antennas to aim the signal toward the intended audience and increase the effective power in that direction. Use of such antennas for international broadcasting began in the mid-1930s and became prominent by the 1950s. By using antennas which focus most of their energy in one direction, a modern station may achieve the equivalent, in that direction, of tens of millions of watts of radio power.

Digital Audio Broadcasting

Some international broadcasters have become available via Digital Audio Broadcasting (DAB) in Europe in the 1990s, and in a similar limited way in the Americas via in-band FM (IBOC) DAB systems in the US in the 2000s. This is a popular method to reach listeners in cars that would otherwise not be accessible during that part of the day. However, in terms of the global international broadcasting audience the DAB listener base is very small—one can assume that it is less than 2% of the listener base globally.

Television

International broadcasting via 24 hour TV news channels has its origins in North America in the early 1980s. CNN technically was the first 24-hour international news channel as it was made available in Canada soon after launch. The BBC World Service considered setting up a global TV news channel as far back as 1975, but abandoned the idea for internal reasons.

Notwithstanding a large number of international 24-hour television news and information broadcasters, the television percentage of viewers is still fairly small when compared to global radio listener numbers.

The rural populations of Sub-Saharan Africa and South Asia (as well as East Asia) have radio listener bases that are far larger than the largest international TV broadcaster could hope for, yet they could be considered underserved since the end of the Cold War (when these regions had more radio broadcasts targeted at them).

Streaming video sites   Many international television broadcasters (as well as domestic television broadcasters) have set up accounts on streaming video sites like YouTube to allow their news and information broadcasts to be globally distributed. The viewer numbers for these sites may seem huge. Cable, TVRO and terrestrial television broadcasters probably have 100 to 1,000 times larger audiences for their international broadcasting content.

International broadcasters known to maintain their own streaming video sites (not authoritative)

- CNN

- France 24

- Russia Today

- Al Jazeera English (except the United States)

- i24news

- Sky News

RSS feeds and email

Many international broadcasters (television or radio) can reach "unreachable" audiences via email and RSS feeds. This is not at all unusual, as the first commonly agreed international broadcast was a Morse Code telegram transmitted from US President Wilson to the German Kaiser (mid-1918) via a high powered longwave transmitter on the US East Coast (this important event in international broadcasting history was described in depth in the IEEE "The History of International Broadcasting" first volume). As Morse Code is considered to be a data format, with email and RSS merely being refinements of the technology it can be said that international broadcasting has a deep relationship with modern day datacasting.

The reach of RSS and email for international broadcasters is not really known that well, especially considering that emails get forwarded. The numbers for active RSS and email audiences are probably 5 to 20 times larger than for streaming video. It may take into the 2010s to get meaningful numbers with respect to the size of these audiences for assorted technical reasons related to the RSS and email technologies.

Email and RSS feeds can traverse telecommunications barriers that streaming video cannot, thus the larger expected audience numbers. The global economic downturn of 2008-2009 will probably increase the email and RSS audience sizes as fewer people will be able to afford high speed internet connections in North America, Western Europe and the Asia-Pacific regions.

1.1.4   Listeners

An international broadcaster may have the technical means of reaching a foreign audience, but unless the foreign audience has a reason to listen, the effectiveness of the broadcaster is in question.

One of the most common foreign audiences consists of expatriates, who cannot listen to radio or watch television programs from home. Another common audience is radio hobbyists, who attempt to listen to as many countries as possible and obtain verification cards or letters (*QSLs*). These audiences send letters and in response few radio stations write them back. These kind of Listeners often take part in weekly and monthly quizzes and contests started by many radio stations. A third audience consists of journalists, government officials, and key businesspersons, who exert a disproportionate influence on a state's foreign or economic policy.

A fourth, but less publicized audience, consists of intelligence officers and agents who monitor broadcasts for both open-source intelligence clues to the broadcasting state's policies and for hidden messages to foreign agents operating in the receiving country. The BBC started its monitoring service in Caversham, Reading in 1936 (now BBC Monitoring). In the United States, the DNI Open Source Center (formerly the Central Intelligence Agency's Foreign Broadcast Information Service) provides the same service. Copies of OSC/FBIS reports can be found in many U.S. libraries that serve as government depositories. In addition, a number of hobbyists listen and report "spook" transmissions.

Without these four audiences, international broadcasters face difficulty in getting funding. In 2001, for example, the BBC World Service stopped transmitting shortwave broadcasts to North America, and other international broadcasters, such as YLE Radio Finland, stopped certain foreign-language programs.

However, international broadcasting has been successful when a country does not provide programming wanted by a wide segment of the population. In the 1960s, when there was no BBC service playing rock and roll, Radio Television Luxembourg (RTL) broadcast rock and roll, including bands such as the Beatles, into the United Kingdom. Similar programming came from an unlicensed, or "pirate" station, Radio Caroline, which broadcast from a ship in the international waters of the North Sea.

1.1.5 Restricting reception

In many cases, governments do not want their citizens listening to international broadcasters. In Nazi Germany, a major propaganda campaign, backed by law and prison sentences, attempted to discourage Germans from listening to such stations. The practice was made illegal in 1939.[13] In addition, the German government sold a cheap, 76-Reichsmark "People's Receiver", as well as an even cheaper 35-Reichsmark receiver,[13] that could not pick up distant signals well.[14]

The idea was copied by Stalin's Soviet Union, which had a nearly identical copy manufactured in the Tesla factory in Czechoslovakia.[14] In North Korea, all receivers are sold with fixed frequencies, tuned to local stations.

The most common method of preventing reception is jamming, or broadcasting a signal on the same frequencies as the international broadcaster. Germany jammed the BBC European service during the Second World War. Russian and Eastern European jammers were aimed against Radio Free Europe, other Western broadcasters, and against Chinese broadcasters during the nadir of Sino-Soviet relations. In 2002, the Cuban government jammed the Voice of America's Radio Martí program and the Chinese government jammed Radio Free Asia, Voice of America, Radio

Taiwan International as well broadcasts made by adherents of Falun Gong, .

North Korea restricts most people to a single fixed frequency mediumwave receiver; those who met political requirements and whose work absolutely required familiarity with events abroad were allowed shortwave receivers.[15] Another method of reaching people with government radio programming, but not foreign programming, is the use of radio broadcasting by direct broadcasting to loudspeakers.[16] David Jackson, director of the Voice of America, noted "The North Korean government doesn't jam us, but they try to keep people from listening through intimidation or worse. But people figure out ways to listen despite the odds. They're very resourceful."[17]

Yet another method of preventing reception involves moving a domestic station to the frequency used by the international broadcaster. During the Batista government of Cuba, and during the Castro years, Cuban medium-wave stations broadcast on the frequencies of popular South Florida stations. In October 2002, Iraq changed frequencies of two stations to block the Voice of America's Radio Sawa program.

Jamming can be defeated by using very efficient transmitting antennas, carefully choosing the transmitted frequency, changing transmitted frequency often, using single sideband, and properly aiming the receiving antenna.

For a list of international broadcasters, see List of international broadcasters.

1.1.6 See also

- List of international television channels
- List of shortwave radio broadcasters
- Shortwave
- Shortwave bands
- FTA receiver
- Medium wave - MW broadcasts generally don't travel as far as shortwave broadcasts, but MW is still used for international broadcasting, particularly to neighboring countries
- Euronews

1.1.7 References

[1] Wood 2000: 56

[2] U.S. Government Printing Office. *International Law Documents: Neutrality, Conduct and Conclusion of Hostilities.* 1919, p. 55

[3] John Bray (2002). *Innovation and the Communications Revolution: From the Victorian Pioneers to Broadband Internet.* IET. pp. 73–75.

[4] *History of Radio Netherlands*

[5] "Daventry Calling - 2: Station G5SW Chelmsford".

[6] BBC World Service. *World Service timeline.*

[7] Wood 2000: 49

[8] Wood 2000:57

[9] Levillain 2002: 1600

[10] Wood 2000: 51

[11] Wood 2000: 58

[12] Horwitz 2001: 287

[13] Hughes and Mann 2002: 93

[14] Graef 2005: 36

[15] Martin 2006: 495

[16] Goetz, Philip W. *The New Encyclopædia Britannica*, 1991 edition, ISBN 0-85229-400-X, p 315

[17] Jackson, David. "The Future of Radio II". *World Radio TV Handbook*, 2007 edition. 2007, Billboard Books. ISBN 0-8230-5997-9. p 38.

Sources

Graef 2005
Graef, Robert. *Bicycling to Amersfoort: A World War II Memoir.* 2005, iUniverse. ISBN 0-595-34621-9

Horwitz 2001
Horwitz, Robert Britt. *Communication and Democratic Reform in South Africa.* 2001, Cambridge University Press ISBN 0-521-79166-9.

Hughes and Mann 2002
Hughes, Matthew, and Chris Mann. *Inside Hitler's Germany: Life Under the Third Reich.* 2002, Brassey's. ISBN 1-57488-503-0

Levillain 2002
Levillain, Philippe. *The Papacy: An Encyclopedia.* Translated by John O'Malley. Routledge, 2002. ISBN 0-415-92228-3

Martin 2006
Martin, Bradley K. *Under the Loving Care of the Fatherly Leader: North Korea and the Kim Dynasty.* 2006, Macmillan. ISBN 0-312-32221-6

Wood 2000
Wood, James. *History of International Broadcasting.* 2000, IET. ISBN 0-85296-920-1

1.1.8   External links

- Hard-Core-DX - serious information about shortwave/AM radio stations

- American Radio Relay League (ARRL), Newington, Connecticut.

- englishradio.co.uk Cataloguing and reviewing every English-language radio station

- Easy-to-construct "interference-reducing" antennas for shortwave portables: U.S. International Broadcasting Bureau and K3MT (the "Villard antenna")

- *World Radio TV Handbook* The bible of international broadcasting

- *RCI Action Committee* Union group created to protect Radio Canada International's international broadcasting mandate and funding.

- *AIB | Association for International Broadcasting* The non-governmental, not-for-profit industry association for international TV and radio

1.2   Shortwave radio

For other uses, see Shortwave (disambiguation).

Shortwave radio is radio transmission using shortwave

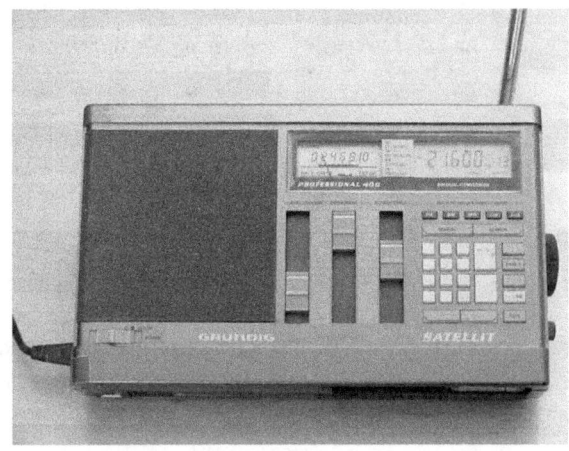

A solid-state, digital shortwave receiver

frequencies, generally 1.6–30 MHz (187.4–10.0 m), just above the medium wave AM broadcast band.

Shortwave radio is used for long distance communication by means of *skywave* or *skip* propagation, in which the radio waves are reflected or refracted back to Earth from the ionosphere, allowing communication around the curve of the Earth. Shortwave radio is used for broadcasting of voice and music, and long-distance communication to ships and aircraft, or to remote areas out of reach of wired communication or other radio services. Additionally, it is used for two-way international communication by amateur radio enthusiasts for hobby, educational and emergency purposes.

1.2.1 Frequency classifications

The widest popular definition of the shortwave frequency interval is the ITU Region 1 (EU+Africa+Russia...) definition, and is the span 1.6–30 MHz, just above the medium wave band, which ends approximately at 1.6 MHz.

There are also other definitions of the shortwave frequency interval:

- 1.71 to 30 MHz in ITU Region 2 (North and South America...)

- 1.8 (160 meter radio amateur band start) to 30 MHz

- 2.3 (120 meter band start) to 30 MHz

- 2.3 (120 meter band start) to 26.1 MHz (11 meter band end)[1][2]

- In Germany and perhaps Austria the ITU Region 1 shortwave frequency interval can be subdivided in:

  - *de:Grenzwelle* ("border waves"): 1.605-3.8 MHz and *de:Kurzwelle* (shortwaves) 3.8-30 MHz[3]

  - *Grenzwelle*: 1.605-4 MHz and *Kurzwelle* (shortwaves) 4-30 MHz

- In Germany these shortwave frequency intervals has also been seen used:

  - 3-30 MHz[4][5] – e.g. some accept that "high frequency" is the same as "short wave". In reality, the definition of the "shortwave" frequency band is a mess, and therefore the "shortwave frequencies" can not be exactly equal "high frequencies".

  - the above other definitions[6]

Shortwave radio received its name because the wavelengths in this band are shorter than 200 m (1500 kHz) which marked the original upper limit of the medium frequency band first used for radio communications. The broadcast medium wave band now extends above the 200 m/1500 kHz limit, and the amateur radio 1.8 MHz – 2.0 MHz band (known as the "top band") is the lowest-frequency band considered to be 'shortwave'.

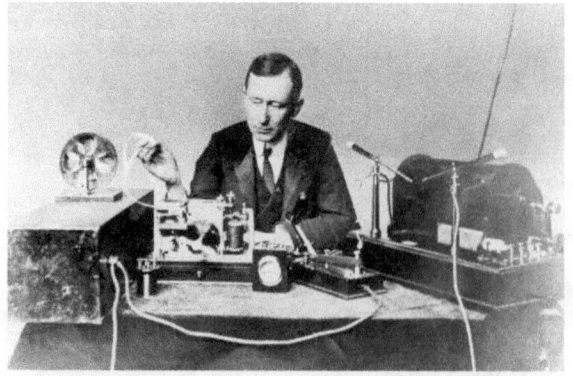

Radio Amateurs carried out the first short wave transmissions over a long distance before Guglielmo Marconi.

1.2.2 History

Development

Early radio telegraphy had used long wave transmissions. The drawbacks to this system included a very limited spectrum available for long distance communication, and the very expensive transmitters, receivers and gigantic antennas that were required. It was also difficult to beam the radio wave directionally with long wave, resulting in a major loss of power over long distances. Prior to the 1920s, the shortwave frequencies above 2 MHz were regarded as useless for long distance communication and were designated in many countries for amateur use.

Guglielmo Marconi, pioneer of radio, commissioned his assistant Charles Samuel Franklin to carry out a large scale study into the transmission characteristics of short wavelength waves and to determine their suitability for long distance transmissions. Franklin rigged up a large antenna at Poldhu Wireless Station, Cornwall, running on 25 kW of power. In June and July 1923, wireless transmissions were completed during nights on 97 meters from Poldhu to Marconi's yacht *Elettra* in the Cape Verde Islands.[7]

In September 1924, Marconi transmitted daytime and nighttime on 32 meters from Poldhu to his yacht in Beirut. Franklin went on to refine the directional transmission, by inventing the curtain array aerial system.[8][9] In July 1924, Marconi entered into contracts with the British General Post Office (GPO) to install high speed shortwave telegraphy circuits from London to Australia, India, South Africa and Canada as the main element of the Imperial Wireless Chain. The UK-to-Canada shortwave "Beam Wireless Service" went into commercial operation on 25 October 1926. Beam Wireless Services from the UK to Australia, South Africa and India went into service in 1927.[7]

Shortwave communications began to grow rapidly in the 1920s,[10] similar to the internet in the late 20th century. By 1928, more than half of long distance communications had moved from transoceanic cables and longwave wireless services to shortwave and the overall volume of transoceanic shortwave communications had vastly increased. Shortwave also ended the need for multimillion-dollar investments in new transoceanic telegraph cables and massive longwave wireless stations, although some existing transoceanic telegraph cables and commercial longwave communications stations remained in use until the 1960s.

The cable companies began to lose large sums of money in 1927, and a serious financial crisis threatened the viability of cable companies that were vital to strategic British interests. The British government convened the Imperial Wireless and Cable Conference[11] in 1928 "to examine the situation that had arisen as a result of the competition of Beam Wireless with the Cable Services". It recommended and received Government approval for all overseas cable and wireless resources of the Empire to be merged into one system controlled by a newly formed company in 1929, Imperial and International Communications Ltd. The name of the company was changed to Cable and Wireless Ltd. in 1934.

Amateur use of shortwave propagation

Amateur radio operators also discovered that long-distance communication was possible on shortwave bands. Early long-distance services used surface wave propagation at very low frequencies,[12] which are attenuated along the path. Longer distances and higher frequencies using this method meant more signal attenuation. This, and the difficulties of generating and detecting higher frequencies, made discovery of shortwave propagation difficult for commercial services.

Radio amateurs may have conducted the first successful transatlantic tests[13] in December 1921, operating in the 200 meter mediumwave band (1500 kHz)—the shortest wavelength then available to amateurs. In 1922 hundreds of North American amateurs were heard in Europe at 200 meters and at least 20 North American amateurs heard amateur signals from Europe. The first two-way communications between North American and Hawaiian amateurs began in 1922 at 200 meters. Although operation on wavelengths shorter than 200 meters was technically illegal (but tolerated as the authorities mistakenly believed at first that such frequencies were useless for commercial or military use), amateurs began to experiment with those wavelengths using newly available vacuum tubes shortly after World War I.

Extreme interference at the upper edge of the 150-200 meter band—the official wavelengths allocated to amateurs by the Second National Radio Conference[14] in 1923—forced amateurs to shift to shorter and shorter wavelengths; how-

ever, amateurs were limited by regulation to wavelengths longer than 150 meters (2 MHz). A few fortunate amateurs who obtained special permission for experimental communications below 150 meters completed hundreds of long distance two way contacts on 100 meters (3 MHz) in 1923 including the first transatlantic two way contacts.[15]

By 1924 many additional specially licensed amateurs were routinely making transoceanic contacts at distances of 6,000 miles (~9,600 km) and more. On 21 September several amateurs in California completed two way contacts with an amateur in New Zealand. On 19 October amateurs in New Zealand and England completed a 90-minute two-way contact nearly halfway around the world. On October 10, the Third National Radio Conference made three shortwave bands available to U.S. amateurs[16] at 80 meters (3.75 MHz), 40 meters (7 MHz) and 20 meters (14 MHz). These were allocated worldwide, while the 10-meter band (28 MHz) was created by the Washington International Radiotelegraph Conference[17] on 25 November 1927. The 15-meter band (21 MHz) was opened to amateurs in the United States on 1 May 1952.

1.2.3   Propagation characteristics

Shortwave radio frequency energy is capable of reaching any location on the Earth as it is influenced by ionospheric reflection back to the earth by the ionosphere, (a phenomenon known as "skywave propagation"). A typical phenomenon of shortwave propagation is the occurrence of a skip zone (see first figure on that page) where reception fails. With a fixed working frequency, large changes in ionospheric conditions may create skip zones at night.

As a result of the multi-layer structure of the ionosphere, propagation often simultaneously occurs on different paths, scattered by the E or F region and with different numbers of hops, a phenomenon that may be disturbed for certain techniques. Particularly for lower frequencies of the shortwave band, absorption of radio frequency energy in the lowest ionospheric layer, the D layer, may impose a serious limit. This is due to collisions of electrons with neutral molecules, absorbing some of a radio frequency's energy and converting it to heat.[18] Predictions of skywave propagation depend on:

- The distance from the transmitter to the target receiver.

- Time of day. During the day, frequencies higher than approximately 12 MHz can travel longer distances than lower ones. At night, this property is reversed.

- With lower frequencies the dependence on the time of the day is mainly due to the lowest ionospheric layer, the D Layer, forming only during the day when photons from the sun break up atoms into ions and free electrons.

- Season. During the winter months of the Northern or Southern hemispheres, the AM/MW broadcast band tends to be more favorable because of longer hours of darkness.

- Solar flares produce a large increase in D region ionization so high, sometimes for periods of several minutes, all skywave propagation is nonexistent.

1.2.4 Types of modulation

Further information: Modulation

Several different types of modulation are used to impress information on a short-wave transmission.

Amplitude modulation is the simplest type and the most commonly used for shortwave broadcasting. The instantaneous amplitude of the carrier is controlled by the amplitude of the signal (speech, or music, for example). At the receiver, a simple detector recovers the desired modulation signal from the carrier.

Single sideband transmission is a form of amplitude modulation but in effect filters the result of modulation. An amplitude-modulated signal has frequency components both above and below the carrier frequency. If one set of these components is eliminated as well as the residual carrier, only the remaining set is transmitted. This saves power in the transmission, as roughly 2/3 of the energy sent by an AM signal is unnecessary to recover the information contained on it. It also saves "bandwidth", allowing about one-half the carrier frequency spacing to be used. The drawback is that the receiver is more complicated, since it must re-recreate the carrier to recover the signal. Small errors in the detector process can greatly affect the pitch of the received signal, so single side band is not usual for music or general broadcast. Single side band is used for long-range voice communications by ships and aircraft, Citizen's Band, and amateur radio operators. LSB (lower sideband) is generally used below 9 MHz and USB (upper sideband) above 9 MHz.

Vestigal sideband transmits the carrier and one complete side-band, but filters out the redundant side-band. It is a compromise between AM and SSB, allowing simple receivers to be used but requiring almost as much transmitter power as AM. One advantage is that only half the bandwidth of an AM signal is used. It can be heard in the transmission of certain radio time signal stations.

Continuous wave (CW) is on-and-off keying of a carrier, used only for Morse code communications.

Narrow-band frequency modulation (NBFM) is mainly used in the higher HF frequencies (typically above 20 MHz). Because of the larger bandwidth required, NBFM is much more commonly used for VHF communication. Regulations limit the bandwidth of a signal transmitted in the HF bands, and the advantages of frequency modulation are greatest if the FM signal is allowed to have a wider bandwidth. NBFM is limited to short-range SW transmissions due to the multiphasic distortions created by the ionosphere.[19]

Digital Radio Mondiale (DRM) is a digital modulation for use on bands below 30 MHz.

Radioteletype, fax, digital, slow-scan television and other systems use forms of frequency-shift keying or audio subcarriers on a shortwave carrier. These generally require special equipment to decode, such as software on a computer equipped with a sound card.

1.2.5 Uses

Some major uses of the shortwave radio band are:

- International broadcasting primarily by government-sponsored propaganda, international news (for example, the BBC World Service) or cultural stations to foreign audiences: the most common use of all.

- Domestic broadcasting: to widely dispersed populations with few longwave, mediumwave and FM stations serving them; or for specialty political, religious and alternative media networks; or of individual commercial and non-commercial paid broadcasts.

- "Utility" stations transmitting messages not intended for the general public, such as aircraft flying between continents, encrypted diplomatic messages, weather reporting, or ships at sea.

- Clandestine stations. These are stations that broadcast on behalf of various political movements, including rebel or insurrectionist forces, and are normally unauthorised by the government-in-charge of the country in question. Clandestine broadcasts may emanate from transmitters located in rebel-controlled territory or from outside the country entirely, using another country's transmission facilities. Clandestine stations were used during World War II to transmit news from the Allied point of view into Axis-controlled areas. Although the Nazis confiscated many radios and executed their owners, many people continued to listen.

- Numbers Stations These stations regularly appear and disappear all over the shortwave radio band but are unlicenced and untraceable. It is believed that Numbers Stations are operated by government agencies, and are used to communicate with clandestine operatives working within foreign countries. However, no definitive proof of such use has emerged. Because the vast majority of these broadcasts contain nothing but the recitation of blocks of numbers, in various languages, with occasional bursts of music, they have become known colloquially as "Number Stations". Perhaps the most noted Number Station is the "Lincolnshire Poacher", named after the 18th century English folk song, which is transmitted just before the sequences of numbers.

- Amateur radio operators.

- Time signal and radio clock stations: In North America, WWV radio and WWVH radio transmit at these frequencies: 2500 kHz, 5000 kHz, 10000 kHz, and 15000 kHz; and WWV also transmits on 20000 kHz. The CHU radio station in Canada transmits on the following frequencies: 3330 kHz, 7850 kHz, and 14670 kHz. Other similar radio clock stations transmit on various shortwave and longwave frequencies around the world. The shortwave transmissions are primarily intended for human reception, while the longwave stations are generally used for automatic synchronization of watches and clocks.

- Over-the-horizon radar: From 1976 to 1989, the Soviet Union's Russian Woodpecker over-the-horizon radar system blotted out numerous shortwave broadcasts daily.

The term DXing, in the context of listening to radio signals of any user of the shortwave band, is the activity of monitoring distant stations. In the context of amateur radio operators, the term "DXing" refers to the two-way communications with a distant station, using shortwave radio frequencies.

The Asia-Pacific Telecommunity estimates that there are approximately 600,000,000 shortwave broadcast-radio receivers in use in 2002.[20] WWCR claims that there are 1.5 billion shortwave receivers worldwide.[21]

1.2.6 Shortwave broadcasting

See International broadcasting for details on the history and practice of broadcasting to foreign audiences.

See Shortwave relay station for the actual kinds of integrated technologies used to bring high power signals to listeners.

Frequency allocations

The World Radiocommunication Conference (WRC), organized under the auspices of the International Telecommunication Union, allocates bands for various services in conferences every few years. The last WRC took place in 2007.

At WRC-97 in 1997, the following bands were allocated for international broadcasting. AM shortwave broadcasting channels are allocated with a 5 kHz separation for traditional analog audio broadcasting.

Although countries generally follow the table above, there may be small differences between countries or regions. For example, in the official bandplan of the Netherlands,[22] the 49 m band starts at 5.95 MHz, the 41 m band ends at 7.45 MHz, the 11 m band starts at 25.67 MHz, and the 120, 90 and 60 m bands are absent altogether. Additionally, international broadcasters sometimes operate outside the normal WRC-allocated bands or use off-channel frequencies. This is done for practical reasons, or to attract attention in crowded bands (60m, 49m, 40m, 41m, 31m, 25m).

The new digital audio broadcasting format for shortwave DRM operates 10 kHz or 20 kHz channels. There are some ongoing discussions with respect to specific band allocation for DRM, as it mainly transmitted in 10 kHz format.

The power used by shortwave transmitters ranges from less than one watt for some experimental and amateur radio transmissions to 500 kilowatts and higher for intercontinental broadcasters and over-the-horizon radar. Shortwave transmitting centers often use specialized antenna designs (like the ALLISS antenna technology) to concentrate radio energy at the target area.

Advantages

Shortwave does possess a number of advantages over newer technologies, including the following:

- Difficulty of censoring programming by authorities in restrictive countries: unlike their relative ease in monitoring the Internet, government authorities face technical difficulties monitoring which stations (sites) are being listened to (accessed). For example, during the Russian coup against President Mikhail Gorbachev, when his access to communications was limited, Gorbachev was able to stay informed by means of the BBC World Service on shortwave.[23]

- Low-cost shortwave radios are widely available in all but the most repressive countries in the world. Simple shortwave regenerative receivers can be easily built with a few parts.

- In many countries (particularly in most developing nations and in the Eastern bloc during the Cold War era) ownership of shortwave receivers has been and continues to be widespread[24] (in many of these countries some domestic stations also used shortwave).

- Many newer shortwave receivers are portable and can be battery-operated, making them useful in difficult circumstances. Newer technology includes hand-cranked radios which provide power without batteries.

- Shortwave radios can be used in situations where Internet or satellite communications service is temporarily or long-term unavailable (or unaffordable).

- Shortwave radio travels much farther than broadcast FM (88-108 MHz). Shortwave broadcasts can be easily transmitted over a distance of several thousands of kilometers, including from one continent to another.

- Particularly in tropical regions, SW is somewhat less prone to interference from thunderstorms than medium wave radio, and is able to cover a large geographic area with relatively low power (and hence cost). Therefore, in many of these countries it is widely used for domestic broadcasting.

- Very little infrastructure is required for long-distance two-way communications using shortwave radio. All one needs is a pair of transceivers, each with an antenna, and a source of energy (such as a battery, a portable generator, or the electrical grid). This makes shortwave radio one of the most robust means of communications, which can be disrupted only by interference or bad ionospheric conditions. Modern digital transmission modes such as MFSK and Olivia are even more robust, allowing successful reception of signals well below the noise floor of a conventional receiver.

Disadvantages

Shortwave radio's benefits are sometimes regarded as being outweighed by its drawbacks, including:

- In most Western countries, shortwave radio ownership is usually limited to true enthusiasts, since most new standard radios do not receive the shortwave band. Therefore, Western audiences are limited.

- In the developed, world, shortwave reception is very difficult in urban areas because of excessive noise from switched mode power adapters, fluorescent or led light sources, internet modems and routers, computers and many, many other sources of radio interference.

1.2.7 Shortwave listening

A pennant sent to overseas listeners by Radio Budapest in the late 1980s

Main article: Shortwave listening

Many hobbyists listen to shortwave broadcasters without operating their own transmitters. In some cases, the goal is to hear as many stations from as many countries as possible *(DXing)*; others listen to specialized shortwave utility, or "ute", transmissions such as maritime, naval, aviation, or military signals. Others focus on intelligence signals from numbers stations,stations which transmit strange broadcast usually for intelligence operations, or the two way communications by amateur radio operators. Some short wave listeners behave analogously to "lurkers" on the Internet, in that they listen only and never make any attempt to send out their own signals. Other listeners participate in clubs, or actively send and receive QSL cards, or become involved with amateur radio and start transmitting on their own.

Many listeners tune the shortwave bands for the programmes of stations broadcasting to a general audience (such as Radio Taiwan International, Voice of Russia, China Radio International, Radio Canada International, Voice of America, Radio France Internationale, BBC World Service, Radio Australia, Radio Netherlands, Voice of Korea, Radio Free Sarawak etc.). Today, through the evolution of the Internet, the hobbyist can listen to short-wave signals via remotely controlled shortwave receivers around the world, even without owning a shortwave radio. Many international broadcasters (such as Radio Canada International , the BBC and Radio Australia) offer live streaming audio on their websites. Shortwave listeners, or SWLs, can obtain QSL cards from broadcasters, utility stations or amateur radio operators as trophies of the hobby. Some stations even give out special certificates, pennants, stickers and other tokens and promotional materials to shortwave listeners.

1.2.8 Amateur radio

Main article: Amateur radio

The practice of operating a shortwave radio transmitter for non-commercial two-way communications is known as amateur radio. Licenses are granted by authorized government agencies.

Amateur radio operators have made many technical advancements in the field of radio, and make themselves available to transmit emergency communications when normal communications channels fail. Some amateurs practice operating *off the power grid* so as to be prepared for power loss. Many amateur radio operators started out as Shortwave Listeners (SWLs) and actively encourage SWLs to become amateur radio operators.

1.2.9 Utility stations

Main article: Utility station

Utility stations are stations that do not intentionally broadcast to the general public (although their signals can be received by anybody with appropriate equipment). There are shortwave bands allocated to the use of merchant shipping, marine weather, and ship-to-shore stations; for aviation weather and air-to-ground communications; for military communications; for long-distance governmental purposes, and for other non-broadcast communications. Many radio hobbyists specialize in listening to "ute" broadcasts, which often originate from geographic locations without known shortwave broadcasters.

1.2.10 Unusual signals

The short wave bands are also used by unlicensed individuals who may want mostly short-range "party line" like communications. Two examples are the use of HF for communication between fishing boats in many areas of the world,

and the unlicensed use of the 11-meter band, which is effectively permitted in some areas of the world. Unlicensed operators, called "pirates", can cause signal interference to licensed stations. Many third-world countries have shops selling HF transmitter radios to any customer without regard to license or operator knowledge. As of 2012, there were virtually no national or international efforts to control such pirate operations.

The short wave bands are also used for various experiments, some continuing for years. In 2011, signals traceable to China regularly sent powerful HF transmissions scanning wide ranges of HF frequencies, perhaps to determine the maximum usable frequency (MUF) or other variables.

Numbers stations are broadcasts on shortwave radio that are coded into groups of numbers. Their content is generally encrypted and their purpose remains a mystery.

1.2.11 Shortwave broadcasts and music

Some musicians have been attracted to the unique aural characteristics of shortwave radio which—due to the nature of amplitude modulation, varying propagation conditions, and the presence of interference—generally has lower fidelity than local broadcasts (particularly via FM stations). Shortwave transmissions often have bursts of distortion, and "hollow" sounding loss of clarity at certain aural frequencies, altering the harmonics of natural sound and creating at times a strange "spacey" quality due to echoes and phase distortion. Evocations of shortwave reception distortions have been incorporated into rock and classical compositions, by means of delays or feedback loops, equalizers, or even playing shortwave radios as live instruments. Snippets of broadcasts have been mixed into electronic sound collages and live musical instruments, by means of analogue tape loops or digital samples. Sometimes the sounds of instruments and existing musical recordings are altered by remixing or equalizing, with various distortions added, to replicate the garbled effects of shortwave radio reception.

The first attempts by serious composers to incorporate radio effects into music may be those of the Russian physicist and musician Léon Theremin, who perfected a form of radio oscillator as a musical instrument in 1928 (regenerative circuits in radios of the time were prone to breaking into oscillation, adding various tonal harmonics to music and speech); and in the same year, the development of a French instrument called the Ondes Martenot by its inventor Maurice Martenot, a French cellist and former wireless telegrapher. A notable chamber piece by Mexican composer Silvestre Revueltas—*Ocho x radio*, 1933—features a complex texture of pseudo-mariachi musics, overlapping and cross-fading as if heard from distant stations: quite similar to shortwave radio signal propagation disturbance. John

Cage used actual radios (of unspecified wavelength) live on several occasions, starting in 1942 with *Credo in Us*, while Karlheinz Stockhausen used shortwave radio and effects in works including *Hymnen* (1966–67), *Kurzwellen* (1968)—adapted for the Beethoven Bicentennial in *Opus 1970* with filtered and distorted snippets of Beethoven pieces—*Spiral* (1968), *Pole*, *Expo* (both 1969–70), and *Michaelion* (1997).

Holger Czukay, a student of Stockhausen, was one of the first to use shortwave in a rock music context. In 1975, German electronic music band Kraftwerk recorded a full length concept album around simulated radiowave and shortwave sounds, entitled *Radio-Activity*. Among others, The The whose Radio Cineola monthly broadcasts draw heavily on shortwave radio sound,[25] The B-52s, Shearwater, Tom Robinson, Peter Gabriel, Pukka Orchestra, AMM, John Duncan, Orchestral Manoeuvres in the Dark (on their *Dazzle Ships* album), Pat Metheny, Aphex Twin, Boards of Canada, PressureWorks, Rush, Able Tasmans, Team Sleep, Underworld, Meat Beat Manifesto, Tim Hecker, Jonny Greenwood of Radiohead, Roger Waters (on *Radio KAOS* album), Wilco, code 000 and Samuel Trim have also used or been inspired by shortwave broadcasts.

1.2.12 Shortwave's future

Further information: The future of shortwave listening

The development of direct broadcasts from satellites has reduced the demand for shortwave receiver hardware, but there are still a great number of shortwave broadcasters. A new digital radio technology, Digital Radio Mondiale (DRM), is expected to improve the quality of shortwave audio from very poor to standards comparable to the FM broadcast band. The future of shortwave radio is threatened by the rise of power line communication (PLC), also known as Broadband over Power Lines (BPL), which uses a data stream transmitted over unshielded power lines. As the BPL frequencies used overlap with shortwave bands, severe distortions can make listening to analog shortwave radio signals near power lines difficult or impossible. However, because shortwave is a cheap and effective way to receive communications in countries with poor infrastructure, it will be around for years to come.

Shortwave use by hobbyists and licensed amateur ham radio operators continues, and after declining interest for a few years due to competing interests in computers and other communication devices, a new resurgence of interest has occurred as evidenced by the increase of new amateur operator licenses issued worldwide. Some hobbyists have combined amateur radio HF with computers for experimental and established data modes that can communicate very close to under the noise floor of receivers - e.g. WSJT, WSPR.

1.2.13 See also

- ALLISS — a very large rotatable antenna system used in international broadcasting

- Amateur radio — also uses the shortwave bands, but with power levels under 2 kW

- International broadcasting

- List of American shortwave broadcasters

- List of shortwave radio broadcasters

- Long wave

- Medium wave

- Shortwave bands — shortwave spectrum allocation

- shortwave relay station — the fundamental way in which programmes are broadcast on shortwave

- SSB — a method of radio signal modulation

1.2.14 References

[1] Inconsistent article: itwissen.info: KW (Kurzwelle) SW (short wave) Quote: "... Der KW-Frequenzbereich (SW) liegt zwischen 3 MHz und 30 MHz [<−1. definition] ... Kurzwelle ist auch eine sendetechnische Bezeichnung für Rundfunk im Frequenzbereich zwischen 2,3 MHz und 26,1 MHz [<−2. definition] ..."

[2] oldtimeradio.de: Kleines Radio-Lexikon Quote: "... Kurzwellen, Kurzwellenbereich ... Wellenbereich, der von den Rundfunksendern (je nach geographischer Lage) von 11 bis 120 m = 26.100 bis 2.300 kHz ..."

[3] darc.de: Amateurfunk Frequenzen Quote: "... Grenzwelle (Kurzwelle) ...", *de:Grenzwelle* Quote: "... Als Grenzwelle wird der Frequenzbereich zwischen 1605 kHz und 3800 kHz bezeichnet, weil er auf der „Grenze" zwischen Mittelwelle und Kurzwelle liegt ..."

[4] Inconsistent article: Die Meterbänder der Kurzwelle Quote: "...Bereich der Kurzwelle von 3.000 kHz bis 30.000 kHz ... [tabel] ... 120 m ... 2.300 kHz ... 2.495 kHz ... Tropenband ..."

[5] Universal-Lexikon: Kurzwellen Quote: "... entsprechend Frequenzen von 30-3 MHz ..."

[6] Grundig Satellit 1000 TR6002, schematic See at the bottom of the schematic, just below the transformer. In the schematic it is written that the first shortwave band starts at 1.6 MHz (just after the band end of MW/AM): "KW1-SW1-OC1 1,6 .... 5,0 MHz"

[7] John Bray (2002). *Innovation and the Communications Revolution: From the Victorian Pioneers to Broadband Internet.* IET. pp. 73–75.

[8] Beauchamp, K. G. (2001). *History of Telegraphy.* IET. p. 234. ISBN 0-85296-792-6. Retrieved 2007-11-23.

[9] Burns, R. W. (1986). *British Television: The Formative Years.* IET. p. 315. ISBN 0-86341-079-0. Retrieved 2007-11-23.

[10] "Full text of "Beyond the ionosphere : fifty years of satellite communication"". Archive.org. Retrieved 2012-08-31.

[11] Cable and Wireless Plc History

[12] Stormfax. Marconi Wireless on Cape Cod

[13] "1921 - Club Station 1BCG and the Transatlantic Tests". Radio Club of America. Retrieved 2009-09-05.

[14] "Radio Service Bulletin No. 72, pp. 9-13". Bureau of Navigation, Department of Commerce. 1923-04-02. Retrieved 2009-09-05.

[15] Archived November 30, 2009 at the Wayback Machine

[16] "Recommendations for Regulation of Radio: October 6-10, 1924". Earlyradiohistory.us. Retrieved 2012-08-31.

[17] http://www.twiar.org/aaarchives/WB008.txt

[18] Karl Rawer:"Wave Propagation in the Ionosphere". Kluwer, Dordrecht 1993 ISBN 0-7923-0775-5

[19] Ian Robertson Sinclair, *Audio and Hi-Fi Handbook,* Newnes, 2000 ISBN 0-7506-4975-5 pp. 195-196

[20] Archived February 10, 2005 at the Wayback Machine

[21] Arlyn T. Anderson. *Changes at the BBC World Service: Documenting the World Service's Move From Shortwave to Web Radio in North America, Australia, and New Zealand,* Journal of Radio Studies 2005, Vol. 12, No. 2, Pages 286-304 doi:10.1207/s15506843jrs1202_8 mentioned in WWCR FAQ

[22] Nationaal Frequentieplan

[23] http://www.w4uvh.net/dxld7078.txt

[24] Habrat, Marek. "Odbiornik "Roksana" (Radio constructor's recollections)". Retrieved 2008-08-05.

[25] Archived December 18, 2011 at the Wayback Machine

- Ulrich L. Rohde, Jerry Whitaker "Communications Receivers, Third Edition" McGraw Hill, New York, NY, 2001, ISBN 0-07-136121-9.

1.2.15 External links

- SWLing.com - A beginner's guide to shortwave radio listening.

- Glenn Hauser's World of Radio website

- Space Weather and Radio Propagation Center View live and historical data and images of space weather and radio propagation.

- Short-wave radio, Snap and crackle goes pop, Life in the old wireless yet The Economist article describing pros and cons of short wave radio since the Cold War.

- "Short-Wave Radio Telephone is Success in Tests" *Popular Mechanics,* July 1931, mid page experiments carried out for the French and British governments

- Que Escuchar en la Onda Corta en Español website

Chapter 2

International Broadcasting & Shortwave Radio Articles

2.1 Agência Brasil

Logotipo

Agência Brasil (**ABR**) is the national public news agency, run by the Brazilian government. It is a part of the public media corporation Empresa Brasil de Comunicação (EBC), created in 2007 to unite two government media enterprises Radiobrás and TVE (TV Educativa). It is publishing contents under CC-BY.

ABr is one of the most important Brazilian news agencies, that feeds thousands of regional newspapers and websites throughout Brazil but also national media outlets like Estadao, O Globo, Folha de S.Paulo, UOL and Terra.

Its headquarters are located in Brazilian capital, Brasília. There are also two regional offices located in São Paulo and Rio de Janeiro.

2.1.1 External links

Media related to Agência Brasil at Wikimedia Commons

- Official website

2.2 Agência Estado

Agência Estado is a Private News Agency in Brazil. It was started by a Brazilian media group Grupo Estado.[1][2]

2.2.1 References

[1] "Brazil County Profile Media". BBC. Retrieved 16 August 2013.

[2] "Agência Estado". AFP. Retrieved 16 August 2013.

2.2.2 External links

- Agência Estado website (Portuguese)

2.3 ALLISS

For other uses, see Alliss (disambiguation).
  ALLISS is a fully rotatable antenna system for high power

ALLISS antenna as viewed underneath

shortwave radio broadcasting in the 6 MHz to 26 MHz range. An ALLISS module is a self-contained shortwave relay station that is used for international broadcasting.

2.3.1 FAQ

ALLISS is a special design case of HRS type antennas. True ALLISS systems have solid radiators (horizontal radiating elements) versus tensioned flexible (open) radiators found with all other variations of ITU HRS type antennas systems. The name is based on a concatenation of two French towns ALLouis and ISSoudun.

2.3.2 Technological ambiguities

There are some factors that separate true ALLISS technology from 'run-of-the-mill' rotatable HRS Type antennas

- Thales pseudo-ALLISS rotatable antenna designs were procured from other antenna manufacturers that Thales acquired by corporate transactions.

- Technically only solid radiators distinguish true ALLISS systems from all other rotatable HRS type antennas.

- Only about 12% (estimate) of all HRS antennas in use globally are rotatable, and of these only 28 of the ALLISS systems have solid radiators.

- One must assume that only about 10% of HRS type antennas are rotatable, but compiled statistics are fragmentary. Only about 20% of rotatable HRS antennas are ALLISS, but this may be a slight overestimate.

- The Transmitter Documentation Project has most but not all stats on shortwave relay station antennas in use or historical.

- The Chinese SARFT is said to contain replicated ALLISS module technology, so to consider ALLISS technology as being exclusively in the domain of Thales is no longer true.

Corporate name changes

Information about ALLISS can also be found associated with Thomson-CSF—the previous name of Thales Group.

2.3.3 Technology FAQ (overview)

ALLISS technology, due to its cost and complexity—is out of reach to most consumers as a consumer product. Cheaper solutions to ALLISS exist in the shortwave broadcasting technology area.

- Any competent and reasonable transmission planning person should look at all other shortwave transmission options before considering ALLISS.

- As a rule of thumb ALLISS systems should only be purchased if 360 degrees of coverage is necessary.

- ALLISS is only used by well funded broadcasting and telecommunications operations that intend to use the modules over their design lifespan of 50–60 years.

Technology FAQ (operation)

ALLISS allows a broadcaster to change the following shortwave transmission parameters at any time:

- direction (azimuths from 0 to 360 degrees, rate: ~1 deg / 6 sec)

- broadcast frequency

- antenna configuration (i.e.: HR 4/4/1 -> HR 6/4/1)

All of these transmission mode changes can take effect in as little as 5 minutes. This flexibility can allow a broadcaster to redirect the entire shortwave transmission network to a strategically important target area in as little as 15 minutes.

ALLISS advantages vs traditional shortwave relay stations

Modular construction

- ALLISS relay stations can be built on a module by module basis.

- An ALLISS module can start broadcasting as soon as construction is completed.

Higher RFI & EMF (electromagnetic) compatibility vs traditional relay stations

- ALLISS modules should be geographically scattered for security and RFI exposure reasons. However, few broadcasters have chosen this option mainly due to poor understanding of the technology.

- Ironically, TDF did not pursue this option at Allouis or Issoudun — a technological blow to French security.

Each ALLISS module is fully automated, so there is no need for technical staff. When there are 2-5 ALLISS modules scattered over several hundred square kilometers, a three-person support staff is enough to keep the modules in operation year round (provided these modules are visited monthly for repair and maintenance).

With conventionally designed HRS type antennas shortwave relay stations and their obligatory transmitter hall, switch matrix, coaxial or open feeder line systems and multiple antennas (~90% of shortwave relay stations are built this way) much larger staffs are required.

Cost per module

Around US$10 million.

- Some modules have been rumored to cost as much as US$15 million.

- With 4 different module versions cost per module can vary by as much as US$5 million.

- At least 30% of the cost of each module is related to the still exotic metallurgy and metalworking requirements needed to construct each module.

- Because of the costly and complex metallurgy construction requirements, ALLISS technology is 'off limits' to many developing nations including even a few advanced nations in the developed world.

Versions of ALLISS modules

According to the current Thales brochure on ALLISS, there are 6 different versions of the ALLISS system. These versions are sorted by date of initial installation.

- 1995: France at Issoudun and Allouis
  - *Low Band* Modes (HR) : 4/4, 4/3, 4/2, 2/4, 2/3, 2/2
  - *High Band* Modes (HR) : 4/6, 4/4, 4/2, 2/6, 2/4
  - Band coverage: 5.9 MHz to 26.1 MHz (10 modules)
  - Band coverage: 5.9 MHz to 17.9 MHz (2 modules)

- 1997: Germany at Nauen-A
  - *Low Band* Modes (HR) : 4/4, 2/2
  - *High Band* Modes (HR) : 4/4
  - 2 systems installed

- 1997: Germany at Nauen-B

- *Low Band* Modes (HR) : 4/4
- *High Band* Modes (HR) : 4/4
- 2 systems installed

- 2002: Oman at Al-Seela
  - *Low Band* Modes (HR) : 4/4, 2/4, 4/2, 2/2
  - *High Band* Modes (HR) : 4/4, 2/4, 4/2, 2/2
  - 2 systems installed
  - Band coverage: 5.9 MHz to 17.9 MHz

- 2003: China
  - *Low Band* Modes (HR) : 4/4, 2/4, 4/2, 2/2
  - *High Band* Modes (HR) : 4/4, 2/4, 4/2, 2/2
  - Band coverage: 5.9 MHz to 26.1 MHz

- 2009 : Kuwait at the Voice of Kuwait shortwave relay station at Kabd operated by the Ministry of Information.
  - Image of the HR 2/2 ALLISS rotatable antenna similar Thales model HP-RCA 2/2 and a full ALLISS system. The second and third rotatable systems are due South and South Southwest of the current ALLISS and have been in operation since the first Gulf War. http://maps.google.de/maps?f=q& source=s_q&hl=de&geocode=&q=kuwait& sll=51.151786,10.415039&sspn=20.453081, 56.733398&ie=UTF8&ll=29.153047,47. 762251&spn=0.003467,0.006925&t=h&z=18
  - This model appears to be similar to those sold to China. There may be some export requirements that may keep HR 6/4/x models from being exported to politically sensitive regions.

- 2009: Cuba—The SARFT is said to have received a contract via the Chinese Foreign Affairs Department to replicate ALLISS modules in Cuba at an undisclosed or undetermined location. This is according to Glen Hauser's *World of Radio* transmission of 25 June 2009. There is no mapping evidence to indicate that construction has begun.

Transmitter

Typically ALLISS modules possess a 500 kW polyphase shortwave transmitter.

- Digital 'AM' type transmitters are preferred for their compactness.

- Many 'Push-pull' (Class-B) transmitters may be too large for some ALLISS installations.

- Essentially all 300 kW and 500 kW PDM, PSM, polyphase (4 x PDM) transmitters are preferred for structural reasons.

- It is not customary to install a 300 kW transmitter in an ALLISS module, but such installations are possible.

- TDF's Montsinnery Relay Station has 2 ALLISS modules installed, but without an installed shortwave transmitter. This same design arrangement is used by the BBC World Service Oman Relay Station Al-Seela.

Antennas (high band)

Three HRS array antennas types are available for broadcasting in the traditional shortwave broadcasting bands.

For tropical and lower frequency shortwave broadcasting

- HR 4/2/1 (using low band antenna)

- HR 2/4/1 (using low band antenna)

- HR 2/2/1

For traditional shortwave broadcasting

- HR 4/2/1

- HR 4/4/1

For highly directional shortwave broadcasting

- HR 6/4/1

- HR 6/2/1

The HRS 6/4/1 is not available for use in the 26 MHz band.

Antennas (low band)

One Low Band antenna exists for Tropical Band broadcasting. It takes up the entire back side of the ALLISS module. This Low Band antenna counterbalances the primary transmission antennas used in traditional shortwave broadcasting.

Relay stations with ALLISS modules

Documentation format — Nation : Broadcaster : City(Modules, Date Sold)

- France : TDF : Issoudun (12 modules, 1993 and 1997)

- Germany : DW : Nauen (4 modules, 1997)

- French Guiana : TDF : Montsinery (2 modules, 1997)

- Oman : BBCWS & VT Merlin : (2 modules, 2002)

- China : SARFT : Qiqihari : (12 modules, 2005)

Total number of modules sold since 1989: 32

2.3.4   Notable sites

- RFI at Issoudun

- Volga ALLISS Module

- Ganges ALLISS Module

- Former RFI Issoudun Relay station feeders and curtain arrays

- Former RFI Issoudun Relay curtain arrays

The International broadcasting center of TDF (Télédiffusion de France) is at Issoudun/Ste Aoustrille. Issoudun is currently utilized by TDF for shortwave transmissions. The site uses 12 rotary ALLISS antennas fed by 12 transmitters of 500 kW each to transmit shortwave broadcasts by Radio France International (RFI), along with other broadcast services.

2.3.5   See also

General category

- shortwave

- international broadcasting

- shortwave relay station

- Cities : Allouis & Issoudun

Applicable related technologies

- VOACAP can simulate all ITU HRS antenna types

- HRS type antennas

Broadcasters using ALLISS modules

- BBC World Service

- Deutsche Welle

- China Radio International

- Radio France International

2.3.6 External links

Technology portals (non-Thales)

- http://HireMe.geek.nz/ALLISS.html

2.4 Amateur radio

"Ham radio" redirects here. For other uses, see Ham radio (disambiguation).

Amateur radio (also called **ham radio**) describes

An example of an amateur radio station with four transceivers, amplifiers, and a computer for logging and for digital modes. On the wall are examples of various amateur radio awards, certificates, and a reception report card (QSL card) from a foreign amateur station.

the use of radio frequency spectra for purposes of non-commercial exchange of messages, wireless experimentation, self-training, private recreation and emergency communication. The term "amateur" is used to specify "a duly authorised person interested in radioelectric practice with a purely personal aim and without pecuniary interest;"[1] (either direct monetary or other similar reward) and to differentiate it from commercial broadcasting, public safety (such as police and fire), or professional two-way radio services (such as maritime, aviation, taxis, etc.).

The amateur radio service (*amateur service* and *amateur-satellite service*) is established by the International Telecommunication Union (ITU) through the International Telecommunication Regulations. National governments regulate technical and operational characteristics of transmissions and issue individual stations licenses with an identifying call sign. Prospective amateur operators are tested for their understanding of key concepts in electronics and the host government's radio regulations. Radio amateurs use a variety of voice, text, image, and data communications modes and have access to frequency allocations throughout the RF spectrum to enable communication across a city, region, country, continent, the world, or even into space.

Amateur radio is officially represented and coordinated by the International Amateur Radio Union (IARU), which is organized in three regions and has as its members the national amateur radio societies which exist in most countries. According to an estimate made in 2011 by the American Radio Relay League, two million people throughout the world are regularly involved with amateur radio.[2] About 830,000 amateur radio stations are located in IARU Region 2 (the Americas) followed by IARU Region 3 (South and East Asia and the Pacific Ocean) with about 750,000 stations. A significantly smaller number, about 400,000, are located in IARU Region 1 (Europe, Middle East, CIS, Africa).

2.4.1 History

Main article: History of amateur radio

The origins of amateur radio can be traced to the late

An amateur radio station in the United Kingdom. Multiple transceivers are employed for different bands and modes. Computers are used for control, datamodes, SDR and logging.

19th century, but amateur radio as practiced today began in the early 20th century. The *First Annual Official Wireless Blue Book of the Wireless Association of America*, produced in 1909, contains a list of amateur radio stations.[3] This radio callbook lists wireless telegraph stations in Canada and the United States, including 89 amateur radio stations. As with radio in general, amateur radio was associated with various amateur experimenters and hobbyists. Amateur radio enthusiasts have significantly contributed to science, engineering, industry, and social services. Research by amateur operators has founded new industries,[4] built economies,[5] empowered nations,[6] and saved lives in times of emergency.[7][8] Ham radio can also be used in the classroom to teach English, map skills, geography, math, science and computer skills.[9]

Ham radio

Main article: Etymology of ham radio

The term "ham radio" was first a pejorative that mocked amateur radio operators with a 19th-century term for being bad at something, like "ham-fisted" or "ham actor". It had already been used for bad wired telegraph operators.

Subsequently, the community adopted it as a welcome moniker, much like the "Know-Nothing Party", or other groups and movements throughout history. Other, more entertaining explanations have grown up throughout the years, but they are apocryphal.

2.4.2 Activities and practices

The many facets of amateur radio attract practitioners with a wide range of interests. Many amateurs begin with a fascination of radio communication and then combine other personal interests to make pursuit of the hobby rewarding. Some of the focal areas amateurs pursue include radio contesting, radio propagation study, public service communication, technical experimentation, and computer networking.

Amateur radio operators use various modes of transmission to communicate. The two most common modes for voice transmissions are frequency modulation (FM) and single sideband (SSB). FM offers high quality audio signals, while SSB is better at long distance communication when bandwidth is restricted.[10]

Radiotelegraphy using Morse code, also known as "CW" from "continuous wave", is the wireless extension of land line (wired) telegraphy developed by Samuel Morse and dates to the earliest days of radio. Although computer-based (digital) modes and methods have largely replaced CW for commercial and military applications, many amateur radio operators still enjoy using the CW mode— particularly on the shortwave bands and for experimental work, such as earth-moon-earth communication, because of its inherent signal-to-noise ratio advantages. Morse, us-

ing internationally agreed message encodings such as the Q code, enables communication between amateurs who speak different languages. It is also popular with homebrewers and in particular with "QRP" or very-low-power enthusiasts, as CW-only transmitters are simpler to construct, and the human ear-brain signal processing system can pull weak CW signals out of the noise where voice signals would be totally inaudible. A similar "legacy" mode popular with home constructors is amplitude modulation (AM), pursued by many vintage amateur radio enthusiasts and aficionados of vacuum tube technology.

Demonstrating a proficiency in Morse code was for many years a requirement to obtain an amateur license to transmit on frequencies below 30 MHz. Following changes in international regulations in 2003, countries are no longer required to demand proficiency.[11] The United States Federal Communications Commission, for example, phased out this requirement for all license classes on February 23, 2007.[12][13]

Modern personal computers have encouraged the use of digital modes such as radioteletype (RTTY) which previously required cumbersome mechanical equipment.[14] Hams led the development of packet radio in the 1970s, which has employed protocols such as AX.25 and TCP/IP. Specialized digital modes such as PSK31 allow real-time, low-power communications on the shortwave bands. Echolink using Voice over IP technology has enabled amateurs to communicate through local Internet-connected repeaters and radio nodes,[15] while IRLP has allowed the linking of repeaters to provide greater coverage area. Automatic link establishment (ALE) has enabled continuous amateur radio networks to operate on the high frequency bands with global coverage. Other modes, such as FSK441 using software such as WSJT, are used for weak signal modes including meteor scatter and moonbounce communications.

Fast scan amateur television has gained popularity as hobbyists adapt inexpensive consumer video electronics like camcorders and video cards in PCs. Because of the wide bandwidth and stable signals required, amateur television is typically found in the 70 cm (420 MHz–450 MHz) frequency range, though there is also limited use on 33 cm (902 MHz–928 MHz), 23 cm (1240 MHz–1300 MHz) and higher. These requirements also effectively limit the signal range to between 20 and 60 miles (30 km–100 km).

Linked repeater systems, however, can allow transmissions of VHF and higher frequencies across hundreds of miles.[16] Repeaters are usually located on heights of land or tall structures and allow operators to communicate over hundreds of miles using hand-held or mobile transceivers. Repeaters can also be linked together by using other amateur radio bands, landline, or the Internet.

Amateur radio satellites can be accessed, some using a

NASA astronaut Col. Doug Wheelock, KF5BOC, Expedition 24 flight engineer, operates the NA1SS ham radio station in the Zvezda Service Module of the International Space Station. Equipment is a Kenwood TM-D700E transceiver.

The top of a tower supporting a Yagi-Uda antenna and several wire antennas

hand-held transceiver (HT), even, at times, using the factory "rubber duck" antenna.[17] Hams also use the moon, the aurora borealis, and the ionized trails of meteors as reflectors of radio waves.[18] Hams can also contact the International Space Station (ISS) because many astronauts and cosmonauts are licensed as amateur radio operators.[19][20]

Amateur radio operators use their amateur radio station to make contacts with individual hams as well as participating in round table discussion groups or "rag chew sessions" on the air. Some join in regularly scheduled on-air meetings with other amateur radio operators, called "nets" (as in "networks"), which are moderated by a station referred to as "Net Control".[21] Nets can allow operators to learn procedures for emergencies, be an informal round table, or cover specific interests shared by a group.

Amateur radio operators, using battery- or generator-powered equipment, often provide essential communications services when regular channels are unavailable due to natural disaster or other disruptive events.

Many amateur radio operators participate in radio contests, during which an individual or team of operators typically seek to contact and exchange information with as many other amateur radio stations as possible in a given period of time. In addition to contests, a number of Amateur radio operating award schemes exist, sometimes suffixed with "on the Air", such as Summits on the Air, Islands on the Air, Worked All States and Jamboree on the Air.

2.4.3 Licensing

All countries that license citizens to use amateur radio require operators to display knowledge and understanding of key concepts, usually by passing an exam; however some au-

A handheld VHF/UHF transceiver

thorities also recognize certain educational or professional qualifications (such as a degree in electrical engineering) in lieu.[22] In response, hams receive operating privileges in larger segments of the radio frequency spectrum using a wide variety of communication techniques with higher power levels permitted compared to unlicensed personal radio services such as CB radio, Family Radio Service or PMR446 that require type-approved equipment restricted in frequency, range, and power.

Amateur licensing is a routine civil administrative matter in many countries. Amateurs therein must pass an examination to demonstrate technical knowledge, operating competence and awareness of legal and regulatory requirements in order to avoid interference with other amateurs and other radio services. A series of exams are often available, each progressively more challenging and granting more privileges: greater frequency availability, higher power output, permitted experimentation, and in some countries, distinctive call signs. Some countries, such as the United Kingdom and Australia, have begun requiring a practical training course in addition to the written exams in order to obtain a beginner's license, which they call a Foundation License.

Amateur radio licensing in the United States exemplifies the way in which some countries award different levels of amateur radio licenses based on technical knowledge: three sequential levels of licensing exams (Technician Class, General Class and Amateur Extra Class) are currently offered, which allow operators who pass them access to larger portions of the Amateur Radio spectrum and more desirable (shorter) call signs.

In some countries, an amateur radio license is necessary in order to purchase or possess amateur radio equipment.[23] An amateur radio license is only valid in the country in which it is issued or in another country that has a reciprocal licensing agreement with the issuing country.

Both the requirements for and privileges granted to a licensee vary from country to country, but generally follow the international regulations and standards established by the International Telecommunication Union[24] and World Radio Conferences.

In most countries, an individual will be assigned a call sign with their license. In some countries, a separate "station license" is required for any station used by an amateur radio operator. Amateur radio licenses may also be granted to organizations or clubs. Some countries only allow ham radio operators to operate club stations. Others, such as Syria and Cuba restrict all operation by foreigners to club stations only. Radio transmission permits are closely controlled by nations' governments because clandestine uses of radio can be made, and, because radio waves propagate beyond national boundaries, radio is an international matter.

Licensing requirements

Prospective amateur radio operators are examined on understanding of the key concepts of electronics, radio equipment, antennas, radio propagation, RF safety, and the radio regulations of the government granting the license. These examinations are sets of questions typically posed in either a short answer or multiple-choice format. Examinations can be administered by bureaucrats, non-paid certified examiners, or previously licensed amateur radio operators.

The ease with which an individual can acquire an amateur radio license varies from country to country. In some countries, examinations may be offered only once or twice a year in the national capital and can be inordinately bureaucratic (for example in India) or challenging because some amateurs must undergo difficult security approval (as in Iran). Currently only Yemen and North Korea do not issue amateur radio licenses to their citizens, although in both cases a limited number of foreign visitors have been permitted to obtain amateur licenses in the past decade. Some developing countries, especially those in Africa, Asia, and Latin America, require the payment of annual license fees that can be prohibitively expensive for most of their citizens. A few small countries may not have a national licensing process and may instead require prospective amateur radio operators to take the licensing examinations of a foreign country. In countries with the largest numbers of amateur radio licensees, such as Japan, the United States, Thailand, Canada, and most of the countries in Europe, there are frequent license examinations opportunities in major cities.

Granting a separate license to a club or organization generally requires that an individual with a current and valid amateur radio license who is in good standing with the telecommunications authority assumes responsibility for any operations conducted under the club license or club call sign. A few countries may issue special licenses to novices or beginners that do not assign the individual a call sign but instead require the newly licensed individual to operate from stations licensed to a club or organization for a period of time before a higher class of license can be acquired.

Reciprocal licensing

Further information: Amateur radio international operation
A reciprocal licensing agreement between two countries allows bearers of an amateur radio license in one country under certain conditions to legally operate an amateur radio station in the other country without having to obtain an amateur radio license from the country being visited, or the bearer of a valid license in one country can receive a separate license and a call sign in another country, both of which have a mutually-agreed reciprocal licensing approvals. Reciprocal licensing requirements vary from country to country. Some countries have bilateral or multilateral reciprocal operating agreements allowing hams to operate within their borders with a single set of requirements. Some countries lack reciprocal licensing systems.

When traveling abroad, visiting amateur operators must follow the rules of the country in which they wish to oper-

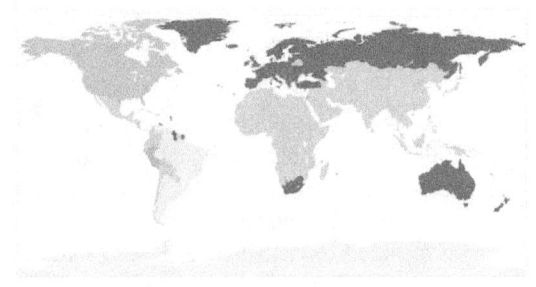

Reciprocal Agreements by Country

CEPT Member Nations
IARP Member Nations
Members of CEPT and IARP
USA and Canada Treaty, CEPT and IARP

ate. Some countries have reciprocal international operating agreements allowing hams from other countries to operate within their borders with just their home country license. Other host countries require that the visiting ham apply for a formal permit, or even a new host country-issued license, in advance.

The reciprocal recognition of licenses frequently not only depends on the involved licensing authorities, but also on the nationality of the bearer. As an example, in the US, foreign licenses are only recognized if the bearer does not have US citizenship and holds no US license (which may differ in terms of operating privileges and restrictions). Conversely, a US citizen may operate under reciprocal agreements in Canada, but not a non-US citizen holding a US license.

Newcomers

Many people start their involvement in amateur radio by finding a local club. Clubs often provide information about licensing, local operating practices, and technical advice. Newcomers also often study independently by purchasing books or other materials, sometimes with the help of a mentor, teacher, or friend. Established amateurs who help newcomers are often referred to as "Elmers", as coined by Rodney Newkirk, W9BRD,[25] within the ham community.[26][27] In addition, many countries have national amateur radio societies which encourage newcomers and work with government communications regulation authorities for the benefit of all radio amateurs. The oldest of these societies is the Wireless Institute of Australia, formed in 1910; other notable societies are the Radio Society of Great Britain, the American Radio Relay League, Radio Amateurs of Canada, Bangladesh NGOs Network for Radio and Communication, the New Zealand Association of Radio Transmitters and South African Radio League. (*See*

Category:Amateur radio organizations)

Call signs

Further information: Amateur radio call signs

An amateur radio operator uses a call sign on the air to legally identify the operator or station.[28] In some countries, the call sign assigned to the station must always be used, whereas in other countries, the call sign of either the operator or the station may be used.[29] In certain jurisdictions, an operator may also select a "vanity" call sign although these must also conform to the issuing government's allocation and structure used for Amateur Radio call signs.[30] Some jurisdictions require a fee to obtain such a vanity call sign; in others, such as the UK, a fee is not required and the vanity call sign may be selected when the license is applied for. The FCC in the U.S. discontinued its fee for vanity call sign applications in September 2015.[31]

Call sign structure as prescribed by the ITU, consists of three parts which break down as follows, using the call sign ZS1NAT as an example:

1. ZS – Shows the country from which the call sign originates and may also indicate the license class. (This call sign is licensed in South Africa. CEPT Class is no longer "encoded" in South African callsigns. Where specific classes of amateur radio license exist, the call signs may be assigned by class, but the specifics vary by issuing country.)

2. 1 – Gives the subdivision of the country or territory indicated in the first part (this one refers to the Western Cape).

3. NAT – The final part is unique to the holder of the license, identifying that station specifically.

Many countries do not follow the ITU convention for the numeral. In the United Kingdom the original calls G0xxx, G2xxx, G3xxx, G4xxx, were Full (A) License Holders along with the last M0xxx full call signs issued by the City & Guilds examination authority in December 2003. Additional full licenses were originally granted in respect of (B) Licensees with G1xxx, G6xxx, G7xxx, G8xxx and 1991 onward with M1xxx calls. The newer three level Intermediate licensees are 2E1xxx and 2E0xx and basic Foundation license holders are granted a M6xxx call sign.[32] In the United States, for non-Vanity licenses, the numeral indicates the geographical district the holder resided in when the license was issued. Prior to 1978, US hams were required to obtain a new call sign if they moved out of their geographic district.

Also, for smaller entities, a numeral may be part of the country identification. For example, VP2xxx is in the British West Indies (subdivided into VP2Exx Anguilla, VP2Mxx Montserrat, and VP2Vxx British Virgin Islands), VP5xxx is in the Turks and Caicos Islands, VP6xxx is on Pitcairn Island, VP8xxx is in the Falklands, and VP9xxx is in Bermuda.

Online callbooks or callsign databases can be browsed or searched to find out who holds a specific callsign.[33] Non-exhaustive lists of famous people who hold or have held amateur radio callsigns have also been compiled and published.[34]

Many jurisdictions issue specialty vehicle registration plates to licensed amateur radio operators often in order to facilitate their movement during an emergency.[35][36] The fees for application and renewal are usually less than the standard rate for specialty plates.[35][37]

Privileges

In most administrations, unlike other RF spectrum users, radio amateurs may build or modify transmitting equipment for their own use within the amateur spectrum without the need to obtain government certification of the equipment.[38][39] Licensed amateurs can also use any frequency in their bands (rather than being allocated fixed frequencies or channels) and can operate medium to high-powered equipment on a wide range of frequencies[40] so long as they meet certain technical parameters including occupied bandwidth, power, and maintenance of spurious emission.

Radio amateurs have access to frequency allocations throughout the RF spectrum, usually allowing choice of an effective frequency for communications across a local, regional, or worldwide path. The shortwave bands, or HF, are suitable for worldwide communication, and the VHF and UHF bands normally provide local or regional communication, while the microwave bands have enough space, or bandwidth, for amateur television transmissions and high-speed computer networks.

In most countries, an amateur radio license grants permission to the license holder to own, modify, and operate equipment that is not certified by a governmental regulatory agency. This encourages amateur radio operators to experiment with home-constructed or modified equipment. The use of such equipment must still satisfy national and international standards on spurious emissions.

The amount of output power an amateur radio licensee may legally use varies from country to country. Although allowable power levels are moderate by commercial standards, they are sufficient to enable global communication.

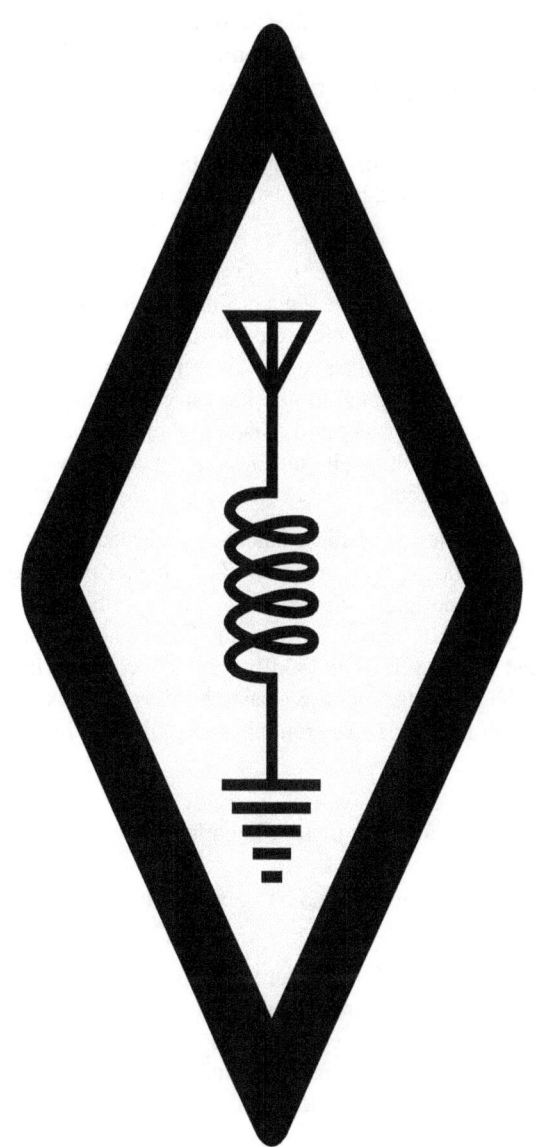

The international symbol for amateur radio, included in the logos of many IARU member societies. The diamond holds a circuit diagram featuring components common to every radio: an antenna, inductor and ground.

Power limits vary from country to country and between license classes within a country. For example, the peak envelope power limits for the highest available license classes in a few selected countries are: 2.25 kW in Canada,[41] 1.5 kW in the United States, 1.0 kW in Belgium, Luxembourg, Switzerland and New Zealand, 750 W in Germany, 500 W in Italy, 400 W in Australia, India and the United Kingdom, and 150 W in Oman. Lower license classes usually have lower power limits; for example, the lowest license class in the UK (Foundation licence) has a limit of 10 W. Amateur radio operators are encouraged both by regulations and tradition of respectful use of the spectrum to use as little power as possible to accomplish the communication.[42] This is to minimise interference or EMC to any other device.

Output power limits may also depend on the mode of transmission. In Australia, for example, 400 W may be used for SSB transmissions, but FM and other modes are limited to 120 W.

The point at which power output is measured may also affect transmissions. The United Kingdom measures at the point the antenna is connected to the signal feed cable, which means the radio system may transmit more than 400 W to overcome signal loss in the cable; conversely, Germany measures power at the output of the final amplification stage, which results in a loss in radiated power with longer cable feeds.

Certain countries permit amateur radio licence holders to hold a Notice of Variation that allows higher power to be used than normally allowed for certain specific purposes. E.g. in the UK some amateur radio licence holders are allowed to transmit using (33dBw) 2.0 kW for experiments entailing using the moon as a passive radio reflector (known as Earth-Moon-Earth communication) (EME).

Band plans and frequency allocations

Main article: Amateur radio frequency allocations

The International Telecommunication Union (ITU) governs the allocation of communications frequencies worldwide, with participation by each nation's communications regulation authority. National communications regulators have some liberty to restrict access to these bandplan frequencies or to award additional allocations as long as radio services in other countries do not suffer interference. In some countries, specific emission types are restricted to certain parts of the radio spectrum, and in most other countries, International Amateur Radio Union (IARU) member societies adopt voluntary plans to ensure the most effective use of spectrum.

In a few cases, a national telecommunication agency may also allow hams to use frequencies outside of the internationally allocated amateur radio bands. In Trinidad and Tobago, hams are allowed to use a repeater which is located on 148.800 MHz. This repeater is used and maintained by the National Emergency Management Agency (NEMA), but may be used by radio amateurs in times of emergency or during normal times to test their capability and conduct emergency drills. This repeater can also be used by non-ham NEMA staff and REACT members. In Australia and New Zealand ham operators are authorized to use one of the UHF TV channels. In the U.S., amateur radio oper-

ators providing essential communication needs in connection with the immediate safety of human life and immediate protection of property when normal communication systems are not available may use any frequency including those of other radio services such as police and fire and in cases of disaster in Alaska may use the statewide emergency frequency of 5167.5 kHz with restrictions upon emissions.[43]

Similarly, amateurs in the United States may apply to be registered with the Military Auxiliary Radio System (MARS). Once approved and trained, these amateurs also operate on US government military frequencies to provide contingency communications and morale message traffic support to the military services.

2.4.4 Modes of communication

Amateurs use a variety of voice, text, image, and data communications modes over radio. Generally new modes can be tested in the amateur radio service, although national regulations may require disclosure of a new mode to permit radio licensing authorities to monitor the transmissions. Encryption, for example, is not generally permitted in the Amateur Radio service except for the special purpose of satellite vehicle control uplinks. The following is a partial list of the modes of communication used, where the mode includes both modulation types and operating protocols.

Voice

- Amplitude modulation (AM)

- Double Sideband Suppressed Carrier (DSB-SC)

- Independent Sideband (ISB)

- Single Sideband (SSB)

- Amplitude Modulation Equivalent (AME)

- Frequency modulation (FM)

- Phase modulation (PM)

Image

- Amateur Television (ATV), also known as *Fast Scan television*

- Slow-Scan Television (SSTV)

- Facsimile

Text and data

Most amateur digital modes are transmitted by inserting audio into the microphone input of a radio and using an analog scheme, such as amplitude modulation (AM), frequency modulation (FM), or single-sideband modulation (SSB).

- Continuous Wave (CW), usually used for Morse code

- Automatic Link Establishment (ALE)

- AMateur Teleprinting Over Radio (AMTOR)

- D-STAR

- Echolink

- Hellschreiber, also referred to as either *Feld-Hell*, or *Hell*

- Discrete multi-tone modulation modes such as Multi Tone 63 (MT63)

- Multiple Frequency-Shift Keying (MFSK) modes such as

    - FSK441, JT6M, JT65, and

    - Olivia MFSK

- Packet radio (AX.25)

    - Automatic Packet Reporting System (APRS)

- PACTOR

- Phase-Shift Keying

    - 31 baud binary phase shift keying: PSK31

    - 31 baud quadrature phase shift keying: QPSK31

    - 63 baud binary phase shift keying: PSK63

    - 63 baud quadrature phase shift keying: QPSK63

- Spread spectrum (SS)

- Radioteletype (RTTY)

Modes by activity

The following "modes" use no one specific modulation scheme but rather are classified by the activity of the communication.

- Earth-Moon-Earth (EME)

- Internet Radio Linking Project (IRLP)

- Low Transmitter Power (QRP)

- Satellite (OSCAR - Orbiting Satellite Carrying Amateur Radio)

2.4.5   See also

- DX Century Club

- Hamfest

- List of amateur radio magazines

- List of amateur radio organizations

- Maritime mobile amateur radio

- Prosigns for Morse code

- Piracy in amateur and two-way radio

- Worked All Continents

2.4.6   References

[1] "International Radiotelegraph Convention (Washington, 1927)" (PDF). Retrieved 26 April 2015.

[2] Sumner, David (August 2011). "How Many Hams?". *QST* (American Radio Relay League). p. 9.

[3] Gernsback, H (May 1909). *First Annual Official Wireless Blue Book of the Wireless Association of America* (PDF). New York: Modern Electrics Publication. Retrieved 2009-06-19.

[4] ""THE INFLUENCE OF AMATEUR RADIO ON THE DEVELOPMENT OF THE COMMERCIAL MARKET FOR QUARTZ PIEZOELECTRIC RESONATORS IN THE UNITED STATES." (1996) By Patrick R. J. Brown, Hewlett Packard Company, Spokane Division". Bliley.net. Retrieved 2012-11-22.

[5] "Inventor of IC 'chip', Nobel Prize Winner Jack S. Kilby Credits Amateur Radio for His Start in Electronics". Nobelprize.org. 2005-06-20. Retrieved 2012-11-22.

[6] *Role of Amateur Radio in Development Communication of Bangladesh.* Information & Communication Technology for Development. By Bazlur Rahman

[7] Jim Taylor. "Canadian Amateur Radio Bulletin, "Amateur Radio "Saved Lives" in South Asia" (2004-12-29)". Hfradio.net. Archived from the original on March 6, 2012. Retrieved 2012-11-22.

[8] "What is Ham Radio?". *ARRL.org*. Archived from the original on 4 May 2010. Retrieved 2010-06-01.

[9] Weaver, Bruce D. (January 2003). "On the Air Learning". *Teaching Pre K-8* **33** (4): 50–51. ISSN 0891-4508.

[10] "Ham Radio Frequently Asked Questions". *ARRL.org*. Archived from the original on 6 May 2010. Retrieved 2010-05-23.

[11] "FCC Report and Order 06-178A1" (PDF). Federal Communications Commission. 2006-12-19. p. 7. Retrieved 2007-05-16.

[12] Federal Communications Commission (2007-01-24). "47 CFR Part 97" (PDF). *Federal Register* (Washington, D.C.: Government Printing Office) **72** (15): 3081–3082. Retrieved 2007-12-18.

[13] "FCC to Drop Morse Testing for *All* Amateur License Classes". *ARRL.org* via *UnwiredAdventures.com*. 2006-12-15. Retrieved 2010-05-17.

[14] "KH6BB USS Missouri Radio Room Photos". *KH6BB USS Missouri Battleship Radio Room, kh6bb.org*. Retrieved 2010-05-23.

[15] Valdes, Robert (2001-05-09). "HowStuffWorks: Use of VoIP in Amateur Radio". Communication.howstuffworks.com. Retrieved 2012-11-22.

[16] Taggart, Ralph E (April 1993). "An Introduction to Amateur Television" (PDF). *QST* via *ARRL.org*. pp. 19–23. Archived from the original (PDF) on June 5, 2007.

[17] Holmstead, Stephen (30 December 1994). "Amateur Satellite FAQ". The Radio Amateur Satellite Corporation. Retrieved 14 March 2010.

[18] Taylor, Joe (December 2001). "*WSJT*: New Software for VHF Meteor-Scatter Communication". *QST* via *ARRL.org*. pp. 36–41. Archived from the original (PDF) on January 28, 2010.

[19] "ARISS: Amateur Radio on the International Space Station". *ARRL.org*. Archived from the original on 11 January 2007. Retrieved 2007-01-10.

[20] Jurrens, Gerald. "Astronaut (and Former Astronaut) Hams". *gjurrens* at *Tellurian.com*. Archived from the original on 30 December 2006. Retrieved 2007-01-10.

[21] Haag, Jerry. "Principles of Amateur Radio Net Control". *SCC-AREA-RACES.org*. Retrieved 2007-01-10.

[22] brweb (2000-05-01). ""International Telecommunication Union", Minimum Qualifications For Radio Amateurs". Itu.int. Retrieved 2012-11-22.

[23] "Amateur radio licensing in Thailand – sect. Equipment license". The Radio Amateur Society of Thailand 7 August 2010. Retrieved 13 February 2011.

[24] "Amateur and Amateur-satellite service". International Telecommunication Union. Archived from the original on 22 August 2010. Retrieved 2010-08-16.

[25] "285 TechConnect Radio Club". Na0tc.org. Retrieved 2012-11-22.

[26] "ARRL Mentor Program". *ARRL.org*. Archived from the original on 2007-10-14.

[27] Wilson, Mark J; Reed, Dana G (2006). *The ARRL Handbook for Radio Communications 2007* (84th ed.). Newington, CT: American Radio Relay League. ISBN 0-87259-976-0.

[28] "Amateur Radio (Intermediate) License (A) or (B) Terms, Provisions and Limitations Booklet BR68/I".

[29] "Amateur Radio (Intermediate) License (A) or (B) Terms, Provisions and Limitations Booklet BR68/I". *Ofcom.org.uk*. Retrieved 2007-06-02.

[30] "Common Filing Task: Obtaining Vanity Call Sign". *FCC.gov*. Retrieved 2007-06-02.

[31] "Vanity Call Sign Fees". *ARRL.org*. Retrieved 2015-09-28.

[32] "UK Amateur Radio Call Signs (callsigns)". *Electronics and Radio Today*. 2010. Archived from the original on 30 April 2011. Retrieved 21 March 2011.

[33] "License Search". *Universal Licensing System*. US Federal Communications Commission. Archived from the original on 22 August 2010. Retrieved 29 August 2010.

[34] "Famous Radio Amateurs 'Hams' & Call Signs". Bedworth Lions Club. Retrieved 29 August 2010.

[35] "ARRL Web: Amateur Radio License Plate Fees". Archived from the original on 2007-08-04.

[36] "Ham Radio Callsign License Plates (Canada)". Archived from the original on 7 December 2008. Retrieved 2008-12-04.

[37] "ICBC – HAM radio plates". Archived from the original on 19 October 2008. Retrieved 2008-12-03.

[38] OFTA, Equipment for Amateur Station: *Radio amateurs are free to choose any radio equipment designed for the amateur service. Radio amateurs may also design and build their own equipment provided that the requirements and limitations specified in the Amateur Station Licence and Schedules thereto are complied with.* Archived October 14, 2007 at the Wayback Machine

[39] "FCC.gov, About Amateur Stations. 'They design, construct, modify, and repair their stations. The FCC equipment authorization program does not generally apply to amateur station transmitters.'". Wireless.fcc.gov. 2002-02-19. Retrieved 2012-11-22.

[40] "Australian Radio Amateur FAQ". *AMPR.org*. June 24, 2006. Archived from the original on July 18, 2008.

[41] Industry Canada (September 2007). "RBR-4 – Standards for the Operation of Radio Stations in the Amateur Radio Service, s. 10.2". Government of Canada. Retrieved 21 January 2013.

[42] "FCC Part 97 : Sec. 97.313 Transmitter power standards". *W5YI.org*. Retrieved 2010-08-27.

[43] "FCC Part 97 : Sec. 97.401 and 97.403 Emergency Communications". Retrieved 2012-06-21.

General References

Australia

- Wireless Institute of Australia (2005). *The Foundation Licence Manual: Your Entry into Amateur Radio*. Wireless Institute of Australia, November, 2005. ISBN 0-9758342-0-7

Canada

- Cleveland-Iliffe, John, and Smith, Geoffrey Read (1995). *The Canadian Amateur Study Guide for the Basic Qualification*. Fifth Edition, Second Printing. Ottawa, Ontario, Canada: Radio Amateurs of Canada. ISBN 1-895400-08-2

India

- Amateur radio licensing in India. Retrieved Aug. 13, 2007.

United Kingdom

- Betts, Alan (2001). *Foundation Licence – Now!*. London, United Kingdom: Radio Society of Great Britain, December, 2001. ISBN 1-872309-80-1

United States

- Straw, R. Dean, Reed, Dana G., Carman, R. Jan, and Wolfgang, Larry D. (ed.) (2003). *Now You're Talking!*. Fifth Edition. Newington, Connecticut, U.S.: American Radio Relay League, May, 2003. ISBN 0-87259-881-0
- American Radio Relay League (2003). *The ARRL FCC Rule Book: Complete Guide to the FCC Regulations*. 13th Edition. Newington, Connecticut, U.S.: American Radio Relay League, August, 2003. ISBN 0-87259-900-0

- Silver, H. Ward (2004). *Ham Radio For Dummies*. John Wiley and Sons, Ltd., April, 2004. ISBN 0-7645-5987-7

2.4.7   Further reading

- Bergquist, Carl J (May 2001). *Ham Radio Operator's Guide* (2nd ed.). Indianapolis: Prompt Publications. ISBN 0-7906-1238-0.

- Dennison, Mike; Fielding, John, eds. (2009). *Radio Communication Handbook* (10th ed.). Bedford, England: Radio Society of Great Britain. ISBN 978-1-905086-54-2.

- Haring, Kristen (2007). *Ham Radio's Technical Culture*. Cambridge, MA: MIT Press. ISBN 0-262-08355-8.

- Poole, Ian D (October 2001). *HF Amateur Radio*. Potters Bar, Hertfordshire, England: Radio Society of Great Britain. ISBN 1-872309-75-5.

- Rohde, Ulrich L; Whitaker, Jerry C (2001). *Communications Receivers: DSP, Software Radios, and Design* (3rd ed.). New York City: McGraw-Hill. ISBN 0-07-136121-9.

- *The ARRL Handbook for Radio Communications 2010* (87th ed.). Newington, CT: American Radio Relay League. November 2009. ISBN 0-87259-144-1.

2.4.8   External links

- Amateur Radio at DMOZ

2.5   Arne Skoog

Arne Skoog was born in 1913 in the northern Swedish province of Jämtland. He died on 7 June, 1999 aged 86.

2.5.1   Sweden Calling DXers

In 1948, Arne, then a young engineer at Radio Sweden, founded Sweden Calling DXers as a way of keeping listeners in touch with news in the world of international radio. Shortwave came into its own during the Second World War, and after the war shortwave listening became a popular hobby. But hobbyists needed access to the latest news, and Arne felt the best place for them to find out was on a shortwave station.

Many hobbyists say they remember Arne presenting Sweden Calling DXers, which is curious since he never did. Arne wrote the scripts, which were read by colleagues, first by the English Service, and then in all of Radio Sweden's languages, except, ironically, Swedish.

Initially Arne gathered all his own news for the program, but soon listeners were writing in with material for the program. Almost immediately a written version of the program scripts was being mailed out to everyone who contributed. At its height there were between 1500 and 2000 names on the list, and they constituted a "Who's Who" of the shortwave community.

When Arne retired in 1978, on the program's 30th anniversary, George Wood, who wasn't even born in 1948, took over. But changing times and new Radio Sweden management led to radical changes with the introduction of satellite broadcasting in the 1980s.

The mailing list was discontinued, and the program became a twice a month English-only program called MediaScan, which concentrated more and more on satellite DXing and general Scandinavian media news. The raido program was completely discontinued in 2001, and continues as a sporadically updated page on the Radio Sweden website.

2.5.2 Arne and the DX Community

Arne Skoog was also one of the founders of the Swedish DX Federation, and later the European DX Council. He rejected, however, the sometimes expressed claim that he was "the father of all Swedish DXers".

In his retirement, Arne turned to his second pastime, violin-making, for which he was quite respected. He taught many courses in the skill. He died on June 7, 1999, in his beloved home province of Jämtland.

2.6 Association for International Broadcasting

The **Association for International Broadcasting - AIB -** is the industry association that represents and supports international television and radio broadcasters.

Founded in 1993, the AIB has developed into a truly global organisation whose membership extends from New Zealand west through to the USA. The AIB provides its members with market intelligence, lobbying, networking and marketing support. It publishes the international media magazine, *The Channel*, that has a regular subscriber base of more than 6,600 senior executives in broadcasting and electronic media organisations in over 120 countries. The

Logo of the Association for International Broadcasting.

AIB also produces regular electronic news letters that reach the desktops of more than 26,000 people worldwide.

The AIB has an immense collection of data about broadcasting and electronic media covering territories throughout the world. As part of its role to support the broadcasting industry, the AIB publishes its *Global Broadcasting Sourcebook* with extracts of its data concerning broadcasters, regulators and key personnel.

The AIB runs a festival that celebrates the best in factual TV and radio broadcasting. Called the AIBs, this annual festival attracts entries from broadcasters and independent production companies on every continent.

The AIB is a non-governmental, not-for-profit organisation with its headquarters in the United Kingdom. It is governed by an Executive Council of six members elected from the AIB's membership including representatives of BBC Global News, Bloomberg Television, Audiovisuel exterieur de France, RT channel, Deutsche Welle and Voice of Nigeria. The AIB's permanent staff is led by Chief Executive Simon Spanswick.

2.6.1 External links

• Official site

• Official awards site

2.7 British DX Club

The British DX Club (abbreviated form "BDXC") is an association of radio hobbyists, based in the United Kingdom. It caters mainly for, though not exclusively, DXers and Short Wave listeners. It was founded in 1974 and was originally known as the "Twickenham DX Club" (after the Middlesex, UK, town where it was originally based), but relaunched in 1979 as the British DX Club. The name change was made to reflect its growing national and international membership which currently stands at around 500. To distinguish it from similar organisations in other countries,the club's abbreviated name for international use is BDXC-UK.

BDXC publishes a monthly magazine "Communication", which is registered with the British Library (ISSN 0958-2142). The main contents of a typical magazine consist of a mixture of DX "loggings", details of radio station frequencies and programmes, letters and articles of general interest on broadcasting topics. These topics include:

UK News, Webwatch, DX News, Propagation, DRM News, Mediumwave Report, Collectors Corner (vintage Radio's etc.), The Audio Circle, Beyond The Horizon, Mediumwave Logbook, Tropical Logbook, HF Logbook, Alternative Airwaves (Pirate Radio), and members contributions.

In 1976, the club launched a monthly audio magazine on cassette tape, BDXC Audio Circle (formerly known as BDXC Tape Circle). This is now also distributed on Compact Disc and as an MP3 downloadable file for subscribing members only.

The BDXC also has an email news group for club members to distribute news and DX 'catches' to subscribed members. The Audio Circle also has its own email group.

BDXC and its members are important contributors to each edition of the World Radio Television Handbook (WRTH) an annual publication listing broadcasting times and frequencies of most of the world's national and international broadcasters. Information on other UK DX clubs is also included in WRTH.

2.7.1   External links

- www.bdxc.org.uk Official British DX Club website.

- www.wrth.co.uk World Radio TV Handbook Website

- www.worldofradio.com Glenn Hauser's World Of Radio

2.8   Digital Radio Mondiale

Digital Radio Mondiale (abbreviated **DRM**; *mondiale* being Italian and French for "worldwide") is a set of digital audio broadcasting technologies designed to work over the bands currently used for analogue radio broadcasting including AM broadcasting, particularly shortwave, and FM

Official DRM logo

broadcasting. DRM is more spectrally efficient than AM and FM, allowing more stations, at higher quality, into a given amount of bandwidth, using various MPEG-4 audio coding formats.

Digital Radio Mondiale is also the name of the international non-profit consortium that has designed the platform and is now promoting its introduction. Radio France Internationale, TéléDiffusion de France, BBC World Service, Deutsche Welle, Voice of America, Telefunken (now Transradio) and Thomcast (now Thomson Broadcast) took part at the formation of the DRM consortium.

The principle of DRM is that bandwidth is the limited element, and computer processing power is cheap; modern CPU-intensive audio compression techniques enable more efficient use of available bandwidth, at the expense of processing resources.

2.8.1   Features

DRM can deliver FM-comparable sound quality on frequencies below 30 MHz (long wave, medium wave and short wave), which allow for very-long-distance signal propagation. The modes for these lower frequencies are often collectively known under the term "DRM30". In the VHF bands, the term "DRM+" is used. DRM+ is able to use available broadcast spectrum between 30 and 300 MHz; generally this means band I (47 to 68 MHz), band II (87.5 to 108 MHz) and band III (174 to 230 MHz). DRM has been designed to be able to re-use portions of existing analogue transmitter facilities such as antennas, feeders, and, especially for DRM30, the transmitters themselves, avoiding major new investment. DRM is robust against the fading and interference which often plague conventional broadcasting in these frequency ranges.

The encoding and decoding can be performed with digital

signal processing, so that a cheap embedded computer with a conventional transmitter and receiver can perform the rather complex encoding and decoding.

As a digital medium, DRM can transmit other data besides the audio channels (datacasting) — as well as RDS-type metadata or program-associated data as Digital Audio Broadcasting (DAB) does. DRM services can be operated in many different network configurations, from a traditional AM one-service one-transmitter model to a multi-service (up to four) multi-transmitter model, either as a single-frequency network (SFN) or multi-frequency network (MFN). Hybrid operation, where the same transmitter delivers both analogue and DRM services simultaneously is also possible.

DRM incorporates technology known as Emergency Warning Features that can override other programming and activates radios which are in standby in order to receive emergency broadcasts.

2.8.2 Status

The technical standard is available free-of-charge from the ETSI,[1] and the ITU has approved its use in most of the world. Approval for ITU region 2 (North and South America and the Pacific) is pending amendments to existing international agreements. The inaugural broadcast took place on June 16, 2003, in Geneva, Switzerland, at the ITU's World Radio Conference.

Current broadcasters include All India Radio, BBC World Service, Deutschlandradio, biteXpress, HCJB, Deutsche Welle, Radio Netherlands Worldwide, RTÉ Radio (RTÉ), Radio Exterior de España, RAI, Kuwait Radio, Radio New Zealand International, Vatican Radio, Voice of Russia and Radio Romania International.[2]

Until now DRM receivers have typically used a personal computer. A few manufacturers have introduced DRM receivers which have thus far remained niche products due to limited choice of broadcasts. It is expected that the transition of national broadcasters to digital services on DRM, notably All India Radio, will stimulate the production of a new generation of affordable, and efficient receivers.

Chengdu NewStar Electronics is offering the DR111 from May 2012 on which meets the minimum requirements for DRM receivers specified by the DRM consortium and is sold worldwide.[3]

The General Overseas Service of All India Radio broadcasts daily in DRM to Western Europe on 9,950 kHz at 17:45 to 22:30 UTC.[4] All India Radio is in the process of replacing and refurbishing many of its domestic AM transmitters with DRM. The project which began in 2012 is scheduled to complete during 2015.[5]

The British Broadcasting Corporation BBC has trialed the technology in the United Kingdom by broadcasting BBC Radio Devon in the Plymouth area in the MF band. The trial lasted for a year (April 2007 – April 2008).[6] The BBC also trialed DRM+ in the FM band in 2010 in the Fife area of Scotland, including the city of Edinburgh. In this trial, a previously used 10 kW (ERP) FM transmitter was replaced with a 1 kW DRM+ transmitter in two different modes and the coverage compared with FM [7] Digital Radio Mondiale was included in the 2007 Ofcom consultation on the future of radio in the United Kingdom for the AM medium wave band.[8]

RTÉ has also run single and multiple programme overnight tests during a similar period on the 252 kHz LW transmitter in Trim, Co.Meath, Ireland which was upgraded to support DRM after Atlantic 252 closed.

International regulation

On 28 September 2006, the Australian spectrum regulator, the Australian Communications and Media Authority, announced that it had "placed an embargo on frequency bands potentially suitable for use by broadcasting services using Digital Radio Mondiale until spectrum planning can be completed" "those bands being "5,950–6,200; 7,100–7,300; 9,500–9,900; 11,650–12,050; 13,600–13,800; 15,100–15,600; 17,550–17,900; 21,450–21,850 and 25,670–26,100 kHz.[9]

The United States Federal Communications Commission states in 47 C.F.R. 73.758 that: "For digitally modulated emissions, the Digital Radio Mondiale (DRM) standard shall be employed." Part 73, section 758 is for HF broadcasting only.

2.8.3 Technological Overview

Audio source coding

Useful bitrates for DRM30 range from 6.1 kbit/s (Mode D) to 34.8 kbit/s (Mode A) for a 10 kHz bandwidth (±5 kHz around the central frequency). It is possible to achieve bit rates up to 72 kbit/s (Mode A) by using a standard 20 kHz (±10 kHz) wide channel.[10] (For comparison, pure digital HD Radio can broadcast 20 kbit/s using channels 10 kHz wide and up to 60 kbit/s using 20 kHz channels.)[11] Useful bitrate depends also on other parameters, such as:

- desired robustness to errors (error coding)
- power needed (modulation scheme)
- robustness in regard to propagation conditions (multipath propagation, doppler effect), etc.

When DRM was originally designed, it was clear that the most robust modes offered insufficient capacity for the then state-of-the-art audio coding format MPEG-4 HE-AAC (High Efficiency Advanced Audio Coding). Therefore the standard launched with a choice of three different audio coding systems (source coding) depending on the bitrate:

- MPEG-4 HE-AAC (High Efficiency Advanced Audio Coding). AAC is a perceptual coder suited for voice and music and the High Efficiency is an optional extension for reconstruction of high frequencies (SBR: spectral bandwidth replication) and stereo image (PS: Parametric Stereo). 24 kHz or 12 kHz sampling frequencies can be used for core AAC (no SBR) which correspond respectively to 48 kHz and 24 kHz when using SBR oversampling.

- MPEG-4 CELP which is a parametric coder suited for voice only (vocoder) but that is robust to errors and needs a small bit rate.

- MPEG-4 HVXC which is also a parametric coder for speech programs that uses an even smaller bitrate than CELP.

However with the development of MPEG-4 xHE-AAC, which is an implementation of MPEG Unified Speech and Audio Coding, the DRM standard was updated and the two speech-only coding formats, CELP and HVXC, were replaced. USAC is designed to combine the properties of a speech and a general audio coding according to bandwidth constraints and so is able to handle all kinds of programme material. Given that there were few CELP and HVXC broadcasts on-air, the decision to drop the speech-only coding formats has passed without issue.

Many broadcasters still use the HE-AAC coding format because it still offers an acceptable audio quality, somewhat comparable to FM broadcast at bitrates above about 15 kbit/s. However, it is anticipated that in future, most broadcasters will adopt xHE-AAC.

Additionally, as of v2.1, the popular Dream software can broadcast using the Opus coding format. Whilst not within the current DRM standard the inclusion of this codec is provided for experimentation. Aside from perceived technical advantages over the MPEG family such as low latency (delay between coding and decoding), this codec provides an open source (therefore free to use) alternative to the proprietary MPEG family whose use is permitted at the discretion of the patent holders. Equipment manufacturers currently pay royalties for incorporating the MPEG codecs.

Bandwidth

DRM broadcasting can be done using a choice of different bandwidths:

- 4.5 kHz. Gives the ability for the broadcaster to do a simulcast and use the lower-sideband area of a 9 kHz raster channel for AM, with a 4.5 kHz DRM signal occupying the area traditionally taken by the upper-sideband.[12] However the resulting bit rate and audio quality is not good.

- 5 kHz. Gives the ability for the broadcaster to do a simulcast and use the lower-sideband area of a 10 kHz raster channel for AM, with a 5 kHz DRM signal occupying the area traditionally taken by the upper-sideband. However the resulting bit rate and audio quality is marginal (7.1–16.7 kbit/s for 5 kHz). This technique could be used on the short wave bands throughout the world.

- 9 kHz. Occupies half the standard bandwidth of a region–1 long wave or medium wave broadcast channel.

- 10 kHz. Occupies half the standard bandwidth of a region–2 broadcast channel. could be used to simulcast with analogue audio channel restricted to NRSC5. Occupies a full worldwide short wave broadcast channel (giving 14.8–34.8 kbit/s)

- 18 kHz. Occupies full bandwidth of region–1 long wave or medium wave channels according to the existing frequency plan. This offers better audio quality.

- 20 kHz. Occupies full bandwidth of region–2 or 3 AM channel according to the existing frequency plan. This offers highest audio quality of the DRM30 standard (giving 30.6–72 kbit/s).

- 100 kHz for DRM+. This bandwidth can be used in band I, II, and III and DRM+ can transmit four different programs in this bandwidth.

Modulation

The modulation used for DRM is coded orthogonal frequency division multiplexing (COFDM), where every carrier is modulated with quadrature amplitude modulation (QAM) with a selectable error coding.

The choice of transmission parameters depends on signal robustness wanted and propagation conditions. Transmission signal is affected by noise, interference, multipath wave propagation and Doppler effect.

It is possible to choose among several error coding schemes and several modulation patterns: 64-QAM, 16-QAM and 4-QAM. OFDM modulation has some parameters that must be adjusted depending on propagation conditions. This is the carrier spacing which will determine the robustness against Doppler effect (which cause frequencies offsets, spread: Doppler spread) and OFDM guard interval which determine robustness against multipath propagation (which cause delay offsets, spread: delay spread). The DRM consortium has determined four different profiles corresponding to typical propagation conditions:

- A: Gaussian channel with very little multipath propagation and Doppler effect. This profile is suited for local or regional broadcasting.

- B: multipath propagation channel. This mode is suited for medium range transmission. It is nowadays frequently used.

- C: similar to mode B, but with better robustness to Doppler (more carrier spacing). This mode is suited for long distance transmission.

- D: similar to mode B, but with a resistance to large delay spread and Doppler spread. This case exists with adverse propagation conditions on very long distance transmissions. The useful bit rate for this profile is decreased.

The trade-off between these profiles stands between robustness, resistance in regards to propagation conditions and useful bit rates for the service. This table presents some values depending on these profiles. The larger the carrier spacing, the more the system is resistant to Doppler effect (Doppler spread). The larger the guard interval, the greater the resistance to long multipath propagation errors (delay spread).

The resulting low-bit rate digital information is modulated using COFDM. It can run in simulcast mode by switching between DRM and AM, and it is also prepared for linking to other alternatives (e.g., DAB or FM services).

DRM has been tested successfully on shortwave, mediumwave (with 9 as well as 10 kHz channel spacing) and longwave.

There is also a lower bandwidth two-way communication version of DRM as a replacement for SSB communications on HF[13] - note that it is *not* compatible with the official DRM specification. It may be possible in some future time for the 4.5 kHz bandwidth DRM version used by the Amateur Radio community to be merged with the existing DRM specification.

The Dream software will receive the commercial versions and also limited transmission mode using the FAAC AAC encoder.

Error coding

Error coding can be chosen to be more or less robust.

This table shows an example of useful bitrates depending on protection classes

- OFDM propagation profiles (A or B)

- carrier modulation (16QAM or 64QAM)

- and channel bandwidth (9 or 10 kHz)

The lower the protection class the higher the level of error correction.

2.8.4 DRM+

While the initial DRM standard covered the broadcasting bands below 30 MHz, the DRM consortium voted in March 2005 to begin the process of extending the system to the VHF bands up to 108 MHz.[14]

On 31 August 2009, DRM+ (Mode E) became an official broadcasting standard with the publication of the technical specification by the European Telecommunications Standards Institute; this is effectively a new release of the whole DRM spec with the additional mode permitting operation above 30 MHz up to 174 MHz.[15]

Wider bandwidth channels are used, which allows radio stations to use higher bit rates, thus providing higher audio quality. A 100 kHz DRM+ channel has sufficient capacity to carry one low-definition 0.7 megabit/s wide mobile TV channel: it would be feasible to distribute mobile TV over DRM+ rather than DMB or DVB-H.

DRM+ has been successfully tested in all the VHF bands, and this gives the DRM system the widest frequency usage; it can be used in band I, II and III. DRM+ can coexist with DAB in band III.[16] but also the present FM-band can be utilized. The ITU has published three recommendations on DRM+, known in the documents as Digital System G. This indicate the introduction of the full DRM system (DRM 30 and DRM+). ITU-R Rec. BS.1114 is the ITU recommendation for sound broadcasting in the frequency range 30 MHz to 3 GHz. DAB, HD-Radio and ISDB-T were already recommended in this document as Digital Systems A, C and F respectively.

In 2011, the paneuropean organisation Community Media Forum Europe [17] has recommended to the European Commission that DRM+ should rather be used for small scale broadcasting (local radio, community radio) than DAB/DAB+.

2.8.5 See also

- AMSS AM signalling system
- Digital Audio Broadcasting (DAB)
- Digital Multimedia Broadcasting (DMB)
- DVB-H (Digital Video Broadcasting - Handhelds)
- DVB-T (Digital Video Broadcasting - Terrestrial)
- ETSI Satellite Digital Radio (SDR)
- HD Radio, American system for digital radio
- ISDB-Tsb, Japanese system for digital radio.
- Cliff effect, which affects digital communications such as radio
- In-band on-channel

2.8.6 References

[1] DRM System specification

[2] DRM Broadcast Schedule

[3] "DR111 DRM Radio". Chengdu NewStar Electronics | 成都新星电子股份有限公司:. 2014. Retrieved 2014-04-15.

[4] http://allindiaradio.gov.in/Services/Digital%20Transmission/Pages/simple.aspx

[5] http://www.drm.org/?page_id=2494

[6] Digital medium wave trial report (BBC)

[7] BBC Research White Paper WHP199

[8] The Future of Radio (Ofcom, 2007)

[9] article: ACMA embargoes spectrum to plan for Digital Radio Mondiale

[10] "DRM Introduction and Implementation Guide" (PDF; 6.7 MB). DRM. p. 22.

[11] *dead link*] The Structure and Generation of Robust Waveforms for AM In-Band On-Channel Digital Broadcasting PDF

[12] "See section 5: "DRM/AM single channel simulcast"" (PDF).

[13] WinDRM - software for Audio and Fast Data over HF SSB

[14] [http://www.drm.org/uploads/files/drm_plus_pres.pdf DRM+ Presentation], DRM.org, accessed 2009-02-02

[15] ETSI ES 201 980 V3.1.1

[16] Symposium about the DRM+ field trial in VHF band III

[17] Community Media Forum Europe

2.8.7 External links

- Digital Radio Mondiale (DRM) - official homepage
- How to receive DRM on the long-, medium- and short-wave bands
- Diorama DRM receiver. An open source DRM receiver written by the Institute of Telecommunications of the University Kaiserslautern (Germany))
- WinDRM DRM software for amateur radio users
- Dream - an open-source software DRM Receiver
- gr-drm GNU Radio transmitter implementation
- DRM Software DRM software collection

2.9 Duga radar

*Panorama receive antennas of radar "Duga" in **Chernobyl 2***

Duga Russian: Дуга−3 was a Soviet over-the-horizon (OTH) radar system used as part of the Soviet ABM early-warning network. The system operated from July 1976 to December 1989. Two operational Duga radars were deployed, one near Chernobyl and Chernihiv in what was then called the Ukrainian SSR (present-day Ukraine), the other in eastern Siberia.

The Duga systems were extremely powerful, over 10 MW in some cases, and broadcast in the shortwave radio bands. They appeared without warning, sounding like a sharp, repetitive tapping noise at 10 Hz,[1] which led to it being nicknamed by shortwave listeners the **Russian Woodpecker**. The random frequency hops disrupted legitimate

Duga-3 array within the Chernobyl Exclusion Zone. The array of pairs of cylindrical/conical cages on the right are the driven elements, fed at the facing points with a form of ladder line suspended from stand-off platforms at top right. A backplane axel reflector of small wires can just be seen left of center, most clearly at the bottom of the image.

Steel structure of Duga-3 from the bottom

broadcast, amateur radio, commercial aviation communications, utility transmissions, and resulted in thousands of complaints by many countries worldwide. The signal became such a nuisance that some receivers such as amateur radios and televisions actually began including 'Woodpecker Blankers' in their design.

The unclaimed signal was a source for much speculation, giving rise to theories such as Soviet mind control and weather control experiments. However, many experts and amateur radio hobbyists quickly realized it to be an over-the-horizon radar system. NATO military intelligence had already photographed the system and given it the NATO reporting name of either *STEEL WORK* or *STEEL YARD*. This theory was publicly confirmed after the fall of the Soviet Union.

2.9.1 History

Genesis

The Soviets had been working on early warning radar for their anti-ballistic missile systems through the 1960s, but most of these had been line-of-sight systems that were useful for raid analysis and interception only. None of these systems had the capability to provide early warning of a launch, within seconds or minutes of a launch, which would give the defences time to study the attack and plan a response. At the time, the Soviet early-warning satellite network was not well developed, and there were questions about their ability to operate in a hostile environment including anti-satellite efforts. An over-the-horizon radar sited in the USSR would not have any of these problems, and work on such a system for this associated role started in the late 1960s.

The first experimental system, Duga-1, was built outside Mykolaiv in Ukraine, successfully detecting rocket launches from Baikonur Cosmodrome at 2,500 kilometers. This was followed by the prototype Duga-2, built on the same site, which was able to track launches from the far east and submarines in the Pacific Ocean as the missiles flew towards Novaya Zemlya. Both of these radar systems were aimed east and were fairly low power, but with the concept proven, work began on an operational system. The new Duga-3 systems used a transmitter and receiver separated by about 60 km.[2]

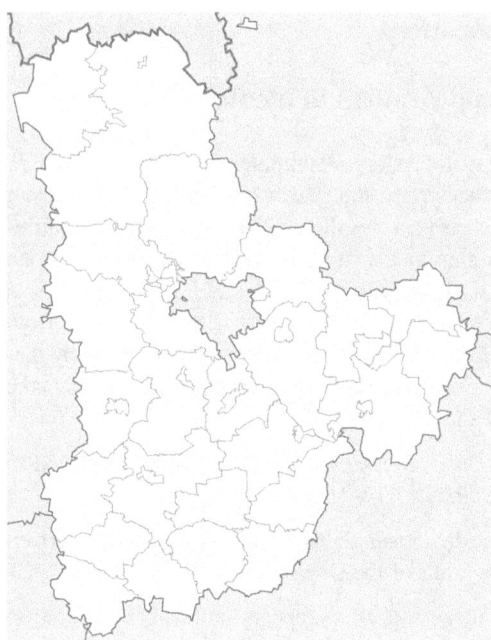

Transmitter

Receiver

Russian Woodpecker in Kiev Oblast

Russian Woodpecker

Starting in 1976 a new and powerful radio signal was detected worldwide, and quickly dubbed the Woodpecker by amateur radio operators. Transmission power on some woodpecker transmitters was estimated to be as high as 10 MW equivalent isotropically radiated power.[3][4]

Triangulation quickly revealed the signals came from Ukraine, at the time called Ukrainian Soviet Socialist Republic (part of USSR). Confusion due to small differences in the reports being made from various sources led to the site being alternately located near Kyiv, Minsk, Chernobyl, Gomel or Chernihiv. All of these reports were describing the same deployment, with the transmitter only a few kilometers southwest of Chernobyl (south of Minsk, northwest of Kyiv) and the receiver about 50 km northeast of Chernobyl (just west of Chernihiv, south of Gomel). At one time there was speculation that several transmitters were in use.[3]

The radar system was given the code 5H32-West by the Soviets, and was set up in two closed towns, Liubech-1 held the two transmitters and Chernobyl-2 the receivers.[4] Unknown to civilian observers at the time, NATO was aware of the new installation. A second installation was built near Komsomolsk-on-Amur, in Bolshya Kartel and Lian, but did

not become active for some time.

NATO Reporting Name

The NATO Reporting Name for the Duga-3 is often quoted as STEEL YARD. Many online and several print references use this name. However some sources also use the term STEEL WORK (or STEEL WORKS). As any "official" sources using NATO Reporting Names are likely to be classified deconflicting this will be difficult. The earliest found open source mention of a NATO Reporting Name for this system, a reference publication in print while the system was still active, unambiguously uses the term STEEL WORK.[5]

Civilian identification

Even from the earliest reports it was suspected that the signals were tests of an over-the-horizon radar,[3] and this remained the most popular hypothesis during the Cold War. Several other theories were floated as well, including everything from jamming western broadcasts to submarine communications. The broadcast jamming theory was discarded early on when a monitoring survey showed that Radio Moscow and other pro-Soviet stations were just as badly affected by woodpecker interference as Western stations.

As more information about the signal became available, its purpose as a radar signal became increasingly obvious. In particular, its signal contained a clearly recognizable structure in each pulse, which was eventually identified as a 31-bit pseudo-random binary sequence, with a bit-width of 100 μs resulting in a 3.1 ms pulse.[6] This sequence is usable for a 100 μs chirped pulse amplification system, giving a resolution of 15 km (10 mi) (the distance light travels in 50 μs). When a second Woodpecker appeared, this one located in eastern Russia but also pointed toward the US and covering blank spots in the first system's pattern, this conclusion became inescapable.

In 1988, the Federal Communications Commission conducted a study on the Woodpecker signal. Data analysis showed an inter-pulse period of about 90 ms, a frequency range of 7 to 19 MHz, a bandwidth of 0.02 to 0.8 MHz, and typical transmission time of 7 minutes.

- The signal was observed using three repetition rates: 10 Hz, 16 Hz and 20 Hz.

- The most common rate was 10 Hz, while the 16 Hz and 20 Hz modes were rather rare.

- The pulses transmitted by the woodpecker had a wide bandwidth, typically 40 kHz.

Jamming

The array at Chernobyl, viewed from a distance

To combat this interference, amateur radio operators attempted to "jam" the signal by transmitting synchronized unmodulated continuous wave signals at the same pulse rate as the offending signal. They formed a club called The Russian Woodpecker Hunting Club.[7]

Disappearance

Starting in the late 1980s, even as the U.S. Federal Communications Commission (FCC) was publishing studies of the signal, the signals became less frequent, and in 1989, they disappeared altogether. Although the reasons for the eventual shutdown of the Duga systems have not been made public, the changing strategic balance with the end of the Cold War in the late 1980s likely had a major part to play. Another factor was the success of the US-KS early-warning satellites, which entered preliminary service in the early 1980s, and by this time had grown into a complete network. The satellite system provides immediate, direct and highly secure warnings, whereas any radar-based system is subject to jamming, and the effectiveness of OTH systems is also subject to atmospheric conditions.

According to some reports, the Komsomolsk-na-Amure installation in the Russian Far East was taken off combat alert duty in November 1989, and some of its equipment was subsequently scrapped. The original Duga-3 site lies within the 30 kilometer Zone of Alienation around the Chernobyl power plant. It appears to have been permanently deactivated, since their continued maintenance did not figure in the negotiations between Russia and Ukraine over the active Dnepr early warning radar systems at Mukachevo and Sevastopol. The antenna still stands, however, and has been used by amateurs as a transmission tower (using their own antennas) and has been extensively photographed.

2.9.2 Locations

2.9.3 Appearances in media

The Ukrainian-developed computer game *S.T.A.L.K.E.R.* has a plot focused on the Chernobyl Nuclear Power Plant and the nuclear accident there. The game heavily features actual locations in the area, including the Duga-1 array. The array itself appears in *S.T.A.L.K.E.R.: Clear Sky* in the city of Limansk-13. While the 'Brain Scorcher' from *S.T.A.L.K.E.R.: Shadow of Chernobyl* was inspired by theories that Duga-1 was used for mind control, it does not take the form of the real array.

In *Call of Duty: Black Ops*, the map "Grid" is placed in Pripyat near the DUGA-1 array.

In the movie *Divergent*, the wall around Chicago is derived from photographs of the Duga-1 array.[8]

The 'Russian woodpecker' appears in Justin Scott's novel *The Shipkiller*.

The Duga at Chernobyl was the focus of the 2015 documentary film, *The Russian Woodpecker*, by Chad Gracia. The film includes interviews with the commander of the Duga Vladimir Musiets, as well as the Vice-Commander, the Head of the Data Center, and others involved in building and operating the radar. The documentary, which won numerous awards, also includes drone video footage of the array and handheld video footage of the surroundings as well as a climb to the top by the cinematographer, Artem Ryzhykov.[9]

2.9.4 See also

- Duga-1 and Duga-2
- Cobra Mist
- Jindalee Operational Radar Network
- Numbers station
- UVB-76

2.9.5 References

[1] David L. Wilson (Summer 1985). "The Russian Woodpecker... A Closer Look". *Monitoring Times*. Retrieved 2007-06-15.

[2] Bukharin, Oleg; et al. (2001). Pavel Podvig, ed. *Russian Strategic Nuclear Forces*. Cambridge, Massachusetts: MIT Press.

[3] "Mystery Soviet over-the-horizon tests". *Wireless World*: 53. February 1977. Retrieved 2007-06-15.

[4] Nazaryan, Alexander (18 April 2014). "The Massive Russian Radar Site in the Chernobyl Exclusion Zone". *Newsweek*.

[5] *The International Countermeasures Handbook, 14th Edition.* Englewood, Colorado, USA: Cardiff Publishing. 1989.

[6] J.P. Martinez (April 1982). "Letter from J. P. Martinez". *Wireless World*: 89. Retrieved 2007-06-15.

[7] Dave Finley (7 July 1982). "Radio hams do battle with 'Russian Woodpecker'". *The Miami Herald*. Retrieved 2007-06-15.

[8] The "Woodpecker" moves to fururistic Chicago! //QRZ.com; The Russian Woodpecker = the wall around Chicago in Divergent; Marcel Birgelen : "thing is surrounded by a great wall, which has some eery similarities to the Russian Woodpecker."

[9] The Russian Woodpecker documentary (2015)

2.9.6 Further reading

- Headrick, James M. (1 July 1990). "Looking over the horizon (HF radar)". *IEEE Spectrum* **27** (7): 36–39. doi:10.1109/6.58421.

- Headrick, James M.; Skolnik, Merrill I. (1 January 1974). "Over-the-Horizon radar in the HF band". *Proceedings of the IEEE* **62** (6): 664–673. doi:10.1109/PROC.1974.9506.

- Headrick, James M., Ch. 24: "HF over-the-horizon radar," in: Radar Handbook, 2nd ed., Merrill I. Skolnik, ed. [New York: McGraw-Hill, 1990].

- Kosolov, A. A., ed. Fundamentals of Over-the-Horizon Radar (translated by W. F. Barton) [Norton, Mass.: Artech House, 1987].

- John Pike. "Steel Yard OTH". GlobalSecurity.org. Retrieved 2010-04-08.

2.9.7 External links

- Chernobyl-2. Secret Military Facility in the territory of exclusion zone. Text and photos 2008

- OTH-Radar "Chornobyl - 2" and Center of space-communication

- "Circle" is an auxiliary system for OTH-Radar "Chornobyl - 2"

- The Russian Woodpecker, *Miami Herald*, July 1982.

- Steel Yard OTH, globalsecurity.org

- Some pictures of Chernobyl-2

- 'Duga' photos at englishrussia.com

2.10   Duga-1 and Duga-2

Coordinates: 47°04′30″N 31°39′00″E / 47.075000°N 31.650000°E

Duga-1 and **Duga-2** (Russian: Дуга - "duga" means "arc") were Soviet over-the-horizon radar (OTH) systems. They were developed for the Soviet ABM early-warning network. Both of these radar systems were aimed east and were fairly high power. They were predecessors of the Duga-3 or "Steel Yard" OTH system, which operated from 1976 to 1989.

Duga-1 was the first experimental system.[1][2] It was built outside Mykolaiv in the Ukraine, and successfully detected rocket launches from Baikonur Cosmodrome, about 2,500 kilometers away.

The second prototype, Duga-2, was built on the same site. Duga-2 is able to track launches from the Far East, and from submarines in the Pacific Ocean, as the missiles flies towards Novaya Zemlya in the Arctic Ocean. This huge radar complex was restored recently (2002) after a fire which seriously damaged it.

2.10.1   References

[1] John Pike. "Steel Yard OTH". GlobalSecurity.org. Retrieved 2010-04-08.

[2] A. Karpenko Nevsky Bastion (1999). "ABM AND SPACE DEFENSE". No. 4: 2–47.

2.11   DXing

Not to be confused with DJing, another audio and radio hobby.

This article is about the hobby of receiving & identifying radio or television signals. For the article about the Philippine FM station in General Santos City, see DXER.

DXing is the hobby of receiving and identifying distant radio or television signals, or making two way radio contact with distant stations in amateur radio, citizens' band radio or other two way radio communications. Many DXers also attempt to obtain written verifications of reception or contact, sometimes referred to as "QSLs" or "veries". The name of the hobby comes from DX, telegraphic shorthand for "distance" or "distant".[1]

The practice of DXing arose during the early days of radio broadcasting. Listeners would mail "reception reports" to radio broadcasting stations in hopes of getting a written acknowledgement or a QSL card that served to officially verify they had heard a distant station. Collecting these cards became popular with radio listeners in the 1920s and 1930s, and reception reports were often used by early broadcasters to gauge the effectiveness of their transmissions. Although international shortwave broadcasts are on the decline, DXing remains popular among dedicated shortwave listeners. The pursuit of two-way contact between distant amateur radio operators is also a significant activity within the amateur radio hobby. [2][3]

2.11.1   Types of DXing

AM radio DX

Main article: MW DX

Early radio listeners, often using home made crystal sets and long wire antennas, found radio stations few and far between. With the broadcast bands uncrowded, signals of the most powerful stations could be heard over hundreds of miles, but weaker signals required more precise tuning or better receiving gear.

By the 1950s, and continuing through the mid-1970s, many of the most powerful North American "clear channel" stations such as KDKA, WLW, CKLW, CHUM, WABC, WJR, WLS, WKBW, KFI, KAAY, KSL and a host of border blasters from Mexico pumped out Top 40 music played by popular disc jockeys. As most smaller, local AM radio stations had to sign off at night, the big 50 kW stations had loyal listeners hundreds of miles away.

The popularity of DXing the medium-wave band has diminished as the popular music formats quickly migrated to the clearer, though less propagating, FM radio beginning in the 1970s. Meanwhile, the MW band in the United States was getting more and more crowded with new stations and existing stations receiving FCC authorization to operate at night. In Canada, just the opposite occurred as AM stations began moving to FM beginning in the 1980s and continuing through today.

Outside of the Americas and Australia, most AM radio broadcasting was in the form of synchronous networks of government-operated stations, operating with hundreds, even thousands of kilowatts of power. Still, the lower powered stations and occasional trans-oceanic signal were popular DX targets.[4]

Shortwave DX

Main article: Shortwave listening

Especially during wartime and times of conflict, reception of international broadcasters, whose signals propagate around the world on the shortwave bands has been popular with both casual listeners and DXing hobbyists.

With the rise in popularity of streaming audio over the internet, many international broadcasters (including the BBC and Voice of America) have cut back on their shortwave broadcasts. Missionary Religious broadcasters still make extensive use of shortwave radio to reach less developed countries around the world.

In addition to international broadcasters, the shortwave bands also are home to military communications, RTTY, amateur radio, pirate radio, and the mysterious broadcasts of numbers stations. Many of these signals are transmitted in single side band mode, which requires the use of specialized receivers more suitable to DXing than to casual listening.[5]

VHF DXing

Main article: TV-FM DX

Though sporadic in nature, signals on the FM broadcast and VHF television bands - especially those stations at the lower end of these bands - can "skip" for hundreds, even thousands of miles. American FM stations have been occasionally received in Western Europe, though no reports exist of European FM signals propagating to North America.

Police, fire, and military communications on the VHF bands are also DX'ed to some extent on multi-band radio scanners, though they are mainly listened to strictly on a local basis. One difficulty is in identifying the exact origins of communications of this nature, as opposed to commercial broadcasters which must identify themselves at the top of each hour, and can often be identified through mentions of sponsors, slogans, etc. throughout their programming.

Amateur radio DX

Main article: Amateur radio

Amateur radio operators who specialize in making two way radio contact with other amateurs in distant countries are also referred to as "DXers". On the HF (also known as shortwave) amateur bands, DX stations are those in foreign countries. On the VHF/UHF amateur bands, DX stations can be within the same country or continent, since making a long-distance VHF contact, without the help of a satellite, can be very difficult. DXers collect QSL cards as proof of contact and can earn special certificates and awards from amateur radio organizations.[6]

In addition, many clubs offer awards for communicating with a certain number of DX stations. For example, the ARRL offers the DX Century Club award, or DXCC. The basic certificate is awarded for working and confirming at least 100 entities on the ARRL DXCC List. [7] For award purposes, other areas than just political countries can be classified as "DX countries". For example, the French protectorate of Reunion Island in the Indian Ocean is counted as a DX country, even though it is a region of France. The rules for determining what is a DX country can be quite complex and to avoid potential confusion, radio amateurs often use the term *entity* instead of country. In addition to entities, some awards are based on island groups in the world's oceans. On the VHF/UHF bands, many radio amateurs pursue awards based on Maidenhead grid locators.

In order to give other amateurs a chance to confirm contacts at new or exotic locations, amateurs have mounted DXpeditions to countries or regions that have no permanent base of amateur radio operators. [6] There are also frequent contests where radio amateurs operate their stations on certain dates for a fixed period of time to try to communicate with as many DX stations as possible.

QSL card from Voice of America

2.11.2   DX Clubs

Many radio enthusiasts are members of DX clubs. There are many DX clubs in many countries around the world. They are useful places to find information about up-to-date news relating to international radio. Many people also enjoy social events, which can form a large part of the enjoyment that people can get out of the radio hobby.

2.11.3   QSL cards

Main article: QSL card
 One of the interesting sides of DXing as a hobby is collecting QSL cards (acknowledgement cards from the broadcaster) confirming the listener's reception report (sometimes called SINPO report, see next section).

Usually a QSL card will have a picture on one side and the reception data on the other. Most of the broadcasters will use pictures and messages indicating their country's culture or technological life.

2.11.4   SINPO report

SINPO stands for the following qualities, graded on a scale of 1 to 5, where '1' means the quality was very bad and '5' very good.

S - Signal strength
I - Interference with other stations or broadcasters
N - Noise ratio in the received signal
P - Propagation (ups and downs of the reception)
O - Overall merit

Although this is a subjective measure, with practise the grading becomes more consistent, and a particular broadcast may be assessed by several listeners from the same area, in which case the broadcaster could assess correspondence between reports.

After listening to a broadcast, the listener writes a report with SINPO values, typically including his geographical location (called QTH in amateur radio terminology) in longitude and latitude, the types of receiver and antennae used, the frequency the transmission was heard on, a brief description of the programme listened to, their opinion about it, suggestions if any, and so on.

The listener can send the report to the broadcaster either by post or email, and request verification (QSL) from them.

Variants of this report are: a) the SIO report which omits the Noise and Propagation, b) grading on a scale of 1 to 3 (instead of 1 to 5) and c) the SINFO report where the F stands for fading.

2.11.5 DX Communication

DX communication is communication over great distances using the ionosphere to refract the transmitted radio beam. The beam returns to the Earth's surface, and may then be reflected back into the ionosphere for a second bounce. Ionospheric refraction is generally only feasible for frequencies below about 50 MHz, and is highly dependent upon atmospheric conditions, the time of day, and the eleven-year sunspot cycle. It is also affected by solar storms and some other solar events, which can alter the Earth's ionosphere by ejecting a shower of charged particles.

The angle of refraction places a minimum on the distance at which the refracted beam will first return to Earth. This distance increases with frequency. As a result, any station employing DX will be surrounded by an annular *dead zone* where they can't hear other stations or be heard by them.

This is the phenomenon that allows short wave radio reception to occur beyond the limits of line of sight. It is utilized by amateur radio enthusiasts (hams), shortwave broadcast stations (such as BBC and Voice of America) and others, and is what allows one to hear AM (MW) stations from areas far from their location. It is one of the backups to failure of long distance communication by satellites, when their operation is affected by electromagnetic storms from the sun.

For example, in clear ionosphere conditions, one can hear France Inter on 711 kHz, far into the UK and as far as Reading, Berkshire. It is also possible to hear Radio Australia from Melbourne as far away as Lansing, Michigan, a distance of some 9835 miles (15,827 kilometers).

2.11.6 DXing equipment

Radio equipment used in DXing ranges from inexpensive portable receivers to deluxe equipment costing thousands of dollars. Using just a simple AM radio, one can easily hear signals from the most powerful stations propagating hundreds of miles at night. Even an inexpensive shortwave radio can receive signals emanating from several countries during any time of day.

Serious hobbyists use more elaborate receivers designed specifically for pulling in distant signals, and often build their own antennas specifically designed for a specific frequency band. There is much discussion and debate in the hobby about the relative merits of lesser priced shortwave receivers vs. their multi-thousand dollar "big brother" radios. In general, a good desktop or "PC Radio" will be able to "hear" just about what a very expensive high-performance receiver can receive. The difference between the two types comes into play during difficult band or reception conditions. The expensive receiver will have more filtering options and usually better adjacent channel interference blocking, sometimes resulting in the difference of being able to receive or not receive a signal under poor conditions. Reception of international broadcasting seldom shows a noticeable difference between the two radios. Car radios are also used for DXing the broadcast bands.

Another recent trend is for the hobbyist to employ multiple radios and antennas connected to a personal computer. Through advanced radio control software, the radios can be automatically ganged together, so that tuning one radio can tune all the others in the group. This DXing technique is sometimes referred to as diversity reception and facilitates easy "A to B" comparison of different antennas and receivers for a given signal. For more details on "PC Radios" or computer controlled shortwave receivers see the discussion in Shortwave listening.

Having a minimum of two Dipole antenna at right angles to each other, for example, one running North-South and one running East-West can produce dramatically different reception patterns. These simple antennas can be made for a few dollars worth of wire and a couple of insulators.

2.11.7 See also

- 802.11 non-standard equipment

2.11.8 References

[1] http://www.dxing.info/introduction.dx *Introduction To DXing*, DXing.info

[2] Jerome S. Berg (30 October 2008). *Listening on the Short Waves, 1945 to Today*. McFarland. pp. 330–. ISBN 978-0-7864-3996-6. Retrieved 12 April 2012.

[3] Susan J. Douglas (25 February 2004). *Listening in: radio and the American imagination*. U of Minnesota Press. pp. 73–. ISBN 978-0-8166-4423-0. Retrieved 12 April 2012.

[4] http://www.dxing.com/amband.htm *AM Band DXing*, DXing.com

[5] http://www.dxing.info/introduction.dx *Introduction To DXing*. DXing.info

[6] Danny Gregory; Paul Sahre (1 April 2003). *Hello world: a life in ham radio.* Princeton Architectural Press. pp. 217–. ISBN 978-1-56898-281-6. Retrieved 4 April 2012.

[7] http://www.arrl.org/awards/dxcc/dxcclist.txt DXCC List - ARRL

2.11.9 External links

- ARRL - American Radio Relay league.

- DXing at DMOZ

- DX News Ham Radio

- The DXZone.com A web site dedicated to the DXing

2.12 Eastern Bloc media and propaganda

Eastern Bloc media and propaganda was controlled directly by each country's Communist party, which controlled the state media, censorship and propaganda organs. State and party ownership of print, television and radio media served as an important manner in which to control information and society in light of Eastern Bloc leaderships viewing even marginal groups of opposition intellectuals as a potential threat to the bases underlying Communist power therein.

Circumvention of dissemination controls occurred to some degree through samizdat and limited reception of western radio and television broadcasts. In addition, some regimes heavily restricted the flow of information from their countries to outside of the Eastern Bloc by heavily regulating the travel of foreigners and segregating approved travellers from the domestic population.

2.12.1 Background

Creation

Bolsheviks took power following the Russian Revolution of 1917. During the Russian Civil War that followed, coinciding with the Red Army's entry into Minsk in 1919, Belarus was declared the Socialist Soviet Republic of Byelorussia. After more conflict, the Byelorussian Soviet Socialist Republic was declared in 1920. With the defeat of the Ukraine in the Polish-Ukrainian War, after the March 1921 Peace of Riga following the Polish-Soviet War, central and eastern Ukraine were annexed into the Soviet Union as the Ukrainian Soviet Socialist Republic. In 1922, the Russian SFSR, Ukraine SSR, Byelorussian SSR and Transcaucasian

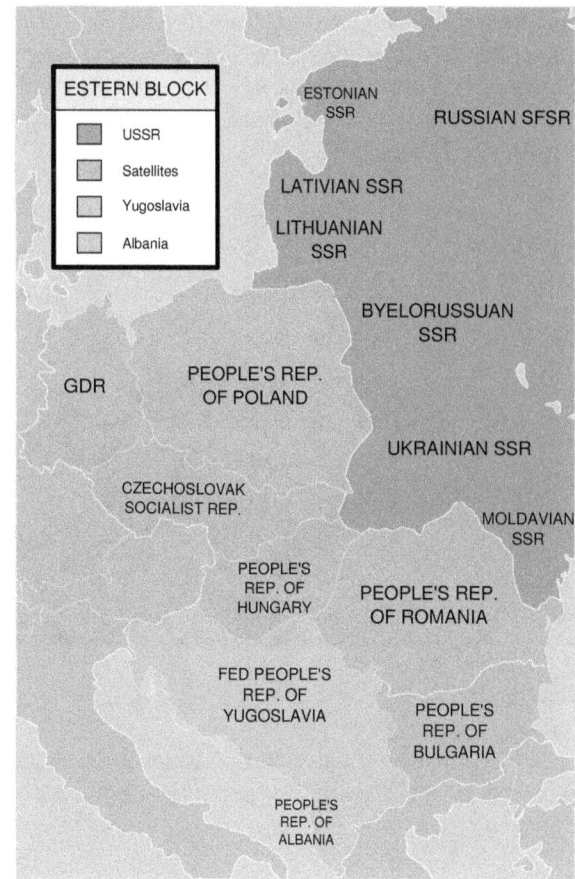

Map of the Eastern Bloc

SFSR were officially merged as republics creating the Union of Soviet Socialist Republics (Soviet Union).

At the end of World War II, all eastern and central European capitals were controlled by the Soviet Union.[1] During the final stages of the war, the Soviet Union began the creation of the Eastern Bloc by occupying several countries as Soviet Socialist Republics that were originally effectively ceded to it by Nazi Germany in the Molotov-Ribbentrop Pact. These included eastern Poland (incorporated into two different SSRs),[2] Latvia (became Latvia SSR),[3][4] Estonia (became Estonian SSR),[3][4] Lithuania (became Lithuania SSR),[3][4] part of eastern Finland (became Karelo-Finnish SSR)[5] and northeastern Romania (became the Moldavian SSR).[6][7]

By 1945, these additional annexed countries totaled approximately 180,000 further square miles, or slightly more than the area of West Germany, East Germany and Austria combined.[8] Other nations were converted into Soviet Satellite states, such as the People's Republic of Poland, the People's Republic of Hungary,[9] the Czechoslovak Socialist Republic,[10] the People's Republic of Romania, the People's Republic of Albania,[11] and later East Germany from the Soviet zone of German occupation.[12] The Federal People's Republic of Yugoslavia was also wrongfully considered part of the Bloc,[13][14] though a Tito-Stalin split occurred in 1948[15] followed by the formation of the Non-Aligned Movement.

Conditions

Berlin Wall 1975

Further information: Eastern Bloc, Eastern Bloc emigration and defection, Eastern Bloc politics and Eastern Bloc economies

Throughout the Eastern Bloc, both in the Soviet Socialist Republic and elsewhere, Russia was given prominence, and referred to as the *naibolee vıdayuşayasya naciya* (the most prominent nation) and the *rukovodyaşy narod* (the leading people).[8] The Soviets encouraged the admiration of everything Russian and the reproduction of their own Communist structural hierarchies in each of the Bloc states.[8]

The defining characteristic of Communism as implemented in the Eastern Bloc was the unique symbiosis of the state with society and the economy, resulting in politics and economics losing their distinctive features as autonomous and distinguishable spheres.[16] Initially, Stalin directed systems that rejected Western institutional characteristics of market economies, democratic governance (dubbed "bourgeois democracy" in Soviet parlance) and the rule of law subduing discretional intervention by the state.[17] The Soviets mandated expropriation and *etatization* of private property.[18]

The Soviet-style "replica regimes" that arose in the Bloc not only reproduced Soviet command economies, but also adopted the methods employed by Joseph Stalin and Soviet secret police to suppress real and potential opposition.[18] Communist regimes in the Eastern Bloc saw even marginal groups of opposition intellectuals as a potential threat because of the basis underlying Communist power.[19] The suppression of dissidence and opposition was a central prerequisite for the security of Communist power within the

Eastern Bloc, though the degree of opposition and dissident suppression varied by country and period.[19]

While over 15 million Eastern Bloc residents migrated westward from 1945 to 1949,[20] emigration was effectively halted in the early 1950s, with the Soviet approach to controlling national movement emulated by most of the rest of the Eastern Bloc.[21] Furthermore, the Eastern Bloc experienced economic mis-development by central planners resulting in those countries following a path of extensive rather than intensive development, and thus lagging far behind their western European counterparts in per capita Gross Domestic Product.[22]

2.12.2   Media and information restrictions

Trybuna Ludu *December 13, 1981 reports Martial law in Poland*

Media and information control

Further information: Deutscher Fernsehfunk, Rundfunk der DDR, Mass media in Communist Czechoslovakia and Soviet Information Bureau

In the Eastern Bloc, the state owned and operated the means of mass communication.[23] The ruling authorities viewed media as a propaganda tool, and widely practiced censorship to exercise almost full control over the information dissemination.[23] The press in Communist countries was an organ of, and completely reliant on, the state.[24] Until the late 1980s, all Eastern Bloc radio and television organizations were state-owned (and tightly controlled), while print media was usually owned by political organizations, mostly by the local Communist party.[25]

Youth newspapers and magazines were owned by youth organizations affiliated with the communist party.[25] The governing body in the Soviet Union was "USSR State Committee for Television and Radio Broadcasting", or USSR Gosteleradio (Государственный комитет по телевидению и радиовещанию СССР, Гостелерадио СССР), which was in charge both of Soviet TV and Radio in the Soviet Union.

The Communist party exercised control over the media and was responsible for censorship.[25] Media served as an important form of control over information and thus of society.[26] Eastern Bloc authorities viewed the dissemination and portrayal of knowledge as vital to the survival of Communism and thus stifled alternative concepts and critiques.[26] Several state Communist Party newspapers were published. Radio was initially the dominant medium, with television being considered low on the priority list when compiling Five-year plans during the industrialisation of the 1950s.

Censorship and squashing of dissent

Nikolai Yezhov, the young man strolling with Joseph Stalin to his right, was shot in 1940. He was edited out from a photo by Soviet censors.[27] Such retouching was a common occurrence during Stalin's reign.
Further information: Censorship in East Germany, Censorship in the People's Republic of Poland, Censorship in the Soviet Union, Censorship of images in the Soviet Union and Anti-Soviet agitation

Strict censorship existed in the Eastern Bloc, though it was at times circumvented by those engaging in samizdat.[28] Censorship institutions in the countries of the Bloc were organized differently.[23] For example, censorship in Poland was clearly identified whereas it was loosely structured, but no less efficient, in Hungary.[23] Strict censorship was introduced in the People's Republic of Albania and Federal People's Republic of Yugoslavia as early as 1944, though it was somewhat relaxed in Yugoslavia after the Tito-Stalin split of 1948.[29] Unlike the rest of the Eastern Bloc, relative freedom existed for three years in Czechoslovakia until Soviet-style censorship was fully applied in 1948,[29] along with the Czechoslovak coup d'état of 1948.

Throughout the Bloc, the various ministries of culture held

a tight rein on writers.[30] Cultural products reflected the propaganda needs of the state[30] and Party-approved censors exercised strict control in the early years.[31] During the Stalinist period, even the weather forecasts were changed if they would have otherwise suggested that the sun might not shine on May Day.[31] Under Nicolae Ceauşescu in Romania, weather reports were doctored so that the temperatures were not seen to rise above or fall below the levels which dictated that work must stop.[31]

In each country, leading bodies of the ruling Communist Part exercised hierarchical control of the censorship system.[29] Each Communist Party maintained a department of its Central Committee apparatus to supervise media.[29] Censors employed auxiliary tools such as: the power to launch or close down any newspaper, radio or television station, licensing of journalists through unions and the power of appointment.[29] Party bureaucrats held all leading editorial positions.[29] One or two representatives of censorship agencies modeled on the Soviet GLAVLIT (Main Administration for the Protection of Official and Military Secrets) worked directly in all editorial offices.[29] No story could be printed or broadcast without their explicit approval.[29]

Initially, East Germany presented unique issues because of rules for the occupying powers in the divided Germany (e.g. regarding media control) that prevented the outright seizure of all media outlets.[32] The Soviet occupation administration (SVAG) directed propaganda and censorship policies to East German censorship organs through its "sector for propaganda and censorship".[33] While the initial SVAG policies did not appear to differ greatly from those in the western occupation zones governing denazification,[34] censorship became one of the most overt instruments used to manipulated political, intellectual and cultural developments in East Germany.[33] Art societies and associations that had existed prior to World War II were dissolved and all new theatres and art societies had to register with SVAG.[35] Art exhibits were put under a blanket ban unless censorship organs approved them in advance.[35]

After East Germany's official establishment, while the original constitution[36] provided that "censorship of the media is not to occur", both official and unofficial censorship occurred, although to a lesser extent during its later years. Thereafter, official East German censorship was supervised and carried out by two governmental organizations, the *Head office for publishing companies and bookselling trade* (Hauptverwaltung Verlage und Buchhandel, HV), and the *Bureau for Copyright* (Büro für Urheberrechte). The HV determined the degree of censorship and the method of publishing and marketing works. The Bureau for Copyright appraised the work, then decided if it or another publication was permitted to be published in East Germany or in foreign country. For theatres, a "repertory commission" was created that consisted of the *Ministerium für Volksbildung* (MfV), the ruling SED party, the applicable theatre union and the East German office for theatrical affairs.[37]

After a long visa procurement process, western visitors driving over the West German border to East Germany had their car strip-searched for prohibited Western "propaganda material."[38] Nevertheless, the East German authorities found it extremely difficult to prevent their citizens listening to Western radio stations and Western TV was available across most of the GDR. Technical and diplomatic considerations meant attempts at jamming Western Stations were (unlike in other Eastern bloc countries) soon abandoned.

In the Soviet Union, in accordance with the official ideology and politics of the Communist Party, Goskomizdat censored all printed matter, Goskino supervised all cinema, Gosteleradio controlled radio and television broadcasting and the First Department in many agencies and institutions, such as the State Statistical Committee (Goskomstat), was responsible for assuring that state secrets and other sensitive information only reached authorized hands. The Soviets destroyed pre-revolutionary and foreign material from libraries, leaving only "special collections" (*spetskhran*), accessible by special permit from the KGB. The Soviet Union also censored images, included removing repressed persons from texts, posters, paintings and photographs.

Prominent individuals   Throughout the Eastern Bloc, artists or those attempting to disseminate dissenting views were repressed, with a few of the more prominent victims including:

- Gheorghe Ursu - a Romanian poet who grew disillusioned with Romanian communist doctrine after 1949,[39] and was repeatedly sanctioned for disobedience[40][41] In 1985, after being beaten for weeks on end by the Romanian Police, he was transported to the Jilava jail hospital, where he died of peritonitis later in the day.[42][43][44][45][46]

- Night of the Murdered Poets - thirteen writers, poets, artists, musicians and actors were secretly executed on orders from Joseph Stalin.

- Imre Nagy - former Hungarian Prime Minister who had supported Hungary's withdrawal from the Warsaw Pact during the Hungarian Revolution of 1956, was later arrested by Soviet authorities after leaving the Yugoslavian embassy, and then secretly tried, found guilty, sentenced to death and was executed by hanging in June, 1958.[47] His trial and execution were made public only after the sentence was carried out.[48]

- Ion Valentin Anestin - His work centered on denouncing Stalin and the Soviet Union, in a series titled *Măcelarul din Piața Roșie* ("The Red Square Butcher") published by the magazine *Gluma*.[49][50][51] Following the start of Soviet occupation of Romania, Anestin was barred from publishing for a five-year period (1944–1949), and ultimately imprisoned.[49][50][51] He died soon after his release.[49][50]

- Nikolai Getman - a Ukrainian artist arrested in 1946 for possessing a caricature of Joseph Stalin his friend had drawn on a cigarette box, Getman was sent to Siberian Gulag camps. He is one of the few artists to record life in the Gulag, where he survived by sketching propaganda for the authorities.

- Vasyl Stus - a Ukrainian author and journalist who wrote a book that was rejected for discrepancies with Soviet ideology, was arrested in 1972, spent five years in prison, arrested again in 1980 for defending members of the Ukrainian Helsinki group, was sentenced to ten years more imprisonment and was subsequently beaten to death in a Soviet forced labor camp.

- Vasile Voiculescu - A Romanian poet who was imprisoned in 1958, at the age of 74, spending four years in prison, where he became ill, and died of cancer a few months after his release.

- Arno Esch - An East German political writer who was imprisoned by the Soviet NKVD in 1949, sentenced to death for "counterrevolutionary activities" and executed at the Lubyanka (KGB) prison in 1951.

- Lena Constante - During repeated interrogations by the Securitate, Constante tried to fend off false accusations of "Titoism" and "treason", but, the victim of constant beatings and torture[52] (much of her hair was torn from the roots),[53] and confronted with Zilber's testimony — which implicated her —, she eventually gave in and admitted to the charges.[54]

- József Dudás - A Hungarian political activist who spoke of a 25-point program ending Soviet repression in Hungary to a crowd during the Hungarian Revolution of 1956, and was executed the next year.

- Enn Tarto - An Estonian dissident who was imprisoned from 1956 to 1960, 1962 to 1967, and again from 1983 to 1988 for anti-Soviet activity.

- Anton Durcovici - A Romania clergyman openly critical of the Communist regime, Durcovici was placed under surveillance in 1947, arrested by the Securitate in 1949 during a congregation visit, died from torture and prison deprivation and was buried in an unmarked grave.[55][56] Communist authorities subsequently attempted to erase all evidence of his stay in prison, and most documents were destroyed.[55]

- Valeriy Marchenko - a Ukrainian poet who was arrested in 1973 and charged with Anti-Soviet agitation and propaganda, jailed for six years with two years exile, jailed again in 1983 for violating Article 62 of the Soviet penal code, Anti-Soviet Agitation and Propaganda and sentenced to ten years imprisonment and five years of exile, after which he became ill, was moved to a hospital after international pressure, where he died.

- Jüri Jaakson - An Estonian businessman and former politician critical of Soviet rule who was executed by the Soviet Union in 1941.

- Mečislovas Reinys - A Lithuanian archbishop critical of Bolshveism who was arrested in 1947 and sentenced to eight years in a Soviet prison, where he died in 1953.

- Metropolitan Ioann (Vasiliy Bodnarchuk) - A Ukrainian arrested in 1949 for purported Ukrainian nationalist rhetoric and sentenced to 20 years of hard labor in copper mines.

Other artists, such as Geo Bogza, used subtle imagery or allegories within their works to criticize regimes. This did not prevent state scrutiny, as with the case of Bogza coming under the scrutiny of the Securitate.[57]

Media entities

Further information: Printed media in the Soviet Union, Television in the Soviet Union, Radio in the Soviet Union and Internet in the Soviet Union

[23]

The major newspapers were traditionally the daily official publications of the Communist Party.[23] Newspapers served as the main party organs of record and provided official political roadmaps for officials and other readers who needed to be informed.[58] In some countries, the press provided a significant source of income for the ruling Communist parties.[58] Radio and television was controlled by the state.[59] The Telegraph Agency of the Soviet Union (TASS) was the central agency for collection and distribution of internal and international news for all Soviet newspapers, radio and television stations. TASS monopolized the supply of political news.[29] It was frequently infiltrated by Soviet intelligence and security agencies, such as the NKVD and GRU. TASS had affiliates in 14 Soviet republics, including the Lithuanian SSR, Latvian SSR, Estonian SSR, Moldavian SSR. Ukrainian SSR and Byelorussian SSR.

Despite outward similarities in press policy, large differences existed in the roles and functions of the mass media in Eastern Bloc countries.[60] Where the press was allowed more freedom, such as in Poland, Hungary, and Yugoslavia, a national subtext and a significant element of entertainment flourished.[60] In some cases, newspapers and magazines served as the most visible part of liberalizing forces, such as in Poland in 1956 and 1980–81, in Hungary in 1956, and in Czechoslovakia in 1968.[60]

In many instances toward the end of the Eastern Bloc's existence, the ruling Communist parties' messages in the press increasingly diverged from reality, which contributed to the declining faith of the public in Communist rule.[60] At the same time, some press in the Eastern Bloc became more open in the 1980s in countries such as Poland, Hungary and Czechoslovakia.[60] In Yugoslavia, the press after Tito's reign turned increasingly nationalistic.[60] Only in Romania and Albania did the press remain under tight dictatorial control right up until the end of the Eastern Bloc.[60]

In East Germany, where initial control could be less overt because of shared allied occupation rules, the Soviet SVG set up the Deutsche Verwaltung für Volksbildung (DVV) in the fall of 1945.[33] The SVAG and DVV controlled and approved all publication licenses needed to publish newspapers, books, journals and other materials.[61] Those agencies also provided the top publishing priorities and would apportion paper used for printing to the various publications in accordance with those priorities.[61] The SVAG initially licensed some private publishers which required the employment of a greater number of censors.[62]

2.12.3 Control of information flow out of the Eastern Bloc

Beginning in 1935, Joseph Stalin effectively sealed off outside access to the Soviet Socialist Republics (and until his death), effectively permitting no foreign travel inside the Soviet Union such that outsiders did not know of the political processes that had taken place therein.[63] During this period, and even for 25 years after Stalin's death, the few diplomats and foreign correspondents that were permitted inside the Soviet Union were usually restricted to within a few miles of Moscow, their phones were tapped, their residences were restricted to foreigner-only locations and they were constantly followed by Soviet authorities.[63] Dissenters who approached such foreigners were arrested.[64] For many years after World War II, even the best informed foreigners did not know the number of arrest or executed Soviet citizens, or how poorly the Soviet economy had performed.[64]

Similarly, the regimes in Romania carefully controlled foreign visitors in order to restrict the flow of information coming out of (and into) Romania.[65] Accordingly, activities in Romania remained, until the late 1960s, largely unknown to the outside world.[65] As a result, until 1990, very little information regarding labour camps and prisons in Romania appeared in the West.[65] When such information appeared, it was usually in Romanian émigré publications.[65] Romania's Securitate secret police were able to suppress information leaking to the west about resistance to the regime.[66] Stalinist Albania, which had become increasingly paranoid and isolated after de-Stalinization and the death of Mao Zedong,[67] restricted visitors to 6,000 per year, and segregated those few that traveled to Albania.[68]

2.12.4 Propaganda efforts

Further information: Communist propaganda, Propaganda in the People's Republic of Poland and Propaganda in the Soviet Union

Communist leaders in the Eastern Bloc openly discussed the existence of propaganda efforts. Communist propaganda goals and techniques were tuned according to the target audience. The most broad classification of targets was:[69]

- Domestic propaganda

- External propaganda

- Propaganda of Communist supporters outside the Communist states

Communist Party documents reveal a more detailed classification of specific targets (workers, peasants, youth, women, etc.).[69]

Because the Communist Party was portrayed under Marxist-Leninist theory as the protagonist of history pushing toward the inevitable end result of historical materialism as a "vanguard of the working class", Party leaders were claimed to be as infallible and inevitable as the purported historical end itself.[70] Propaganda often worked itself beyond agit prop plays into traditional productions, such as in Hungary after the Tito-Stalin split, where the director of the National Theatre produced a version of *Macbeth* in which the villainous king was revealed as none other than hated (in the Eastern Bloc) Yugoslavian Leader Josip Broz Tito.[71] Regarding economic woes, debilitating wage cuts following economic stagnation were referred to as "blows in the face of imperialism", while forced loans were called "voluntary contributions to the building of socialism".[72]

Communist theoretician Nikolai Bukharin in his *The ABC of Communism* wrote:[73]

The State propaganda of communism becomes in the long run a means for the eradication of the last traces of bourgeois propaganda dating from the old régime; and it is a powerful instrument for the creation of a new ideology, of new modes of thought, of a new outlook on the world.

Penetration of West German TV reception (grey) in East Germany for ARD (regional channels NDR, HR, BR and SFB). Areas with no reception (black) were jokingly referred to as "Valley of the Clueless" (Tal der Ahnungslosen) while ARD was said to stand for "Außer (except) Rügen und Dresden".

Some propaganda would "retell" the western news, such as the East German television program Der schwarze Kanal ("The Black Channel"), which contained bowdlerized programs from West Germany with added Communist commentary.[74] The name "Black channel" was a play on words deriving from the term German plumbers used for a sewer. The program was meant to counter ideas received by some from West German television because the geography of the divided Germany meant that West German television signals (particularly ARD) could be received in most of East Germany, except in parts of Eastern Saxony around Dresden, which consequently earned the latter the nickname "valley of the clueless"[75] (despite the fact that some Western radio was still available there).

Eastern Bloc leaders, including even Joseph Stalin, could become personally involved in dissemination. For example, in January 1948, the U.S. State Department published a collection of documents titled *Nazi-Soviet Relations, 1939–*

1941: Documents from the Archives of The German Foreign Office, which contained documents recovered from the Foreign Office of Nazi Germany[76][77] revealing Soviet conversations with Germany regarding the Molotov-Ribbentrop Pact, including its secret protocol dividing eastern Europe,[78][79] the 1939 German-Soviet Commercial Agreement,[78][80] and discussions of the Soviet Union potentially becoming the fourth Axis Power.[81]

In response, one month later, the Soviet Information Bureau published *Falsifiers of History*.[76][82] Stalin personally edited the book, rewriting entire chapters by hand.[82] The book claimed, for instance, that American bankers and industrialists provided capital for the growth of German war industries, while deliberately encouraging Hitler to expand eastward.[76][78] The book also included the claim that, during the Pact's operation, Stalin rejected Hitler's offer to share in a division of the world, without mentioning the Soviet offers to join the Axis.[83] Historical studies, official accounts, memoirs and textbooks published in the Soviet Union used that depiction of events until the Soviet Union's dissolution.[83]

The book referred to "the American falsifiers and their British and French associates",[84] claimed "[a]s far back as in 1937 it became perfectly clear that a big war was being hatched by Hitler with the direct connivance of Great Britain and France",[85] blasted "the claptrap of the slanderers"[86] and stated "[n]aturally, the falsifiers of history and slanderers are called falsifiers and slanderers precisely because they do not entertain any respect for facts. They prefer to gossip and slander."[87]

In East Germany, the Soviet SVAG and DVV initially controlled all publication priorities.[61] In the initial months of 1946, the Soviets were unsure how to merge propaganda and censorship efforts in East Germany.[61] The SVAG engaged in a broad propaganda campaign that moved beyond customary political propaganda to engage in the practice at unions, women's organizations and youth organizations.[61]

2.12.5 Bypassing censorship

Clandestine information passing

Main articles: Samizdat, Magnitizdat and Polish underground press

Samizdat was the clandestine copying and distribution of government-suppressed literature or other media in Eastern bloc countries. Copies were often made in small quantities of handwritten or typed documents, while recipients were expected to make additional copies. Samizdat traders used underground literature for self-analysis and self-expression under the heavy censorship of the Eastern Bloc.[88] The

practice was fraught with danger as harsh punishments were meted out to people caught possessing or copying censored materials. Former Soviet dissident Vladimir Bukovsky defined it as follows: "I myself create it, edit it, censor it, publish it, distribute it, and [may] get imprisoned for it."[89] One of the longest-running and well-known samizdat publications was the information bulletin "Хроника текущих событий" (Khronika Tekushchikh Sobitiy; *Chronicle of Current Events*),[90] which contained anonymously published pieces dedicated to the defense of human rights in the USSR. Several people were arrested in connection with the *Chronicle*, including Natalya Gorbanevskaya, Yuri Shikhanovich, Pyotr Yakir, Victor Krasin, Sergei Kovalev, Alexander Lavut, Tatyana Velikanova, among others.

Magnitizdat (in Russian магнитиздат) is the process of recopying and self distributing live audio tape recordings in the Soviet Union that were not available commercially. The process of magnitizdat was less risky than publishing literature via samizdat, since any person in the USSR was permitted to own a private reel-to-reel tape recorder, while paper duplication equipment was under control of the state. "Tamizdat" refers to literature published abroad (там, tam, meaning "there"), often from smuggled manuscripts.

Western role in propaganda war

Further information: British Broadcasting Corporation, Voice of America and Radio Free Europe

Western countries invested heavily in powerful transmitters which enabled broadcasters to be heard in the Eastern Bloc, despite attempts by authorities to jam such signals. In 1947, VOA started broadcasting in Russian with the intent to counter Soviet propaganda directed against American leaders and policies, and disseminate pro western propaganda directed against Soviet leaders and policies.[91] These included Radio Free Europe (RFE)), RIAS (Berlin) the Voice of America (VOA), Deutsche Welle, Radio France International and the British Broadcasting Corporation (BBC).[59] The Soviet Union responded by attempting aggressive, electronic jamming of VOA (and some other Western) broadcasts in 1949.[91] The BBC World Service similarly broadcast language-specific programming to countries behind the Iron Curtain.

RFE was developed out of a belief that the Cold War would eventually be fought by political rather than military means.[92] In January 1950, it obtained a transmitter base at Lampertheim, West Germany and on July 4 of the same year, RFE completed its first broadcast aimed at Czechoslovakia[93] Broadcasts were often banned in Eastern Europe and Communist authorities used sophisticated jamming techniques in an attempt to prevent citizens from listening to them.[94] In late 1950, RFE began to assemble a full-fledged foreign broadcast staff, and became more than just a "mouthpiece for exiles" who had fled Eastern Bloc countries.[95] While RFE was cleared of charges that it gave Hungarian listeners false hope during the Hungarian Revolution of 1956, its Broadcast Analysis Division was established to ensure that broadcasts were accurate and professional while maintaining the journalists' former autonomy.[96]

A 1960 study concluded that RFE possessed considerably more listeners than the BBC or VOA.[97] The study concluded that the BBC was regarded as the most objective and the VOA had suffered a notable decline since it stopped critical broadcasts on the communist world after the Hungarian Revolution of 1956, focusing instead on world news, American culture and jazz.[97]

2.12.6 Notes

[1] Wettig 2008, p. 69

[2] Roberts 2006, p. 43

[3] Wettig 2008, p. 21

[4] Senn, Alfred Erich, *Lithuania 1940 : revolution from above*, Amsterdam, New York, Rodopi, 2007 ISBN 978-90-420-2225-6

[5] Kennedy-Pipe, Caroline, *Stalin's Cold War*, New York: Manchester University Press, 1995, ISBN 0-7190-4201-1

[6] Roberts 2006, p. 55

[7] Shirer 1990, p. 794

[8] Graubard 1991, p. 150

[9] Granville, Johanna, *The First Domino: International Decision Making during the Hungarian Crisis of 1956*, Texas A&M University Press, 2004. ISBN 1-58544-298-4

[10] Grenville 2005, pp. 370–71

[11] Cook 2001, p. 17

[12] Wettig 2008, pp. 96–100

[13] Crampton 1997, pp. 216–7

[14] *Eastern bloc*, *The American Heritage New Dictionary of Cultural Literacy*, Third Edition. Houghton Mifflin Company, 2005.

[15] Wettig 2008, p. 156

[16] Hardt & Kaufman 1995, p. 11

[17] Hardt & Kaufman 1995, p. 12

[18] Roht-Arriaza 1995, p. 83

[19] Pollack & Wielgohs 2004, p. xiv

[20] Böcker 1998, pp. 207–9

[21] Dowty 1989, p. 114

[22] Hardt & Kaufman 1995, pp. 15–17

[23] Frucht 2003, p. 489

[24] O'Neil 1997, p. 15

[25] O'Neil 1997, p. 125

[26] O'Neil 1997, p. 1

[27] The Commissar vanishes (The Newseum)

[28] Major & Mitter 2004, p. 6

[29] Frucht 2003, p. 127

[30] Major & Mitter 2004, p. 15

[31] Crampton 1997, p. 247

[32] Pike 1997, p. 219

[33] Pike 1997, pp. 217–8

[34] Pike 1997, pp. 220–1

[35] Pike 1997, pp. 231–2

[36] October 7, 1949 Constitution of East Germany, German Document Archive (German)

[37] Pike 1997, p. 236

[38] Philipsen 1993, p. 9

[39] Ursu, Andrei. "Despărţire de Iordan Chimet". Revista 22, *Nr. 849, June 2006* (in Romanian). Archived from the original on 16 October 2007.

[40] (French) Gabriela Blebea Nicolae, "Les défis de l'identité: Étude sur la problématique de l'identité dans la période post-communiste en Roumanie", in *Ethnologies*, Vol. 25, Nr. 1/2003 (hosted by Érudit.org); retrieved November 19, 2007

[41] Goşu, Armand (July 2006). "Cazul Gheorghe Ursu. SRI a ascuns crimele Securităţii". *Revista 22, Nr. 852,* (in Romanian). Archived from the original on 2006-07-18.

[42] (Romanian) Mirela Corlăţan, "Istorii. 'Notele' către Securitate ale disidentului Gheorghe Ursu", in *Cotidianul*, July 10, 2007

[43] Deletant 1995, p. 331

[44] Olaru, p.41-42

[45] (Romanian) George Tărâţă, "Torţionarul Stanică rămâne liber", in *Ziua*, November 9, 2006

[46] Vladimir Tismăneanu, *Fantasies of Salvation: Democracy, Nationalism, and Myth in Post-Communist Europe*, Princeton University Press, Princeton, 1998, p.138. ISBN 0-691-04826-6

[47] Richard Solash, "Hungary: U.S. President To Honor 1956 Uprising", Radio Free Europe, June 20, 2006

[48] *The Counter-revolutionary Conspiracy of Imre Nagy and his Accomplices* White Book, published by the Information Bureau of the Council of Ministers of the Hungarian People's Republic (No date).

[49] (Romanian) Raluca Alexandrescu, "Mai multe începuturi de drum" ("Several Road Starts"), interview with Ioana Zlotescu, in *Observator Cultural*

[50] (Romanian) Nicole Sima, "Cred în Moş Crăciun!" ("I Believe in Santa Claus!"), memoir hosted by LiterNet

[51] Vida, Mariana (August 4, 2006). "Un caricaturist uitat: Ion Valentin Anestin ("A Forgotten Caricaturist: Ion Valentin Anestin"". *Ziarul Financiar* (in Romanian). Archived from the original on 2007-09-27.

[52] (Romanian) Ruxandra Cesereanu, "Reprezentanţii represiunii: anchetatorul rafinat, torţionarul sadic şi bufonul balcanizat" ("The Representatives of Repression: The Refined Inquirer, the Sadistic Torturer and the Balkanized Buffoon"), at Memoria.ro

[53] (Romanian) Sanda Golpenţia, "Introducere la *Ultima carte* de Anton Golpenţia (Anchetatorii)" ("Introduction to Anton Golpenţia's *Ultima carte* (The Inquisitors)"), at *Memoria.ro*

[54] Tismăneanu, Vladimir "Memorie...", *Stalinism...*, p.294

[55] (Italian) Antonio Borrelli, *Servo di Dio Anton Durcovici. Vescovo e martire*

[56] (Romanian) *Procesul Comunismului. Episcopii Romano-Catolici*

[57] Chiorean, Claudia Talaşman, "Promovarea mitului Erei Noi în perioada 1989–2000 prin *România Literară*", p.138-139

[58] Frucht 2003, p. 639

[59] Frucht 2003, p. 490

[60] Frucht 2003, p. 640

[61] Pike 1997, pp. 225–6

[62] Pike 1997, pp. 227–8

[63] Laqueur 1994, p. 22

[64] Laqueur 1994, p. 23

[65] Deletant 1995, p. ix

[66] Deletant 1995, p. xiv

[67] Olsen 2000, p. 19

[68] Turnock 1997, p. 48

[69] John C. Clews (1964) *Communist Propaganda techniques*, printed in the USA by *Praeger* and in Great Britain

[70] David Satter. *Age of Delirium: The Decline and Fall of the Soviet Union*, Yale University Press, 2001, ISBN 0-300-08705-5

[71] Crampton 1997, p. 269

[72] Crampton 1997, p. 273

[73] Nikolai Bukharin, Yevgeny Preobrazhensky *The ABC of Communism* (1969 translation: ISBN 0-14-040005-2), Chapter 10: Communism and Education

[74] Hancock, Dafydd (2001-01-01). "Fade to black". Intertel from Transdiffusion. Retrieved 2006-02-20.

[75] Mitchener, Brandon (1994-11-09). "East Germany Struggles, 5 Years After Wall Fell". International Herald Tribune. Retrieved 2007-05-12.

[76] Henig 2005, p. 67

[77] Department of State 1948, p. preface

[78] Roberts 2002, p. 97

[79] Department of State 1948, p. 78

[80] Department of State 1948, pp. 32–77

[81] Churchill 1953, pp. 512–524

[82] Roberts 2002, p. 96

[83] Nekrich, Ulam & Freeze 1997, pp. 202–205

[84] Soviet Information Bureau 1948, p. 9

[85] Soviet Information Bureau 1948, p. 19

[86] Soviet Information Bureau 1948, p. 45

[87] Soviet Information Bureau 1948, p. 65

[88] (Russian) *History of Dissident Movement in the USSR. The birth of Samizdat* by Ludmila Alekseyeva. Vilnius, 1992

[89] (Russian) "Самиздат: сам сочиняю, сам редактирую, сам цензурирую, сам издаю, сам распространяю, сам и отсиживаю за него." (autobiographical novel *И возвращается ветер...*, *And the Wind returns...* NY, Хроника, 1978, p.126) Also online at

[90] (Russian) Chronicle of Current Events Archive at memo.ru

[91] *Cold War Propaganda* by John B. Whitton, The American Journal of International Law, Vol. 45, No. 1 (Jan., 1951), pp. 151–153

[92] Puddington 2003, p. 7

[93] Mickelson, Sig, "America's Other Voice: the Story of Radio Free Europe and Radio Liberty" (New York: Praeger Publishers, 1983): Mickelson 30.

[94] Puddington 2003, p. 214

[95] Puddington 2003, p. 37

[96] Puddington 2003, p. 117

[97] Puddington 2003, p. 131

2.12.7 References

- Böcker, Anita (1998), *Regulation of Migration: International Experiences*, Het Spinhuis, ISBN 90-5589-095-2

- Churchill, Winston (1953), *The Second World War*, Houghton Mifflin Harcourt, ISBN 0-395-41056-8

- Cook, Bernard A. (2001), *Europe Since 1945: An Encyclopedia*, Taylor & Francis, ISBN 0-8153-4057-5

- Crampton, R. J. (1997), *Eastern Europe in the Twentieth Century and After*, Routledge, ISBN 0-415-16422-2

- Department of State (1948), *Nazi-Soviet Relations, 1939–1941: Documents from the Archives of The German Foreign Office*, Department of State

- Deletant, Dennis (1995), *Ceauşescu and the Securitate: coercion and dissent in Romania, 1965–1989*, M.E. Sharpe, ISBN 1-56324-633-3

- Frucht, Richard C. (2003), *Encyclopedia of Eastern Europe: From the Congress of Vienna to the Fall of Communism*, Taylor & Francis Group, ISBN 0-203-80109-1

- Graubard, Stephen R. (1991), *Eastern Europe, Central Europe, Europe*, Westview Press, ISBN 0-8133-1189-6

- Grenville, John Ashley Soames (2005), *A History of the World from the 20th to the 21st Century*, Routledge, ISBN 0-415-28954-8

- Hardt, John Pearce; Kaufman, Richard F. (1995), *East-Central European Economies in Transition*, M.E. Sharpe, ISBN 1-56324-612-0

- Henig, Ruth Beatrice (2005), *The Origins of the Second World War, 1933-41*, Routledge, ISBN 0-415-33262-1

- Krasnov, Vladislav (1985), *Soviet Defectors: The KGB Wanted List*, Hoover Press, ISBN 0-8179-8231-0

- Laqueur, Walter (1994), *The dream that failed: reflections on the Soviet Union*, Oxford University Press, ISBN 0-19-510282-7

- Lipschitz, Leslie; McDonald, Donogh (1990), *German unification: economic issues*, International Monetary Fund, ISBN 1-55775-200-1

- Loescher, Gil (2001), *The UNHCR and World Politics: A Perilous Path*, Oxford University Press, ISBN 0-19-829716-5

- Major, Patrick; Mitter, Rana (2004), "East is East and West is West?", in Major, Patrick, *Across the Blocs: Exploring Comparative Cold War Cultural and Social History*, Taylor & Francis, Inc., ISBN 978-0-7146-8464-2

- Miller, Roger Gene (2000), *To Save a City: The Berlin Airlift, 1948–1949*, Texas A&M University Press, ISBN 0-89096-967-1

- Mynz, Rainer (1995), *Where Did They All Come From? Typology and Georgraphy of European Mass Migration In the Twentieth Century; EUROPEAN POPULATION CONFERENCE CONGRES EUROPEEN DE DEMOGRAPHE*, United Nations Population Division

- Nekrich, Aleksandr Moiseevich; Ulam, Adam Bruno; Freeze, Gregory L. (1997), *Pariahs, Partners, Predators: German–Soviet Relations, 1922–1941*, Columbia University Press, ISBN 0-231-10676-9

- O'Neil, Patrick (1997), *Post-communism and the Media in Eastern Europe*, Routledge, ISBN 0-7146-4765-9

- Olsen, Neil (2000), *Albania*, Oxfam, ISBN 0-85598-432-5

- Pike, David (1997), "Censortship in Soviet-Occupied Germany", in Gibianskii, Leonid; Naimark, Norman, *The Establishment of Communist Regimes in Eastern Europe, 1944–1949*, Westview Press, ISBN 0-8133-3534-5

- Philipsen, Dirk (1993), *We were the people: voices from East Germany's revolutionary autumn of 1989*, Duke University Press, ISBN 0-8223-1294-8

- Puddington, Arch (2003), *Broadcasting Freedom: The Cold War Triumph of Radio Free Europe and Radio Liberty*, University Press of Kentucky, ISBN 0-8131-9045-2

- Pollack, Detlef; Wielgohs, Jan (2004), *Dissent and Opposition in Communist Eastern Europe: Origins of Civil Society and Democratic Transition*, Ashgate Publishing, Ltd., ISBN 0-7546-3790-5

- Puddington, Arch (2003), *Broadcasting Freedom: The Cold War Triumph of Radio Free Europe and Radio Liberty*, University Press of Kentucky, ISBN 0-8131-9045-2

- Roberts, Geoffrey (2006), *Stalin's Wars: From World War to Cold War, 1939–1953*, Yale University Press, ISBN 0-300-11204-1

- Roberts, Geoffrey (2002), *Stalin, the Pact with Nazi Germany, and the Origins of Postwar Soviet Diplomatic Historiography* **4** (4)

- Roht-Arriaza, Naomi (1995), *Impunity and human rights in international law and practice*, Oxford University Press, ISBN 0-19-508136-6

- Soviet Information Bureau (1948), *Falsifiers of History (Historical Survey)*, Moscow: Foreign Languages Publishing House, 272848

- Turnock, David (1997), *The East European economy in context: communism and transition*, Routledge, ISBN 0-415-08626-4

- Wegner, Bernd (1997), *From Peace to War: Germany, Soviet Russia, and the World, 1939–1941*, Berghahn Books, ISBN 1-57181-882-0

- Weinberg, Gerhard L. (1995), *A World at Arms: A Global History of World War II*, Cambridge University Press, ISBN 0-521-55879-4

- Wettig, Gerhard (2008), *Stalin and the Cold War in Europe*, Rowman & Littlefield, ISBN 0-7425-5542-9

2.12.8   External links

- "Research on the History of Television Programs of the GDR". Archived from the original on 8 December 2008.

- (English) Library of Congress—The U.S. Naval Academy Collection of Soviet & Russian TV

- (Russian) Russian Museum of Radio and TV website

- RFE Czechoslovak Unit Open Society Archives, Budapest

- Translations of propaganda materials from the GDR.

- Advice for East German propagandists on how to deal with the Solidarity movement

- "CNN Cold War Knowledge Bank". Archived from the original on 9 December 2008. - comparison of articles on Cold War topics in *TIME Magazine* and *Pravda* between 1945 and 1991

- Censorship in the Soviet Union and its Cultural and Professional Results for Arts and Art Libraries

- Radio Berlin International final English broadcast - Part 1

- Radio Berlin International final English broadcast - Part 2

- GDR Censporship regarding Literature

- "Literaturzensur in der DDR" (in German). Archived from the original on 9 February 2007.

2.13   Electromagnetic interference

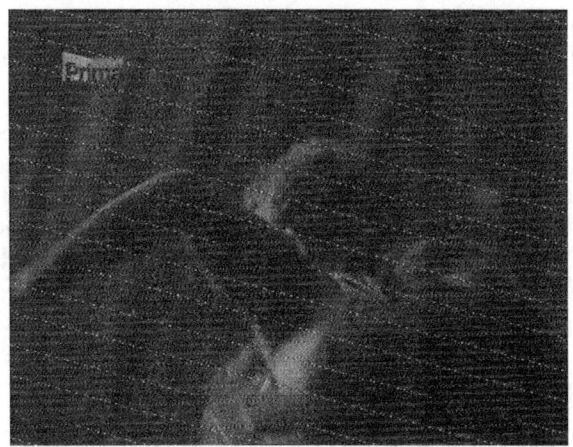

Electromagnetic interference in analog TV signal

Electromagnetic interference (**EMI**), also called **radio-frequency interference** (**RFI**) when in the radio frequency spectrum, is a disturbance generated by an external source that affects an electrical circuit by electromagnetic induction, electrostatic coupling, or conduction.[1] The disturbance may degrade the performance of the circuit or even stop it from functioning. In the case of a data path, these effects can range from an increase in error rate to a total loss of the data.[2] Both man-made and natural sources generate changing electrical currents and voltages that can cause EMI: automobile ignition systems, cell phones, thunder storms, the Sun, and the Northern Lights. EMI frequently affects AM radios. It can also affect cell phones, FM radios, and televisions.

EMI can be used intentionally for radio jamming, as in electronic warfare.

2.13.1   History

Since the earliest days of radio communications, the negative effects of interference from both intentional and unintentional transmissions have been felt and the need to manage the radio frequency spectrum became apparent.

In 1933, a meeting of the International Electrotechnical Commission (IEC) in Paris recommended the International Special Committee on Radio Interference (CISPR) be set up to deal with the emerging problem of EMI. CISPR subsequently produced technical publications covering measurement and test techniques and recommended emission and immunity limits. These have evolved over the decades and form the basis of much of the world's EMC regulations today.

In 1979, legal limits were imposed on electromagnetic emissions from all digital equipment by the FCC in the USA in response to the increased number of digital systems that were interfering with wired and radio communications. Test methods and limits were based on CISPR publications, although similar limits were already enforced in parts of Europe.

In the mid 1980s, the European Union member states adopted a number of "new approach" directives with the intention of standardizing technical requirements for products so that they do not become a barrier to trade within the EC. One of these was the EMC Directive (89/336/EC)[3] and it applies to all equipment placed on the market or taken into service. Its scope covers all apparatus "liable to cause electromagnetic disturbance or the performance of which is liable to be affected by such disturbance".

This was the first time there was a legal requirement on immunity as well as emissions on apparatus intended for the general population. And although there may be additional costs involved for some products to give them a known level of immunity, it increases their perceived quality as they are able to co-exist with apparatus in the active EM environment of modern times and with fewer problems.

Many countries now have similar requirements for products to meet some level of Electromagnetic Compatibility (EMC) regulation.

2.13.2 Types

Electromagnetic interference can be categorized as follows:

- Narrowband EMI or RFI interference typically emanates from intended transmissions, such as radio and TV stations or cell phones.

- Broadband EMI or RFI interference is unintentional radiation from sources such as electric power transmission lines.[4][5][6]

Conducted electromagnetic interference is caused by the physical contact of the conductors as opposed to radiated EMI, which is caused by induction (without physical contact of the conductors). Electromagnetic disturbances in the EM field of a conductor will no longer be confined to the surface of the conductor and will radiate away from it. This persists in all conductors and mutual inductance between two radiated electromagnetic fields will result in EMI.

2.13.3 Susceptibilities of different radio technologies

Interference tends to be more troublesome with older radio technologies such as analogue amplitude modulation, which have no way of distinguishing unwanted in-band signals from the intended signal, and the omnidirectional antennas used with broadcast systems. Newer radio systems incorporate several improvements that enhance the selectivity. In digital radio systems, such as Wi-Fi, error-correction techniques can be used. Spread-spectrum and frequency-hopping techniques can be used with both analogue and digital signalling to improve resistance to interference. A highly directional receiver, such as a parabolic antenna or a diversity receiver, can be used to select one signal in space to the exclusion of others.

The most extreme example of digital spread-spectrum signalling to date is ultra-wideband (UWB), which proposes the use of large sections of the radio spectrum at low amplitudes to transmit high-bandwidth digital data. UWB, if used exclusively, would enable very efficient use of the spectrum, but users of non-UWB technology are not yet prepared to share the spectrum with the new system because of the interference it would cause to their receivers (the regulatory implications of UWB are discussed in the ultra-wideband article).

2.13.4 Interference to consumer devices

In the United States, the 1982 Public Law 97-259 allowed the Federal Communications Commission (FCC) to regulate the susceptibility of consumer electronic equipment.[7][8]

Potential sources of RFI and EMI include:[9] various types of transmitters, doorbell transformers, toaster ovens, electric blankets, ultrasonic pest control devices, electric bug zappers, heating pads, and touch controlled lamps. Multiple CRT computer monitors or televisions sitting too close to one another can sometimes cause a "shimmy" effect in each other, due to the electromagnetic nature of their picture tubes, especially when one of their de-gaussing coils is activated.

Electromagnetic interference at 2.4 GHz can be caused by 802.11b and 802.11g wireless devices, Bluetooth devices, baby monitors and cordless telephones, video senders, and microwave ovens.

Switching loads (inductive, capacitive, and resistive), such as electric motors, transformers, heaters, lamps, ballast, power supplies, etc., all cause electromagnetic interference especially at currents above 2 A. The usual method used for suppressing EMI is by connecting a snubber network, a resistor in series with a capacitor, across a pair of contacts. While this may offer modest EMI reduction at very low currents, snubbers do not work at currents over 2 A with electromechanical contacts.[10][11]

Switched-mode power supplies can be a source of EMI, but have become less of a problem as design techniques have improved, such as integrated power factor correction.

Most countries have legal requirements that mandate electromagnetic compatibility: electronic and electrical hardware must still work correctly when subjected to certain amounts of EMI, and should not emit EMI, which could interfere with other equipment (such as radios).

Radio frequency signal quality has declined throughout the 21st century by roughly one decibel per year as the spectrum becomes increasingly crowded. This has inflicted a Red Queen's race on the mobile phone industry as companies have been forced to put up more cellular towers (at new frequencies) that then cause more interference thereby requiring more investment by the providers and frequent upgrades of mobile phones to match.[12]

2.13.5　Standards

The International Special Committee for Radio Interference or CISPR (French acronym for "Comité International Spécial des Perturbations Radioélectriques"), which is a committee of the International Electrotechnical Commission (IEC) sets international standards for radiated and conducted electromagnetic interference. These are civilian standards for domestic, commercial, industrial and automotive sectors. These standards form the basis of other national or regional standards, most notably the European Norms (EN) written by CENELEC (European committee for electrotechnical standardisation).

2.13.6　EMI in integrated circuits

Main article: Electromagnetic compatibility

Integrated circuits are often a source of EMI, but they must usually couple their energy to larger objects such as heatsinks, circuit board planes and cables to radiate significantly.[13]

On integrated circuits, important means of reducing EMI are: the use of bypass or decoupling capacitors on each active device (connected across the power supply, as close to the device as possible), rise time control of high-speed signals using series resistors,[14] and IC power supply pin filtering. Shielding is usually a last resort after other techniques have failed, because of the added expense of shielding components such as conductive gaskets.

The efficiency of the radiation depends on the height above the ground plane or power plane (at RF, one is as good as the other) and the length of the conductor in relation to the wavelength of the signal component (fundamental frequency, harmonic or transient such as overshoot, undershoot or ringing). At lower frequencies, such as 133 MHz, radiation is almost exclusively via I/O cables; RF noise gets onto the power planes and is coupled to the line drivers via the VCC and GND pins. The RF is then coupled to the cable through the line driver as common-mode noise. Since the noise is common-mode, shielding has very little effect, even with differential pairs. The RF energy is capacitively coupled from the signal pair to the shield and the shield itself does the radiating. One cure for this is to use a braidbreaker or choke to reduce the common-mode signal.

At higher frequencies, usually above 500 MHz, traces get electrically longer and higher above the plane. Two techniques are used at these frequencies: wave shaping with series resistors and embedding the traces between the two planes. If all these measures still leave too much EMI, shielding such as RF gaskets and copper tape can be used. Most digital equipment is designed with metal or conductive-coated plastic cases.

RF immunity and testing

Any unshielded semiconductor (e.g. an integrated circuit) will tend to act as a detector for those radio signals commonly found in the domestic environment (e.g. cell phones).[15] Such a detector can demodulate the high frequency cell phone carrier (e.g., GSM850 and GSM1900, GSM900 and GSM1800) and produce low-frequency (e.g., 217 Hz) demodulated signals.[16] This demodulation manifests itself as unwanted audible buzz in audio appliances such as microphone amplifier, speaker amplifier, car radio, telephones etc. Adding onboard EMI filters or special layout techniques can help in bypassing EMI or improving RF immunity.[17] Some ICs are designed (e.g., LMV831-LMV834, MAX9724) to have integrated RF filters or a special design that helps reduce any demodulation of high-frequency carrier.

Designers often need to carry out special tests for RF immunity of parts to be used in a system. These tests are often done in an anechoic chamber with a controlled RF environment where the test vectors produce a RF field similar to that produced in an actual environment.[16]

2.13.7   RFI in radio astronomy

Interference in radio astronomy, where it is commonly referred to as radio-frequency interference (RFI), is any source of transmission that is within the observed frequency band other than the celestial sources themselves. Because transmitters on and around the Earth can be many times stronger than the astronomical signal of interest, RFI is a major concern for performing radio astronomy. Natural sources of interference, such as lightning and the Sun, are also often referred to as RFI.

Some of the frequency bands that are very important for radio astronomy, such as the 21-cm HI line at 1420 MHz, are protected by regulation. This is called spectrum management. However, modern radio-astronomical observatories such as VLA, LOFAR and ALMA have a very large bandwidth over which they can observe. Because of the limited spectral space at radio frequencies, these frequency bands can not be completely allocated to radio astronomy. Therefore, observatories need to deal with RFI in their observations.

Techniques to deal with RFI range from filters in hardware to advanced algorithms in software. One way to deal with strong transmitters is to filter out the frequency of the source completely. This is for example the case for the LOFAR observatory, which filters out the FM radio stations between 90-110 MHz. It is important to remove such strong sources of interference as soon as possible, because they might "saturate" the highly sensitive receivers (amplifiers and analog-to-digital converters), which means that the received signal is stronger than the receiver can handle. However, filtering out a frequency band implies that these frequencies can never be observed with the instrument.

A common technique to deal with RFI within the observed frequency bandwidth, is to employ RFI detection in software. Such software can find samples in time, frequency or time-frequence space that are contaminated by an interfering source. These samples are subsequently ignored in further analysis of the observed data. This process is often referred to as *data flagging*. Because most transmitters have a small bandwidth and are not continuously present such as lightning or citizens' band (CB) radio devices, most of the data remains available for the astronomical analysis. However, data flagging can not solve issues with continuous broad-band transmitters, such as windmills, digital video or digital audio transmitters.

2.13.8   See also

- Electromagnetic radiation
- Faraday cage
- Power integrity
- Radio receiver
- Signal integrity
- Signal noise
- Twisted pair
- Interference (communication)

2.13.9   References

[1] Based on the "interference" entry of *The Concise Oxford English Dictionary*, 11th edition, online

[2] Sue, M.K. "Radio frequency interference at the geostationary orbit". *NASA*. Jet Propulsion Laboratory. Retrieved 6 October 2011.

[3] "Council Directive 89/336/EEC of 3 May 1989 on the approximation of the laws of the Member States relating to electromagnetic compatibility". EUR-Lex. 3 May 1989. Retrieved 21 January 2014.

[4] "Radio Frequency Interference - And What to Do About It". *Radio-Sky Journal*. Radio-Sky Publishing. March 2001. Retrieved 21 January 2014.

[5] Radio frequency interference / editors, Charles L. Hutchinson, Michael B. Kaczynski ; contributors, Doug DeMaw ... [et al.]. 4th ed. Newington, CT American Radio Relay League c1987.

[6] Radio frequency interference handbook. Compiled and edited by Ralph E. Taylor. Washington Scientific and Technical Information Office, National Aeronautics and Space Administration; [was for sale by the National Technical Information Service, Springfield, Va.] 1971.

[7] Public Law 97-259

[8] Paglin, Max D.; Hobson, James R.; Rosenbloom, Joel (1999), *The Communications Act: A Legislative History of the Major Amendments, 1934-1996*, Pike & Fischer - A BNA Company, p. 210, ISBN 0937275050

[9] "Interference Handbook". Federal Communications Commission. Retrieved 21 January 2014.

[10] "Lab Note #103 *Snubbers - Are They Arc Suppressors?*". Arc Suppression Technologies. April 2011. Retrieved February 5, 2012.

[11] "Lab Note #105 *EMI Reduction - Unsuppressed vs. Suppressed*". Arc Suppression Technologies. April 2011. Retrieved February 5, 2012.

[12] Smith, Tony (7 November 2012). "WTF is... RF-MEMS?". TheRegister.co.uk. Retrieved 21 January 2014.

[13] "Integrated Circuit EMC". Clemson University Vehicular Electronics Laboratory. Retrieved 21 January 2014.

[14] ""Don't "despike" your signal lines, add a resistor instead."". Massmind.org. Retrieved 21 January 2014.

[15] Fiori, Franco (November 2000). "Integrated Circuit Susceptibility to Conducted RF Interference". *Compliance Engineering*. Ce-mag.com. Retrieved 21 January 2014.

[16] Mehta, Arpit (October 2005). "A general measurement technique for determining RF immunity" (PDF). *RF Design*. Retrieved 21 January 2014.

[17] "APPLICATION NOTE 3660: PCB Layout Techniques to Achieve RF Immunity for Audio Amplifiers". Maxim Integrated. 2006-07-04. Retrieved 21 January 2014.

2.13.10   External links

- ARRL, *RFI*

- *Interference Handbook*

- EMC Design Fundamentals

- Clemson's EMC Page (EMI Tools and Information)

- EMC Tutorials

2.14   English-language radio

English-language radio refers to radio stations that broadcast primarily in the English language and are located in countries where English is not an official language or majority language. Often referred to as "English-speaking radio" or "Expat radio" the broadcasts enables expats, vacationers and travelers to listen to radio in their native language while traveling abroad. The idea is that stations broadcast in English to popular holiday destinations such as Pattaya or the French Riviera or places with high expat communities. English language broadcasting also takes the form of military-backed radio such as the American Forces Network. However English-language radio based in foreign countries has to now compete with the introduction of internet radio and satellite technology that has increased listener's access to English-language radio based in "home" counties.

2.14.1   History

The first English-language radio in a foreign country transmission was thought to be in 1925 when Radio Paris broadcast from the Eiffel Tower, a show about fashion design, sponsored by Selfridges of London.

With end of World War II American and British forces began occupation of bases within regions of conquered Axis countries such as Germany and Okinawa for the enjoyment and informative power radio bring to both the troops and their families. Today traditions remain as the American Forces Network and the British Forces Broadcasting Service continue to provide English-language entertainment and information to troops stationed abroad in their respective countries or areas.

Non-military English broadcasting gained momentum with the increase in globalization after World War II. As English-speaking business personnel and expat communities grew because of international trade and investment, so did the demand for English language entertainment. As the number of global English speakers has grown, demand from the local native market has grown as well. Stations that actively reach out to the local community such as International Community Radio Taipei have been on the rise.

2.14.2   Location

The location of most English-language radio stations can be determined by their proximity to large populations of English speakers. Geneva, Switzerland, for example, has many English speakers and expatriates, also the Spanish coastal areas, and thus English radio broadcasts. Radio stations in places such as Baja California and Costa Rica also serve the increasing number of English speakers.

2.14.3   Notable English-language radio stations

- Energy FM http://www.dancemusicradio.net/ Broadcasting "Pure Dance" around the clock on 92.1fm from the Canary islands and worldwide on-line

- Coastline FM http://www.coastline.fm/ Spain's "original" Coastline FM, longest established English radio station covering Costa del Sol on 97.6FM, and broadcasting on the internet, live and local, a lively mix of hits from the past 50 years and today

- 106.8 Ace FM Community radio exclusive the inland areas of Malaga, Spain

- Heart FM Spain 88.2 THE Inland radio station for Andalucia and beyond

- Armed Forces Network

- British Forces Broadcasting Service

- International Community Radio Taipei

- BBC World Service

- XETRA-FM

- Radio X, Brussels

- Riviera Radio, Monaco

2.14.4   See also

- International radio broadcasters

2.14.5   External links

- 106.8 Ace FM Community radio exclusive the inland areas of Malaga, Spain

- Heart FM Spain The radio station for Inland Andalucia and beyond

- ExpatsRadioOnline.com Free online radio from the centre of Europe specifically for expatriates.

2.15   Euronews

Euronews is a multilingual news television channel, headquartered in Lyon, France. Created in 1993, it aims to cover world news from a pan-European perspective. Naguib Sawiris, an Egyptian businessman owns 53% of the station (through *Media Globe Networks*) and is chairman of its supervisory board.

2.15.1   Content

As a rolling-news channel, headlines from both Europe as well as the world are broadcast in thirty-minute intervals. Brief magazine segments typically fill in the remaining schedule, focusing on market data, financial news, sports news, art and culture, science, weather, European politics, and press reviews of the major European newspapers.[2] These item slots will occasionally be preempted by breaking news or live television coverage. Some segments are displayed without commentary under the banner "No Comment", which has been the channel's signature program since its launch.[3]

Euronews headquarters in Lyon.

2.15.2   History and organisation

General

In 1992, following the First Persian Gulf War, during which CNN's position as the preeminent source of 24-hour news programming was cemented, the European Broadcasting Union decided to establish Euronews to present information from a European perspective. The channel's first broadcast was on 1 January 1993 from Lyon. An additional broadcast studio was set up in London in 1996. It was founded by a group of ten European public broadcasters:[4]

- CYBC

- France Télévisions

- RAI

- RTBF

- RTP

- RTVE (*former shareholder*)

- TMC (*former shareholder*)

- YLE

- ERTU

In 1997, the British news broadcaster ITN purchased a 49-percent share of Euronews for £5.1 million from Alcatel-Alsthom.[5] ITN supplies the content of the channel along with the remaining shareholders, which are represented by the SOCEMIE (Société Editrice de la Chaîne Européenne Multilingue d'Information EuroNews) consortium. Euronews SA is the operating company that produces the channel and holds the broadcasting licence. It is co-owned by the 10 founders and:

- ▬ VGTRK
- ☪ TRT
- ◣ ČT
- ˙ ■ PBS
- ▬ SNRT
- ▬ RTVSLO
- ▌▐ RTÉ
- ▬ NTU
- ✚ SRG-SSR
- ▌▌ TVR
- ▬ SVT - MTG
- ◉ ERTT
- ▐ ENTV

The broadcast switched from solely analogue to mainly digital transmission in 1999. In the same year, the Portuguese audio track was added. The Russian audio track appeared in 2001.

In 2003, ITN sold its stake in Euronews as part of its drive to streamline operations and focus on news-gathering rather than channel management.[6]

On 6 February 2006, Ukrainian public broadcaster Natsionalna Telekompanya Ukraïny (NTU) purchased a one-percent interest in SOCEMIE.[7]

On 27 May 2008, Spanish public broadcaster RTVE decided to leave Euronews to promote its international channel TVE Internacional. It also cited legal requirements to maintain low debt levels through careful spending as a factor influencing its decision to leave.[8]

In February 2009, the Turkish public broadcaster TRT became a shareholder in the channel, and joined its supervisory board.[9] TRT purchased 15.70% of the channel's shares and became the fourth main partner after France Télévisions (23.93%), RAI (21.54%), and VGTRK (16.94%).

Language availability

Format

Since 11 January 2011, the channel's video has been broadcast in the 16:9 aspect ratio, which replaced the previous 4:3.

Its audio is monoaural.

Executive Board

CEO : Michael Peters Born in 1971 in Flensburg, Germany, of French and German nationality, Michael Peters graduated from EM Lyon business school with a master's degree in Financial Engineering (1995) and from IAE Lyon III (1992). He began his career at the international auditing firm of Arthur Andersen, in Lyon, France (1995–1998). Michael Peters joined Euronews in 1998 as Finance Manager and then became CFO. He was appointed Deputy Managing Director of Euronews in December 2003.

In May 2005, at the age of 33, and seven years after joining euronews, Michael Peters was appointed Managing Director of Europe's leading news channel. In December 2008, Michael Peters was appointed Managing Director of the Executive Board of Euronews S.A. In December 2011, Michael Peters was appointed Chairman of the Executive Board.

Members of the Executive Board

- Michael Peters, CEO, Chairman of the Executive Board

- Lucian Sârb, Director of News and Programmes

- Cécile Leveaux, Chief Technical Officer

- Olivier de Montchenu, Worldwide Sales Director, Managing Director of Euronews Sales

Management Committee

- Michael Peters, Chairman of the Executive Board

- Lucian Sârb, Director of News and Programmes

- Cécile Leveaux, Chief Technical Officer

- Olivier de Montchenu, Worldwide Sales Director, Managing Director of Euronews Sales

- David Cipel, Chief Financial and Administrative Officer

- Grégory Samak, Director of Broadcasting and Programme Marketing

- Grégoire Olivero de Rubiana, Director of External Relations

- Arnaud Verlhac, Deputy Director Worldwide Distribution

Supervisory Board

Chairman of the Supervisory Board: Paolo Garimberti

A native Italian and educated in Law, Paolo Garimberti has spent his entire career in journalism. He began as a correspondent in Moscow for the daily newspaper La Stampa, then as Office Manager in Rome. In 1986, he moved to La Repubblica as an editorialist, specialising in foreign politics. He also appeared as an expert commentator on RAI's TG3 (television news programme). He was then appointed Manager of RAI's TG2, before going back to La Repubblica as Vice Managing Director, until 2004. During that time, he founded and managed the rolling news website, CNN Italia. He then went on to Espresso Group where he held the positions of Director of International Relations and Development, as well as editorialist. 2009-2012: Paolo Garimberti returned to RAI as its Chairman. In May 2012, he became Chairman of the new museum of the Juventus team at the stadium in Turin. Paolo Garimberti was appointed Chairman of the Supervisory Board of Euronews on 16 December 2011.

Members of the Supervisory Board (natural persons)

- Paolo Garimberti
- Andrey Bystritsky
- Ahmet Koyuncu
- Stéphanie Martin
- Philippe Cayla
- Pier Luigi Malesani

Members of the Supervisory Board (legal entities)

- France Télévisions (France)
- RAI (Italy)
- VGTRK (Russia)
- TRT (Turkey)
- SSR (Switzerland)
- SNRT (Morocco)
- RTP (Portugal)
- RTE (Ireland)
- RTBF (Belgium)
- ERT (Greece)

Acquisition by Egyptian businessman

In February 2015 the channel's executive board has approved the bid by an Egyptian telecom businessman, Naguib Sawiris, to acquire a 53% controlling stake in the media outlet. The deal raised a number of questions over Euronews' future editorial posture and its independence.[10][11][12]

2.15.3 Criticism

Euronews has received criticism for perceived bias towards the European Commission which provides a significant part of the channel's funding.[13][14][15]

2.15.4 Presentation

The channel employs an unusual presentation style: initially, rather than using in-vision presenters, it showed only video footage with recorded voice-overs. This aims to prevent bias. In 2011, however, extended news items featured in-vision reporters, including occasional pieces to camera.

The principal sources of footage come from APTN (Associated Press Television News) and Reuters TV, these being the partner agencies of the European Broadcasting Union. It also draws upon resources from Agence France-Presse, Italian ANSA, Portuguese LUSA, German DPA, Spanish EFE and Russian ITAR-TASS.

2.15.5 Broadcast

The channel is available in 350 million households in 155 countries worldwide. It reaches more than 170 million European households by cable, satellite and terrestrial. It also began to secure availability on multimedia platforms such as IPTV and digital media.[16]

Euronews launched an application for mobile devices (Android, iPhone, and iPad) which is called "Euronews Live". The application is free of charge and is available on Android Market and App Store.[17]

The following countries also broadcast *Euronews* through terrestrial channels:

The channel's programmes are also available by podcast, and it has also maintained a YouTube channel since October 2007.[20]

In 2012, the largest Belarusian state network MTIS stopped broadcasting Euronews for unknown reasons.[21][22]

In 2013, the new commercial channel Planet TV started broadcasting Euronews dubbed in Slovenian after Antenna

TV SL purchased a major stake in the company. Euronews airs after closedown (or sign-off) of Planet TV, but both call sign logos are displayed.

2.15.6　Logos

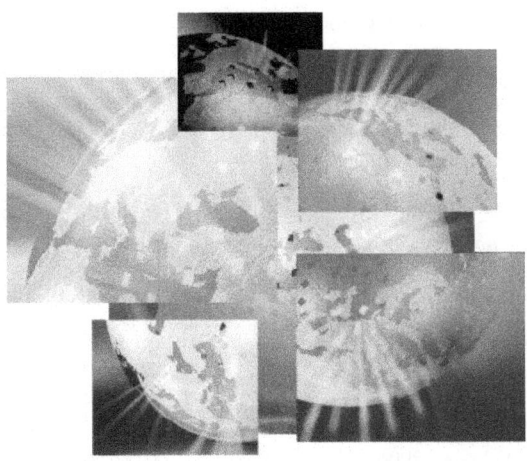

Logo used from 26 October 1998 to 3 June 2008

EuroNews

Logo used from 26 October 1998 to 3 June 2008

The current Euronews logo is the fourth. From 1 January 1993 to 26 October 1998 the logo was in the lower right corner of the screen, between 26 October 1998 and 4 June 2008 it was in the upper left corner of the screen, and since 4 June 2008 it has been in the upper right corner of the screen.

- 1 January 1993 – 8 February 1997: blue lowercase word "euro" in yellow parallelogram and yellow uppercase word "NEWS".

- 8 February 1997 – 26 October 1998: white lowercase word "euro" above and blue lowercase word "news" below.

- 26 October 1998 – 4 June 2008: blue rectangle enclosing white camel case word "EuroNews".

- Since 4 June 2008; white lowercase word "euronews" on a neutral grey background featuring a white circle symbolizing both the world and star circle on the flag of Europe.

2.15.7　Shows and presenters

Programmes

[23]

- *Brussels Bureau* – hosted by James Franey and Efi Koutsokosta

- *Business Line* – hosted by Oleksandra Vakulina and Giacomo Segantini

- *Business Middle East* – hosted by Daleen Hassan

- *Business Planet* – hosted by Serge Rombi

- *Cinema* – hosted by François Menard

- *Europe Weekly*

- *Focus*

- *For Children in War*

- *Futuris*

- *Generation Y*

- *Gravity*

- *Hi-Tech*

- *Innovation*

- *Interview*

- *I-Talk*

- *Learning World*

- *Le Mag*

- *Life*

- *Markets*

- *Metropolitans*

- *Musica*

- *News +*

- *No Comment*

- *On the Frontline*

- *Perspectives*

- *Postcards* – hosted by Seamus Kearney

- *Real Economy*

- *Rendez-Vous*

- *Reporter* – hosted by Hans von der Brelie and Valerie Gauriat
- *Right On*
- *Science*
- *Silent Disasters*
- *Smart Care*
- *Space* – hosted by Claudio Rosimo
- *Speed*
- *Sports United* – hosted by Joe Allen
- *Target*
- *The Corner* – hosted by Cinzia Rizzi
- *The Global Conversation* – hosted by Isabelle Kumar
- *The Network* – hosted by Chris Burns
- *U-Talk*

2.15.8 Captions

Captions are in English, though the names of some countries are not translated (*Deutschland*, not Germany).

2.15.9 See also

- Africanews
- Eurosport
- International broadcasting
- List of international television channels
- List of news channels
- Al Jazeera
- BBC
- CCTV
- CNN
- DW-TV
- France 24
- NHK World
- Press TV
- RT
- TeleSUR
- TouchVision
- State media

2.15.10 References

[1] "Euronews restructures". Informa Telecoms & Media. January 27, 2009. Retrieved 25 October 2012.

[2] *Euronews and Metropolitan Media Ltd* (PDF), Metropolitan Media Ltd, retrieved 20 August 2011

[3] "No comment from EuroNews on YouTube". Advanced Television. 11 October 2007. Retrieved 20 August 2011.

[4] Collins, Richard (1998). *From Satellite to Single Market: New Communication Technology and European Public Service Television*. London: Routledge. p. 130. ISBN 9780415179706.

[5] "ITN ACQUIRES 49% EURONEWS STAKE". *Telecom Paper*. 1 December 1997. Retrieved 5 June 2012.

[6] "ITN Drops Out of Euronews Channel". *Broadcast*.

[7] "NTU Becomes 20th EuroNews Shareholder". DigitalSpy. 5 February 2006.

[8] "TVE abandona EuroNews". *El Mundo* (in Spanish). 5 February 2006.

[9] "Turkey's TRT joins Euronews supervisory board". *World Bulletin*. 15 September 2009. Retrieved 5 June 2012.

[10] "Egyptian Mogul Plans to Buy Controlling Stake in Europe's Answer to CNN". *Hollywood Reporter*. 27 February 2015. Retrieved 25 March 2015.

[11] "Controversial ventures pose questions for Euronews". *EU Observer*. 24 March 2015. Retrieved 25 March 2015.

[12] "Euronews investor Naguib Sawiris: we will resist state interference". *The Guardian*. 27 February 2015. Retrieved 25 March 2015.

[13] "The EU Communication 'propaganda' debate". New Europe. 23 August 2009. Retrieved 20 August 2011.

[14] "Euronews: Channel of Propaganda". EU Democrates. 18 January 2011. Retrieved 20 August 2011.

[15] "EU triples its financial contribution to Euronews". The Parliament. 13 January 2011. Retrieved 20 August 2011.

[16] "Euronews Media Presspack" (PDF). *Euronews*. Retrieved 27 June 2010.

[17] "euronews live apllication". *Euronews*. Retrieved 20 August 2011.

[18] "Programación de Euronews en Extremadura TV" (in Spanish). Extremadura TV. 10 February 2013. Retrieved 10 February 2013.

[19] "Euronews llega a la TDT en España a través de Aragón TV". *Heraldo* (in Spanish). 1 January 2013. Retrieved 27 January 2013.

[20] ""No Comment" sur YouTube ? Affirmatif". *Libération* (in French). 10 August 2007. Retrieved 5 June 2012.

[21] В сети МТИС прекращена трансляция канала "Евроньюс" (in Russian). Naviny. 1 January 2012. Retrieved 5 June 2012.

[22] "В Минске отключают Euronews" (in Russian). Euroradio. 31 December 2011. Retrieved 5 June 2012.

[23] http://www.euronews.com/programs/. Missing or empty |title= (help)

2.15.11   External links

- euronews.com, official international-news website

2.16   European Broadcasting Area

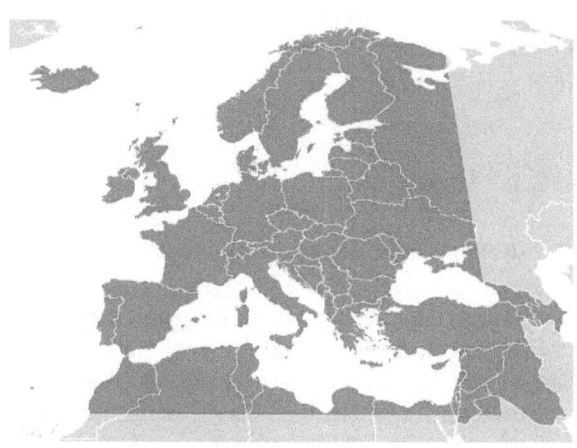

Map showing the European Broadcasting Area in red

The **European Broadcasting Area (EBA)** is defined by the International Telecommunication Union as such:

> The "European Broadcasting Area" is bounded on the west by the western boundary of Region 1, on the east by the meridian 40° East of Greenwich and on the south by the parallel 30° North so as to include the northern part of Saudi Arabia and that part of those countries bordering the Mediterranean within these limits. In addition, Armenia, Azerbaijan, Georgia and those parts of the territories of Iraq, Jordan, Syrian Arab Republic, Turkey and Ukraine lying outside the above limits are included in the European Broadcasting Area.[1]

The EBA includes territory outside Europe, and excludes some territory that is part of the European continent. For example, Armenia, Azerbaijan and Georgia were defined as outside the EBA borders until 2007.[2] After EBA was expanded by the 2007 World Radiocommunication Conference (WRC-07) to include those three countries,[1] the only sovereign undisputed territory belonging to the European continent while remaining outside the EBA borders, is 4% of Kazakhstan. While Kazakhstan is excluded from EBA, it is still eligible to apply for membership of the Council of Europe.[3]

The boundaries of the European Broadcasting Area have their origin in the regions served and linked by telegraphy cables in the 19th and early 20th centuries. The European Broadcasting Area plays a part in the definition of eligibility for active membership in the European Broadcasting Union and thus participation in the Eurovision Song Contest.

2.16.1   List of countries within the EBA

The following territories also rest inside the EBA borders, but can not join EBU due their status as non-sovereign states:

- Abkhazia, claimed as an autonomous republic of Georgia.

- Akrotiri and Dhekelia, dependent territory of United Kingdom.

- Azores, dependent territory of Portugal

- Canary Islands, dependent territory of Spain

- Faroe Islands, dependent territory of Denmark.

- Gibraltar, dependent territory of United Kingdom.

- Guernsey, dependent territory of United Kingdom.

- Isle of Man, dependent territory of United Kingdom.

- Jersey, dependent territory of United Kingdom.

- Kosovo, claimed as an autonomous province of Serbia.

- Kurdistan, claimed as part of Iraq

- Madeira, dependent territory of Portugal

- Nagorno-Karabakh, claimed as part of Azerbaijan.

- Northern Cyprus, claimed as part of Cyprus.

- ▬ Palestine

- ▬ Transnistria, claimed as a territorial unit of Moldova.

- ▬ South Ossetia, claimed as part of Georgia.

2.16.2 Notes

1. ^ *a b c d* Among the 60 EBA countries, only 4 (Liechtenstein, Iraq, Saudi Arabia and Syria) were not active EBU members, as of May 2014.[4]

2. ^ The United Kingdom consists of the countries of England, Northern Ireland, Scotland and Wales. The UK is responsible for the foreign relations and ultimate good governance of the Crown dependencies of Guernsey, the Isle of Man and Jersey, which are otherwise separate. The mainlands of England, Scotland and Wales make up the island of Great Britain (or simply Britain). The two territories: Gibraltar and Akrotiri & Dhekelia have status as British Overseas Territories, and thus also belong within the foreign affairs of United Kingdom.

2.16.3 See also

- International Telecommunication Union region

2.16.4 References

[1] ITU-R Radio Regulations (2012-2015), International Telecommunication Union, available from the Spectrum Management Authority of Jamaica. Armenia, Azerbaijan and Georgia were added to the EBA at the 2007 World Radiocommunication Conference (WRC-07).

[2] "ITU-R Radio Regulations - Articles edition of 2004 (valid in 2004-2007)" (PDF). *International Telecommunication Union*. 2004.

[3] "Situation in Kazakhstan and its Relations with the Council of Europe". *Document 11007: II General information, point 11*. Parliamentary Assembly Council of Europe. 7 July 2006.

[4] "72 active members in 56 countries". EBU. Retrieved 22 May 2014.

2.17 FTA receiver

A free-to-air or **FTA Receiver** is a satellite television receiver designed to receive unencrypted broadcasts. Modern

A Viewsat Xtreme FTA receiver

decoders are typically compliant with the MPEG-2/DVB-S and more recently the MPEG-4/DVB-S2 standard for digital television, while older FTA receivers relied on analog satellite transmissions which have declined rapidly in recent years.

2.17.1 Uses

Mainstream broadcast programming

In some countries, it is common for mainstream broadcasters to broadcast their channels over satellite as FTA.[1] Most notably, in the German-speaking countries, most of the main terrestrial broadcasters, such as ARD Das Erste and ZDF offer FTA satellite broadcasts, as do some of the more recent satellite rivals such as Sat.1 and RTL. The satellites on which these channels broadcast, at Astra's 19.2° east position, are receivable throughout most of Europe.

In the UK, all the original five terrestrial broadcasters, BBC One, BBC Two, ITV, Channel 4, and Five broadcast FTA on digital satellite in some form, including many of their regional variations. However, in some countries, it is not the norm for mainstream channels to broadcast on FTA satellite television.

Ethnic and religious programming

FTA receivers are sold in the United States and Canada for the purpose of viewing unencrypted free-to-air satellite channels, the bulk of which are located on Galaxy 19 (97°W, K_u band).[2] There is also a substantial amount of Christian-based programming available on several satellites over both North America and Europe, such as The God Channel, JCTV, EWTN, and 3ABN.

Educational programming

The PBS Satellite Service offers educational programming on K_u band DVB from the AMC-21 satellite (125°W).[3] As there is no standard MPEG audio on many of these channels, the AC3-only feeds require a Dolby Digital-capable receiver. They are otherwise free. Channels include PBS-HD/PBS-X as well as various secondary programmes normally carried on digital subchannels of PBS terrestrial member stations.

The main PBS New York feed is absent from the free-to-air version of the PBS satellite service to afford local terrestrial member stations a chance to broadcast material before it becomes available on PBS-X or PBS-HD. Typically, PBS-X feeds carried programmes (except news) a day later than the main terrestrial PBS network.

US terrestrial broadcasters

Many of these channels carried programming from major network television affiliates, although these are disappearing, particularly on Ku-band.[4]

Equity Broadcasting used one K_u band (Galaxy 18, 123°W) and one C-band satellite feed as a key part of its Equity C.A.S.H. centralcasting operation; many small UHF local stations were fed from one central point in Little Rock, Arkansas via free-to-air satellite. Most were members of secondary terrestrial networks, including both US English language and Spanish language broadcasters, and content from satellite broadcasts often fed over-the-air digital subchannels of terrestrial stations. Programming such as the Retro Television Network or Retro Jams had been provided at various times; music video broadcasters Mas Música and The Tube were formerly available at 123°W before being taken over (Mas Música is now MTV3) or ceasing operations.

Similarly, unencrypted K_u band satellite television was also used temporarily in the aftermath of 2005's Hurricane Katrina as a means to feed NBC programming into New Orleans from the studios of an out-of-state broadcaster; the feeds contained the content, branding and station identification of the damaged New Orleans station in a form suitable for direct feed to a transmitter (with no further studio processing) in the target market.

Paradoxically, many Equity-owned local UHF stations obtained solid national satellite coverage despite small terrestrial LPTV footprints that barely covered their nominal home communities. In many cases, this brought smaller networks and Spanish-language broadcasting to communities which otherwise would have no free access to this content.

As television market statistics for these stations from firms such as Nielsen Media Research are based on counting viewership within the footprint of the corresponding terrestrial signal, television ratings severely underestimated or failed to estimate the number of households receiving programming such as Univisión from FTA satellite feeds. The liquidation of Equity Broadcasting's station group in mid-2009 greatly reduced the number of US terrestrial stations available from K_u band free-to-air satellite; while a very small handful of uplinked terrestrial stations remain free (mostly on C-band, which requires a much larger antenna) these are from other, independent sources.

Rural and hobby use

Over-the-air digital TV signals do not reach very far outside the city in which they are transmitted. FTA Receivers can be used in rural locations as a fairly reliable source of television without subscribing to cable or a major satellite provider.

Terrestrial broadcasters use some of the nearly 30 North American satellites to transmit their feeds for internal purposes. These unencrypted feeds can then be received by anyone with the proper decoder. Satellite signals are normally receivable well beyond the terrestrial station's coverage area. DXers also use FTA receivers to watch the numerous wildfeeds that are present on many of those satellites.

In theory, a viewer in Glendive, Montana (the smallest North American TV market) could have received what little local CBS and NBC programming is available terrestrially, alongside a K_u band free-to-air dish for additional commercial networks (such as individual ABC and Fox TV affiliates from Equity Broadcasting, formerly at 123°W) and educational programming (PBS Satellite Service at 125°W). Unfortunately, there is no assurance that any individual FTA broadcast will remain available or that those which do remain will continue broadcast in a compatible format - in this example, such a viewer would have lost ABC and Fox in mid-2009 due to Equity's bankruptcy.

Signal piracy

Free-To-Air receivers generally use the same technology standards (such as DVB-S, MPEG-2) as those used by pay-TV networks such as Echostar's Dish Network and BCE's Bell TV. FTA receivers, however, lack the smartcard readers or decryption modules designed for the reception of pay-TV programming, since the receivers are designed only for reception of unencrypted transmissions.

On occasion, where a pay-TV service's encryption system has been very seriously compromised, to the extent that it can be emulated in software and without the presence of a valid access card, hackers have been able to reverse-engineer an FTA receiver's software and add the necessary emulation to allow unauthorized reception of pay TV channels. Manufacturers, importers, and distributors of FTA receivers officially do not condone this practice and some will not sell to or support individuals who they believe will be using their products for this purpose, use of third-party software usually voiding any warranties.

Unlike traditional methods of pirate decryption that involve altered smart cards used with satellite receivers manufactured and distributed by the provider, piracy involving FTA receivers require only an update to the receiver's firmware. Electronic countermeasures that disable access cards may not have the same or any effect on FTA receivers because they are not capable of being updated remotely. The firmware in receivers themselves cannot be overwritten with malicious code via satellite as provider-issued receivers are.

FTA receivers also have the advantage of being able to receive programming from multiple providers plus legitimate free-to-air DVB broadcasts which are not part of any package, a valuable capability which is conspicuously absent from most "package receivers" sold by DBS providers. DVB-S is an international standard and thus the industry-imposed restriction that a Bell TV receiver is not interchangeable with a Dish Network receiver (the same box) and neither are interchangeable with a GlobeCast World TV receiver (also DVB) is an artificial one created by providers and not respected by either pirates or legitimate unencrypted FTA viewers.

Periodically, a provider will change the processes in which its encryption information is sent. When this happens, illegitimate viewing is disrupted. Third-party coders may release an updated altered version of the FTA receiver software on internet forums, sometimes hours to days after the countermeasure is implemented, although some countermeasures have allowed the encryption to remain secure for several months or longer. The receivers, meanwhile, remain able to receive unencrypted DVB-S broadcasts and (for some HDTV models) terrestrial ATSC programming. The same is not true of standard subscription TV receivers, whereby unsubscribing from a pay-TV package causes loss of all channels.

The use of renewable security allows providers to send new smart cards to all subscribers as existing compromised encryption schemes (such as Nagravision 1 and 2) are replaced with new schemes (currently Nagravision 3). This "card swap" process can provide pay-TV operators with more effective control over pirate decryption, but at the expense of replacing smart cards in all existing subscribed receivers. While this approach is used by most providers, deployments tend to be slowed due to cost.

While smart-card piracy often involves individuals who reprogram access cards for others (usually for a price), piracy using FTA receivers involves third-party software that is relatively easy to upload to the receiver and can even be uploaded using a USB device, network, or serial link (a process called "flashing"). Most such firmware is distributed freely on the Internet. Websites that third-party coders use to share this software often have anywhere from 50,000 to over 200,000 registered users.

Another method of pirate decryption that has become popular recently is known as Internet Key Sharing (IKS). This is accomplished by an Ethernet cable hooked to the receiver that allows updated decryption keys to be fed to the unit directly from the internet. The DVB-S common scrambling system and the various conditional access systems are based on the use of a legitimately subscribed smart cards which generates a continuous stream of cryptographic keys usable to decrypt one channel on a receiver. A key-sharing scheme operates by redistributing these keys in real-time to multiple receivers in an unlimited number of locations so that one valid smartcard may serve almost 10000 viewers.

As of June 2009 this was the only active pirate decryption system still in widespread use in North American satellite TV, due to the shutdown of the compromised Nagravision 2 system by providers such as Dish Network and Bell TV.

However, this is limited by the interval of the stream of keys or also called CW (Control Words). Usually, the interval for the renewal of the CWs is ±10 seconds, but other systems (i.e. NDS3) have CW intervals of 5 seconds or less. Each channel usually has a different set of CWs for decryption and thus each box currently watching a specific channel must periodically request the current CWs from the server/smart-card for that specific channel. So arguably, the sharing of the card might not be unlimited. There are some restrictions to this like the frequency of CW changes and also the latency of the network. If the CWs do not arrive in time, there could be a freeze or crackle in the picture.

There are of course other more costly possibilities like having several legitimately subscribed cards each handling a few channels and a CW caching server.

The dependence on an external server also compromises privacy for individual viewers,[5] as well as rendering the system incompatible with many receiver models which lack the ability to connect to an outside network and/or lack the ability to set or modify the various keys or identifiers used in communication between the card and receiver.

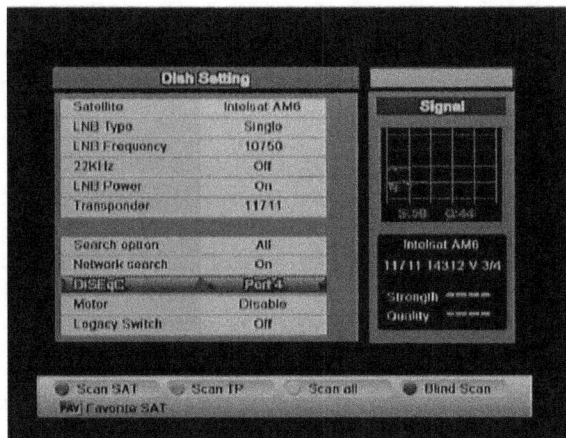

The installation menu

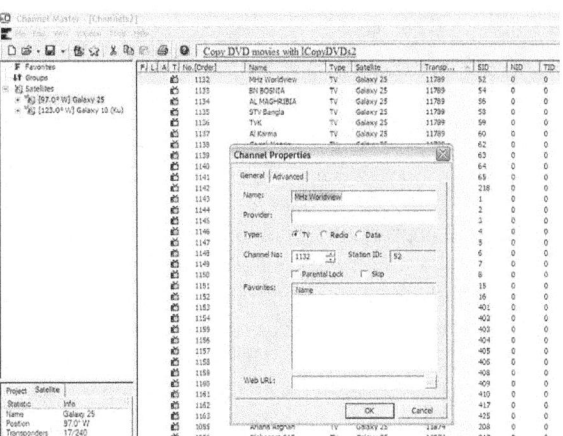

Channel Master editor

2.17.2   Common features

Installation menu

This is the main control panel that allows the user to configure the receiver to interact with LNBs, switches, motors, and other equipment. The user selects the LNB type, local oscillator frequency, appropriate DiSEqC switch port, and motor configuration. If all the settings are correct for the appropriate equipment, a signal bar showing strength and quality will appear. At that point, the receiver can be used to scan the satellite to detect channels.

Blind scan

There are 63 satellites in orbit over the Americas, 57 over Europe and a further 64 over Asia, a significant number of which will be receivable from any one location. Each of these has a different number of active transponders. Each transponder operates at a different frequency and symbol rate. Many FTA receivers are designed to detect any active transponders and any channels on those transponders. Because they are designed to do this without needing to be pre-programmed with the transponder information for each satellite, this process is referred to as a "blind" scan—as opposed to a satellite scan, which scans according to pre-set transponder values.

Channel edit/sort

Once a scan is complete, the channels can often be sorted alphabetically, in satellite/transponder order, or in scrambled/unscrambled order. Additionally, third-party software often allows the option of sorting by the channel's Station Identification (SID) number. This is so that the individual channels can be numbered in a way that mimics the lineup

of Dish Network or Bell TV. Channels can also be renamed or deleted, either in an on-screen menu or with external software.

The most popular software used to configure and sort channels was a database program called Channel Master, which allowed the user to name, number, sort, and delete channels and then save them in a format that can then be written to the receiver. The file created that contains channel information is called a channel list. This channel editor application is not affiliated with the similarly-named antenna manufacturer and appears to have last been updated in 2008. Many older and discontinued receiver models are supported in Channel Master, though most newer and less popular ones are not.

User settings

Most FTA receivers give the user the option of configuring the language, aspect ratio, TV type (NTSC/PAL), and time settings.

Typically, most FTA receivers can accept an MPEG2 video stream in either PAL-compatible (540/704/720 x 576) or NTSC-compatible (640 x 480) image formats and convert it for display on either a PAL or NTSC monitor. There is some loss of image data due to NTSC's lower resolution. Some receivers also support output to SCART, S-Video, HDMI or component video.

Parental control

All FTA receivers contain a parental lock feature.

DiSEqC switch and motor control

Unlike package receivers promoted for use with a limited number of satellites controlled by an individual pay-TV provider, an FTA receiver is designed to be capable of receiving any free signals from all available satellites visible in a given location. To fully exploit this capability, most K_u band FTA receivers will control a DiSEqC motor which can rotate a single dish to view one of any number of multiple satellites.

An alternate approach of pointing a fixed dish (or LNB) at each satellite to be received (then feeding the individual signals into a remotely controlled switch) is compatible both with standard FTA receivers and the more-restricted pay-TV "package receiver". The most common standard for use with FTA receivers is a DiSEqC switch which normally allows automatic selection of signal from four satellites. A simpler two-position remote switch operated by a 22 kHz tone is also occasionally used for North American reception, but this configuration is not compatible with European-style universal LNB's which use the tone internally for band-switching.

A toroidal antenna may be used with multiple LNB's to receive multiple satellites in various locations over a 40° arc. Unlike the single parabola of a standard satellite dish antenna (which is best adapted to focus one target satellite to a single point), the toroidal antenna uses a reflector pair to focus multiple signals to a line.

Individual adjacent or near-adjacent pairs (such as Glorystar on 97°W and 101°W) may be received, due to their close proximity, with two LNB's on what otherwise looks geometrically to be a standard parabolic dish. The outputs from these individual LNB's may then be fed through a switch to a receiver, providing access to all signals on both satellites.

Electronic program guide

An on-screen program schedule can be accessed that also contains descriptive information about a selected program. The availability and quality of programme guide information varies widely between broadcasters (some provide nothing) and the ability of receivers to collect and store guide listings from multiple sources is also variable. Receivers with more memory (or storage on external devices such as hard drives) are often, but not always, better equipped to store and retrieve on-screen programme listings. In some cases, a receiver with both satellite and terrestrial tuners will provide on-screen guide support for one mode of operation but not both.

PVR functions

A few high-end receivers feature the ability to record programs, pause, and review live TV. Often, a hard drive is not included when the unit is purchased, which allows the user to install any desired hard disk drive. Many newer units are equipped with a USB 2.0 port that allows the user to connect a portable hard drive; at least one unit (the Pansat 9200HD) uses external SATA as PVR media storage.

Some receivers, such as TripleDragon or Dream Multimedia's Linux-based Dreambox series, provide local area network interfaces. This allows the use of network-attached storage to provide PVR-like functions (some of these models also include internal hard drives or USB) and allows the unit to be controlled or updated via network.

The use of desktop personal computer cards to deploy DVB-S or terrestrial digital television tuners allows the computer's hard drive and network storage to be used to archive electronic programme guide information and recorded television programming. Most or all of the base PVR functionality becomes available by default at little or no added cost.

MPEG4 and 4:2:2

Most standard FTA receivers support DVB-S, MPEG-2, 480i or 576i SDTV received as unencrypted QPSK from K_u band satellites.

Rarely supported by stand-alone FTA receivers, but likely to be supported by FTA DVB-S tuners for personal computers, are MPEG-4 and MPEG2 4:2:2, variants on the MPEG compression algorithm which provide more compression and more colour resolution, respectively. As personal computers handle much of the video decompression in software, any codec could be easily substituted on the desktop.

High-definition television is also beginning to be supported by a limited-number of high-end receivers; at least one high-end stand-alone receiver (the Quali-TV 1080IR) supports both 4:2:2 and HDTV.

4:2:2 is a version of MPEG-2 compression used in network feeds such as NBC on K_u band (103°W). Some broadcast networks use 4:2:2 encoding for otherwise-unencrypted transmission of sports events to local terrestrial stations, as it provides slightly better colour than the standard 4:2:0 compression.

In some cases support for additional standards (such as DVB-S2, MPEG-4 and 8PSK) will also become necessary to receive a viewable signal.[6] The use of newer means of modulation and compression is likely to become more widespread for high-definition television feeds, in order to partially offset the larger amount of transponder space required to deliver high-definition video to television stations.

Terrestrial DTV

In countries using the DVB-T and DVB-C standards for terrestrial digital television and digital cable, a few higher-end receivers provide an option to install terrestrial DVB tuners either alongside or in place of the stock DVB-S tuner. Dream Multimedia's DreamBox series, for instance, supports this in a few selected models.

In countries using ATSC, inclusion of terrestrial tuners in DVB-S FTA receivers is rare, with one key exception. Some HDTV FTA receivers incorporate terrestrial ATSC tuners. These typically do not support ATSC's unique *major.minor* digital subchannel numbering scheme or the on-screen program guide but are capable of displaying (or timeshifting) local HDTV with no loss in detail. Channels from these receivers are numbered using FTA conventions, by which the first channel found is most often arbitrarily given channel 1 as its virtual channel number.

HDTV

A few high-end receivers feature HDTV. In North America, these often include an ATSC over-the-air digital television tuner and MPEG-4 support. A few HDTV units allow for the addition of a UHF remote control. However, an 8PSK module can be installed in place of the UHF remote and allows the receiver to decode the format used on most Dish Network high definition programming.

These units are superior to DVD recorders for time-shifting HDTV programming, as most DVD units down-convert OTA HDTV signals to standard-definition to match the limitations of the DVD standard. An HDTV FTA receiver with ATSC capability and USB storage can record one channel from a terrestrial or satellite DTV transport stream entirely losslessly, although the on-screen guide for terrestrial reception is often limited and viewing or storage of analog NTSC channels is not supported.

2.17.3 Controversy

Availability of free programming

While significant amounts of programming remain free, there is no assurance to viewers that any individual broadcast currently available free-to-air will remain so. Some will inevitably move to incompatible signal formats (such as MPEG 4:2:2, 8PSK, DVB-S2, or MPEG-4), change from free to encrypted, move to different satellite locations (often across bands, where C band reception requires much larger antennae) or shut down entirely.

Many of the signals are backhaul or "wildfeed" video des-

tined to individual stations, or are feeds to terrestrial transmitters programmed remotely. These were not intentionally created as direct satellite broadcasts to home viewers, but often had been left unencrypted (in the clear) on the assumption that few people were watching. As free-to-air receivers became inexpensive and widely deployed in the 2000s (decade), many of these feeds moved to C band (requiring a huge dish), were encrypted or changed to incompatible modulation or encoding standards which required more advanced receivers, even though the corresponding terrestrial television broadcast may still be free-to-air in its home community.

The onus is on receiver vendors to voluntarily indicate, whenever they use lists of currently available FTA programming for marketing purposes, that free channels frequently may appear, move and disappear, often on a permanent basis, with no advance notice. One North American example was Equity Broadcasting, once a major source of small local terrestrial stations on free satellite television. Equity filed for Chapter 11 bankruptcy on December 9, 2008 and most of Equity's terrestrial stations were sold at auction in mid-2009.[7] As many of the stations (such as New York's WNGS and WNYI) were sold to Daystar and now originate nothing, the corresponding unique free-to-air signals (Galaxy 18, 123°W) are no more.[8] Even where a signal still exists, an incompatible signal format such as that of the NBC feeds (AMC 1 at 103°W, now requires 8PSK, DVB-S2 and HDTV support to receive anything) can remove a channel from virtually all standard FTA receivers.

Receiver obsolescence

Many receivers will provide options for hardware expansion (such as to add 8PSK reception or DVB Common Interface TV subscription cards) and firmware upgrade (either officially or from nominally third-party sources). Most often, once the individual receiver model is discontinued, this support and expandability rapidly disappears from all sources. The migration of existing feeds to formats such as MPEG-4, HDTV, or DVB-S2 (which many current receivers do not support) may also result in viewers losing existing free programming as equipment becomes rapidly obsolete. Unlike digital terrestrial set-top boxes, most standard-definition DVB-S receivers do not down-convert HD programming and thus produce no usable video for these signals.

There have also been incidents where existing receiver designs have been "cloned" or copied by competing manufacturers; a manufacturer will often reduce support for a widely copied receiver design. In some cases, malware has been released, ostensibly in the same format as existing third-party firmware, in an attempt to interfere with the further use of a widely cloned receiver's design.

Legal issues

FTA receivers are ostensibly designed for free-to-air use but can be adapted for other purposes. In some jurisdictions, this dual-use nature can cause problems. Thus, combatting piracy involving FTA receivers has been difficult using legal means.

2.17.4 See also

- free-to-air satellite television

- PBS Satellite Service formerly on AMC-3 (87°W), now on AMC 21 (125°W)

- GlobeCast World TV and ethnic television on Galaxy 19 (97°W)

- Glorystar and religious broadcasting on Galaxy 19 (97°W)

- Equity Broadcasting formerly on Galaxy 18 (123°W), now defunct

- Home2US free-to-air and subscription ethnic programming on AMC4 (101°W)

- Retro Television Network on AMC9 (83°W, C-Band)

- White Springs Television formerly on Galaxy 27 (129°W), no longer on satellite due to uplink failure

- Bell TV, Dish Network and pirate decryption issues surrounding these systems

2.17.5 References

[1] SatcoDX satellite chart

[2] http://www.lyngsat.com/galaxy19.html - the largest source of free ethnic satellite TV in North America

[3] http://www.lyngsat.com/amc21.html carries a wide range of PBS educational programming

[4] Free TV from the United States, Lyngsat satellite TV directory, Christian Lyngemark

[5] A website Satscams.com, owned and operated by NagraStar LLC (a joint venture between EchoStar and Kudelski), contains a long list of boasts of lawsuits and litigation against individual Internet Key Sharing end users and receiver manufacturers such as nFusion, Kbox, Viewtech, Sonicview, Freetech and Panarex.

[6] Lyngsat - AMC 15/18 at 105°W requires 8PSK, C-band, MPEG4, DVB-S2 for NBC reception

[7] Equity Media Holdings Corporation Files Voluntary Petition for Chapter 11 Bankruptcy Protection, MSNBC, Dec. 9, 2008

[8] http://www.lyngsat.com/galaxy18.html - the former home of Equity Broadcasting on FTA DVB-S, now largely vacant

2.17.6 Popular brands

2.17.7 Peripheral equipment

- Satellite dish

- LNB

- DiSEqC

- Universal Satellites Automatic Location System

2.17.8 External links

- Free to air TV channels

- Free to air radio stations

- Channel Master editor (archived, last update October 2008)

- Channel finder for Astra satellites

2.18 Gascony Show

This article is about the Irish broadcaster. For the actor, see John Slattery. For the Colorado politician, see John Henry Slattery. For the Massachusetts politician, see John P. Slattery.

The *Gascony Show* is an English-language radio chat show aired in the southwest of France and a flagship Sunday evening programme broadcast on Radio Coteaux[1] in the Gascony region. The show is presented by Irishman John Slattery.

2.18.1 The show

John Slattery takes over the airwaves at 5pm and finishes at 7pm with a handover to Patrick Martinez. The *Gascony Show* is a music-based programme that includes light-hearted banter and whimsical jokes by the host. Listeners' views and comments are discussed and there are rotating segments such as "Song of The Week" and "Life in the South-West". To keep listeners on their toes, random topics are picked from a big "Bag O Topics."

The Gascony Show *host, John Slattery, in the studio in Saint Blancard*

Podcasts of all previous shows[2] are available on the show's website so that previous shows can be listened to again. Previous shows can also be found on iTunes.[3]

2.18.2   Invited guests

The show runs interviews with regular and/or rotating guests,[4] dealing with subjects of both local interest and of general interest to the expatriate community. The first person to be interviewed on the show was the illustrious English painter, interior designer and writer, Stella Wulf.[5] Stella is the author of "The Lonesome Froom and other Strange Tales," "Rum Rhymes and Vagabond" and "The Song of the Froom." She recounted humorous stories of her years in France and her efforts to get her books published.

Subsequent interviews included music industry notaries such as Joy Askew and her brother Roger Askew.[6] These two truly multi-dimensional singers sang some of their favourite songs on air. Joy, who worked closely with the likes of Peter Gabriel and Joe Jackson, and fronted the all-girl rock band called Bitch, told her story of life as a songwriter and touring musician. The interview proved enormously popular with both English and French speaking listeners in France and it served to create a pattern of story telling and musical treats that would become an underlying trait of later interviews.

A highly commended interview was that with Leaf Fielding,[7] one of the drug-dealers jailed as a result of the famous 1977 drug cartel bust in the UK called Operation Julie. The interview revolved around Leaf's life, his descent into drug dealing, his imprisonment, and his new life in France. Also discussed was Leaf's book *To Live Outside The Law*,[8] that tells the story from his early days and his first experience with LSD, to his amazing falling from grace.

2.18.3   Regular contributors

The show relies heavily on e-mailed material sent in by listeners that is then read out on air, and most of the regular contributors are local expatriates living in the Gascony region.

2.18.4   References

[1] "Emissions thématiques sur le site officiel de radio coteaux, frequences 104.4 et 97.7 FM" By Patrick Martinez, Radio Coteaux

[2] "The Gascony Show Podcasts" Recordings of all previous shows.

[3] "The Gascony Show on iTunes" Recordings of all previous shows.

[4] "The Gascony Show Archives" Recorded shows with guests.

[5] "Stella Wulf, News & Events" Interview, Sunday, 10 July 2011.

[6] "The Gascony Show Channel" Produced by PK Media Promotions.

[7] "Leaf Fielding on the Gascony Show" Interview, Sunday, 11 September 2011.

[8] "To Live Outside The Law" By The Guardian, Thursday, 7 July 2011.

2.18.5   External links

- *Gascony Show* home page

- Radio Coteaux

- Angloinfo Blog on The Gascony Show

2.19   George Wood (Radio Sweden)

George MacClaren Wood III is an American journalist who has worked at Radio Sweden since 1975, He was born in Berkeley, California on August 10, 1949, and grew up in Piedmont, California. He has degrees from the University of California, Santa Barbara and the University of California, Berkeley, and participated in the university's Education Abroad Program to Lund University in Sweden, 1969-1970.

2.19.1 Radio Sweden

George Wood began as a freelance reporter at Radio Sweden in 1975.[1] Following the retirement of Arne Skoog in 1978 he told over the writing and presenting of the program Sweden Calling DXers and its successor MediaScan, until the latter was taken off the air in 2001. In 1994 MediaScan became the first radio program in Sweden and the second in Europe (the first in English) to have its audio posted on the Internet.

He was Radio Sweden's Webmaster since Swedish Radio's first website launched in 1995, while also serving as a journalist for the Radio Swedish English Service. His was one of the voices in the satirical sketches in the program the Saturday Show. He retired from Radio Sweden in August, 2014.

2.19.2 Egyptology

After retirement George Wood enrolled in Egyptology courses at Uppsala University.

2.19.3 Other activities

George Wood was News Director of the university radio station KCSB-FM (at the University of California, Santa Barbara) in 1971-1973, and has also worked at KPFA in Berkeley, California, Earth News Service in San Francisco, and briefly as a freelancer for National Public Radio. Since the late 70's he has been the Stockholm stringer for CBS Radio News. He has written articles for a number of publications including "Satellite Times", "The World Radio TV Handbook", and the "New Age Journal".

2.19.4 References

[1] http://sverigesradio.se/sida/artikel.aspx?programid=2054&artikel=567350

2.19.5 External links

- George Wood's page on the Radio Sweden website

2.20 Glenn Hauser

Glenn Hauser (born April 12, 1945 in Berkeley, California) is an internationally-known American DXer[1][2] and radio host from Enid, Oklahoma. He produces and presents a weekly 30-minute program, *World Of Radio*,[3] heard on a number of non-commercial AM and FM stations throughout the U.S. and worldwide on shortwave.

Hauser began his broadcasting career on Radio Canada International during the late 1970s, providing DX tips on Sunday nights, and his tips also appeared on Radio Nederland's *DX Juke Box* program. He wrote for *Popular Electronics* and *Modern Electronics*, and published *Review of International Broadcasting*.[4]

2.20.1 *World of Radio*

World Of Radio debuted in 1980 on WUOT-FM in Knoxville, Tennessee, moving to shortwave two years later. The half-hour program consists of Hauser reading news about radio around the world in a characteristic monotone. Although *World of Radio* focuses on shortwave news, it covers all aspects of broadcasting. Most items are contributed by listeners to the program or DX publications.[5]

Mundo Radial

Hauser also produced *Mundo Radial*, a Spanish edition of *World of Radio*, from January 2002 to November 2007.[6]

2.20.2 *Review of International Broadcasting*

Hauser introduced *Review of International Broadcasting* in February 1977.[7] The magazine published 154 issues, with columns such as "Listener Insights on Programming," "Radio Equipment Forum," "DX Listening Digest," "The Media Mind" and "Satellite Watch."[7] Contributors included David Newkirk, Loren Cox and Juan Carlos Codina,[7] and *RIB* also featured columns from the BBC, John Norfolk and Alan Roe.[7] It was published monthly during the 1970s and 1980s, later decreasing to quarterly and semiannually before ceasing publication in October 1997.[7] *RIB*'s successor, *DX Listening Digest*, went online in 1999.[7]

2.20.3 Political and religious views

Hauser is a political liberal and an agnostic, which occasionally puts him at odds with the fundamentalist-dominated American shortwave scene which carries *World of Radio*.

2.20.4 References

[1] Berg, Jerome S. (October 2008). *Listening on the short waves, 1945 to today*. McFarland. pp. 222–. ISBN 978-0-7864-3996-6. Retrieved 17 April 2011.

[2] Bennett, Hank; Hardy, David T.; Yoder, Andrew R. (September 1993). *The complete shortwave listener's handbook*. TAB Books. p. 203. ISBN 978-0-8306-4347-9. Retrieved 17 April 2011.

[3] American Radio Relay League (August 1997). *ARRL Operating Manual*. American Radio Relay League. pp. 1–13. ISBN 978-0-87259-614-6. Retrieved 17 April 2011.

[4] Harvey, Sheldon, "Glenn Hauser interview", *CIDX Special Feature #5*, Summer 1999

[5] "DX Listening Digest". Retrieved May 18, 2015.

[6] Mundo Radial

[7] Berg, Jerome S., "Review of International Broadcasting (1977–1997)", 'Listening on the short waves, 1945 to today, pg 221–223

2.20.5 External links

- Glenn Hauser's World of Radio
- Glenn Hauser's website
- World of Radio Podcast via WRN

2.21 Greenwich Time Signal

"The pips" redirects here. For the musical group, see Gladys Knight & the Pips.

The **Greenwich Time Signal (GTS)**, popularly known

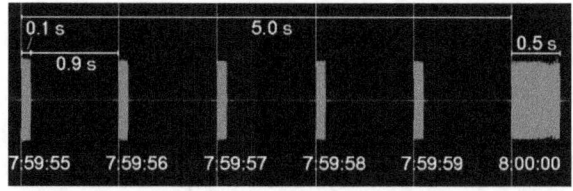

Graph of the six pips

as **the pips**, is a series of six short tones broadcast at one-second intervals by many BBC Radio stations. The pips were introduced in 1924 and have been generated by the BBC since 1990[1] to mark the precise start of each hour. Their utility in calibration is diminishing as digital broadcasting entails time lags.

2.21.1 Structure

There are six pips (short beeps) in total, which occur on the 5 seconds leading up to the hour and on the hour itself. Each pip is a 1 kHz tone (about half way between musical B5 and

C6),[2] the first five of which last a tenth of a second each, while the final pip lasts half a second. The actual moment when the hour changes – the "on-time marker" – is at the very beginning of the last pip.[3]

When a leap second occurs (exactly one second before midnight UTC), it is indicated by a seventh pip. In this case the first pip occurs at 23:59:55 (as usual) and there is a sixth short pip at 23:59:60 (the leap second) followed by the long pip at 00:00:00.[4] The possibility of an extra pip for the leap second thus justifies the final pip being longer than the others, so that it is always clear which pip is on the hour. Before leap seconds were conceived, the final pip was the same length as the others.[5] Although "negative" leap seconds can also be used to make the year shorter, this has never happened in practice.[6][7]

Although normally broadcast only on the hour by BBC domestic radio, BBC World Service use the signal at other times as well. The signal is generated at each quarter-hour and has on occasion been broadcast in error.

2.21.2 Usage

The pips are available to BBC radio stations every 15 minutes but except in rare cases, they are only broadcast on the hour, usually before news bulletins or news programes.

On BBC Radio 4, the pips are broadcast every hour except at 18:00 and 00:00 and at 22:00 on Sundays (at the start of the Westminster Hour) when they are replaced by the Westminster chimes of Big Ben at the Palace of Westminster. No time signal is broadcast at 15:00 on Saturdays and at 10:00 and 11:00 on Sundays.

On BBC Radio 2, the pips are used at 07:00, 08:00, 17:00 and 19:00 on weekdays, at 07:00 and 08:00 on Saturdays and at 08:00 and 09:00 on Sundays.

The pips were used on Radio 1 during *The Chris Moyles Show* at 06:30 just after the news, 09:00 as part of the Tedious Link feature, 10 am (at the end of the show) and often before *Newsbeat*. As most stations only air the pips on the hour, *The Chris Moyles Show* was the only show where the pips were broadcast on the half-hour. Chris Moyles continues to use the pips at the beginning of his show on Radio X. Zane Lowe's Masterpieces, the playing of an album in its entirety, is begun with pips, and they also feature at 19:00 on Fridays to signify the start of the weekend and at 16:00 on Sundays to mark the start of *The Official Chart Show*. *The Weekend Breakfast Show with Dev* begins with the pips at 06:00, and they sometimes feature on the hour at other points during the show, and *Gemma Cairney's Early Breakfast Show* begins with the pips. Dev's previous *Early Breakfast Show* also featured the pips at the beginning, and on the half-hour/hour at other points, particularly at 06:00 before or after the "I'm Here All Week" track. The pips are also used at 19:00 on Saturday evenings at the start of Radio 1's 12-hour simulcast with digital station BBC Radio 1Xtra.

BBC Radio 3 and BBC Radio Five Live do not currently feature the Greenwich Time Signal in their scheduled programming.

The BBC World Service broadcasts the pips every hour.

Pips can also be heard on many BBC Local Radio stations although their use is up to the discretion of individual stations. A rare quarter-hour Greenwich Time Signal can be heard at 05:15 weekdays on Wally Webb's programme on six BBC Local Radio stations in the East of England, as part of his "sychronised cup of tea" feature.

In 1999, pip-like sounds were incorporated into the themes written by composer David Lowe to introduce BBC Television News programmes. They are still used today on BBC One, BBC World News and BBC News.

The BBC does not allow the pips to be broadcast except as a time signal. Radio plays and comedies which use fictional news programmes use various methods to avoid playing the full six pips, ranging from simply fading in the pips to a version played on *On the Hour* in which the sound was made into a small tune between the pips. *The News Quiz* also featured a special Christmas pantomime edition where the pips went missing, and the problem was avoided there by only playing individual pips and not the whole set. The 2012 project *Radio Reunited*, however, did use the pips not as a time signal, but simply to commemorate 90 years of BBC Radio.

2.21.3 Accuracy

The pips for national radio stations and some local radio stations are timed relative to UTC, from an atomic clock in the basement of Broadcasting House synchronised with the National Physical Laboratory's Time from NPL and GPS. On other stations, the pips are generated locally from a GPS-synchronised clock.

The BBC compensates for the time delay in both broadcasting and receiving equipment, as well as the time for the actual transmission. The pips are timed so that they are accurately received on long wave as far as 160 kilometres (100 mi) from the Droitwich AM transmitter, which is the distance to Central London.

As a pre-IRIG and pre-NTP time transfer and transmission system, the pips have been a great technological success. In modern times, however, time can be transferred to systems with CPUs and operating systems by using BCD or some Unix Time variant.

Newer digital broadcasting methods have introduced even

greater problems for the accuracy of use of the pips. On digital platforms such as DVB, DAB, satellite and the Internet, the pips—although *generated* accurately—are not *heard* by the listener exactly on the hour. The encoding and decoding of the digital signal causes a delay, of usually between 2 and 8 seconds. In the case of satellite broadcasting, the travel time of the signal to and from the satellite adds about another 0.25 seconds.

DVB, DAB (Eureka 147 and Digital Radio Mondiale) and FM Radio Data System all support separate time signal transmission subsystems with accuracy equal to or several orders of magnitude better than the pips, so the listener need not worry about decoding the pips to synchronize the clocks on these systems.

2.21.4 History

The machine used to generate the pips in 1970

The pips have been broadcast daily since 5 February 1924,[8] and were the idea of the Astronomer Royal, Sir Frank Watson Dyson, and the head of the BBC, John Reith. The pips were originally controlled by two mechanical clocks located in the Royal Greenwich Observatory that had electrical contacts attached to their pendula. Two clocks were used in case of a breakdown of one. These sent a signal each second to the BBC, which converted them to the audible oscillatory tone broadcast.[8]

The Royal Greenwich Observatory moved to Herstmonceux Castle in 1957 and the GTS equipment followed a few years later in the form of an electronic clock. Reliability was improved by renting two lines for the service between Herstmonceux and the BBC, with a changeover between the two at Broadcasting House if the main line became disconnected.

The tone sent on the lines was inverted: the signal sent to the BBC was a steady 1 kHz tone when no pip was required, and no tone when a pip should be sounded. This let faults on the line be detected immediately by automated monitoring for loss of audio.

The Greenwich Time Signal was the first sound heard in the handover to the London 2012 Olympics during the Beijing 2008 Olympics closing ceremony.[9]

The pips were also broadcast by the BBC Television Service, but this practice was discontinued by the 1960s.

To celebrate the 90th birthday of the pips on 5 February 2014, the *Today* programme broadcast a sequence that included a re-working of the Happy Birthday melody using the GTS as its base sound.[10]

2.21.5 Crashing the pips

The BBC discourages any other sound being broadcast at the same time as the pips; doing so is commonly known as 'crashing the pips'. This was most often referred to on Terry Wogan's show, *Wake Up to Wogan*, although usually only in jest since the actual event happened rarely.[11] Different BBC Radio stations approach this issue differently. Both BBC Radio 1 and Radio 2 generally take a more laidback approach with the pips, usually playing them over the closing seconds of a currently playing song or a jingle 'bed' (background music from a jingle), followed by their respective news jingles. Many BBC local radio stations also play the pips over the station's jingle. BBC Radio 4 is stricter. It is an almost entirely speech-based network; incidents at the end of the *Today* programme regularly cause listeners' complaints.

As a contribution to Comic Relief's 2005 Red Nose Day, the BBC developed a "pips" ring-tone which can be downloaded.[12]

Bill Bailey's BBC Rave includes the BBC News theme, which incorporates a variant of the pips (though not actually broadcast exactly on the hour). The footage can be seen on his DVD *Part Troll*.

In the late 1980s Radio 1 featured the pips played over a station jingle during Jakki Brambles' early show and Simon Mayo's breakfast show. This was not strictly crashing the pips as they were not intended to be, or mistaken for, an accurate time signal.

2.21.6 Technical problems

At 8 am on 17 September 2008, to the surprise of John Humphrys, the day's main presenter on the *Today* programme, and Johnnie Walker, who was standing in for Terry Wogan on Radio 2, the pips went adrift by 6 seconds, and broadcast seven pips rather than six. This was traced

to a problem with the pip generator, which was 'repaired' by switching it off and on again.[13] Part of Humphrys' surprise was probably because of his deliberate avoidance of crashing the pips with the help of an accurate clock in the studio.

A sudden total failure in the generation of the audio pulses that constitute the pips was experienced on 31 May 2011 and silence was unexpectedly broadcast in place of the 17:00 signal. The problem was traced to the power supply of the equipment which converts the signal from the atomic clocks into an audible signal.[14] Whilst repairs were underway the BBC elected to broadcast a "dignified silence" in place of the pips at 19:00.[15] By 19:45 the same day the power supply was repaired[14] and the 20:00 pips were broadcast as normal.[16]

2.21.7 Similar time signals elsewhere

Many radio broadcasters around the world use the Greenwich Time Signal as a means to mark the start of the hour. The pips are both used in domestic and international commercial and public broadcasting. Many radio stations use six tones similar to those used by the BBC World Service; some shorten it to five, four, or three tones.

- Argentina - all news/talk stations (Radio Nacional, Radio Mitre, Radio Continental, Radio 10, Cadena 3, etc.) air the six pips similar to the BBC every hour, and 3 pips for every half-hour similar to Catalonia.

- Australia - pips are used on ABC Radio National and ABC Local Radio at the top of every hour, as well as on Fairfax Media talkback stations- 2UE, 3AW, 4BC and 6PR.

- Belgium - Both RTBF and VRT broadcast six short peeps every hour.

- Canada - the National Research Council Time Signal is broadcast daily on Ici Radio-Canada Première at 12:00 EST/EDT and on CBC Radio One at 13:00 EST/EDT. It is Canada's longest running radio feature and has been broadcast every day since November 5, 1939.

- Finland - on YLE's radio services the pips are broadcast on the hour.

- France - the station France Inter broadcasts four very short pips every hour, which are almost invariably crashed. The last pip, which is as long as the other ones, marks the top of the hour. Some local stations of the France Bleu network also air four pips that are a little longer than Inter's.

- Germany - Deutschlandfunk broadcasts four beeps every hour, the last one being longer than the others; between 05:00 and 18:00 on weekdays they are broadcast every half-hour and often omitted at 21:00 when there is no news programme scheduled. Similarly, Deutschlandradio Kultur uses six pips, the last one being longer than the others.

- Hong Kong - pips are used on RTHK's radio channels for the same purpose and in the same way. The signals, which are provided by the Hong Kong Observatory, are broadcast every half-hour during the day and on the hour at night, immediately before the news headline reports.

- Hungary - the national radio channel Kossuth broadcasts five stereophonic pips at the top of every hour, the fifth being longer than the others.

- Ireland - six pips are broadcast before news bulletins at 07:00, 13:00 and 24:00 on RTÉ Radio 1.

- Israel - Kol Israel's hourly newscasts begin with six tones, with the sixth tone being longer.

- Italy - The National Italian Radio "Rai" uses 6 tones to signal exact time in all its stations. They differ in the timing: Rai 1 tells the hour. Rai 2 tells on the half hour. Rai 3 signals the 45 minute of selected hours. All the signals come from the Istituto Metrologico di Torino, the national study center for measure and time.

- Malaysia - RTM radio stations use the pips hourly before the news broadcast but only the top-of-the-hour pip is sounded. Until late 2012, the time signal is simply a short pip on the 59th second before the hour and a longer pip on the top of the hour. In a news report in The Star on 1 January 1982, the pips were used to sound similar to the BBC's.

- Netherlands - only three pips are used. There used to be six, however it was felt that people would lose count, so now only three are used.

- New Zealand - the equivalent of BBC Radio 4, Radio New Zealand National, plays the six pips at the top of every hour.

- North Korea - the pips are heard on Voice of Korea before its startup at 17:00.

- Sindh The Sindhi language news channel KTN News plays 5 pips at the start of each news hour.

- Spain - the signal is broadcast by almost all radio stations, even by music stations, but depends on the frequency: music stations usually use pips on the hour,

but most of the non-musical stations broadcast the signal every 30 minutes. Los 40 Principales, the most important music radio in Spain, broadcast a different version of GTS: two first pips sound and then a music is added on the background, using the rhythm to create the corporative jingle of the radio. This station in particular uses only 4 pips, typically the two last using two different frequencies (resulting in a modern rhythm). Other musical radios like Máxima FM and M80 Radio, both owned by PRISA, and Europa FM use a similar effect.

  - Catalonia, Spain - dance music station Flaix FM and Hot AC station Ràdio Flaixbac, both owned by the same media group, broadcast every half-hour a very short sequence of two very short tones followed by a longer one, the whole lasting not more than one and a half seconds. Els 40 Principals, the Catalan edition of Spanish radio Los 40 Principales, use the same jingle, using a mix of GTS and corporative music.

- Switzerland - The first channel of the Swiss radio used to play three short peeps, the third being higher than the others. It seems to have disappeared now.

- United States - the pips can be heard on the Middlebury College radio station WRMC.

- former Yugoslavia- JRT broadcast the six pips before the news (on the hour) on radio as well as on television, before the start of the *TV Dnevnik* at 08:00 pm. The broadcast on TV was stopped in 1974 because the *TV Dnevnik* was moved to its current term at the bottom of the hour (07:30 pm)

2.21.8 See also

- Time from NPL

- National Research Council Time Signal - A CBC Radio One indicator for 1300 ET

2.21.9 References

[1] "The 'Six Pips' Time Signal" (pdf). *Eng Inf* (40). Spring 1990.

[2] Scientific pitch notation

[3] "What's the time?". *Astronomy & Time*. Royal Museums Greenwich. Retrieved 5 February 2014.

[4] "Leap second: Keeper of the pips". *BBC News*. 30 December 2008. Retrieved 5 February 2014.

[5] "The comforting tone of the hourly radio pips". *BBC News (Magazine Monitor)*. 5 February 2014. Retrieved 5 February 2014.

[6] "Leap years and leap seconds". Royal Museums Greenwich.

[7] "Adjusting after a 'long' weekend at the Royal Observatory – Precision clocks and the leap second". Royal Museums Greenwich. Retrieved 5 February 2014.

[8] *Sci/Tech - Six pip salute*, BBC News, 5 February 1999, retrieved 2009-04-23

[9] Peter Simpson (25 August 2008). "Baton Passed to London for 2012". *South China Morning Post*. Retrieved 24 September 2011.

[10] http://www.bbc.co.uk/programmes/p01rqqkn

[11] Tom Leonard (28 February 2002). "Pip, pip! Woman of Today is gone tomorrow". *The Daily Telegraph*. Retrieved 18 November 2012.

[12] "The Radio 4 Pips - How you download the pips". *Today Programme*. BBC News. March 2005. Retrieved 18 November 2012.

[13] *'Pips' slip in BBC radio error*, BBC News, 17 September 2008, retrieved 2009-04-23

[14] Denis Nowlan (1 June 2011). "What happened to the Radio 4 pips?". *Radio 4 and 4 Extra Blog*. BBC. Retrieved 2 June 2011.

[15] Harry Wallop (31 May 2011). "Radio 4's pips die". *The Telegraph*. Retrieved 31 May 2011.

[16] "The Pips return from a 3 hour break" *Radio Today* 31 May 2011 Retrieved 31 May 2011

2.21.10 External links

- http://www.miketodd.net/other/gts.htm

- http://www.radionetherlands.nl/features/media/practical/time.html

- http://www.clockco.co.uk/article_info.php?articles_id=15

2.22 Guglielmo Marconi

"Marconi" redirects here. For other uses, see Marconi (disambiguation).

Guglielmo Marconi, 1st Marquis of Marconi (Italian: [ɡuʎˈʎɛlmo marˈkoːni]; 25 April 1874 – 20 July 1937) was an Italian inventor and electrical engineer, known for his pioneering work on long-distance radio transmission[1] and for his development of Marconi's law and a radio telegraph system. He is often credited as the inventor of radio,[2] and he shared the 1909 Nobel Prize in Physics with Karl Ferdinand Braun "in recognition of their contributions to the development of wireless telegraphy".[3][4][5] An entrepreneur, businessman, and founder in Britain in 1897 of The Wireless Telegraph & Signal Company (which became the Marconi Company), Marconi succeeded in making a commercial success of radio by innovating and building on the work of previous experimenters and physicists.[6][7] In 1929 the King of Italy ennobled Marconi as a Marchese (marquis).

2.22.1 Biography

Early years

Marconi was born into the Italian nobility as Guglielmo Giovanni Maria Marconi[8] in Bologna on 25 April 1874, the second son of Giuseppe Marconi (an Italian aristocratic landowner from Porretta Terme) and of his Irish/Scots wife, Annie Jameson (daughter of Andrew Jameson of Daphne Castle in County Wexford, Ireland and granddaughter of John Jameson, founder of whiskey distillers *Jameson & Sons*[9]). Between the ages of two and six Marconi, along with his elder brother Alfonso, was brought up by his mother in the English town of Bedford.[10][11] After returning to Italy he received his early education privately in Bologna in the lab of Augusto Righi, in Florence at the Istituto Cavallero and, later, in Livorno.[12] As a child, according to Robert McHenry, Marconi did not do well in school,[13] though historian Giuliano Corradi in his biography characterizes him as a true genius.[14] Baptized as a Catholic, he had been brought up as a member of the Anglican Church, being married into it (although this marriage was later annulled). Before his marriage to Maria Christina in 1927, Marconi was confirmed in the Catholic faith and became a devout member of the Church.[15]

Radio work

During his early years, Marconi had an interest in science and electricity. One of the scientific developments during this era came from Heinrich Hertz, who, beginning in 1888, demonstrated that one could produce and detect electromagnetic radiation—now generally known as radio waves, at the time more commonly called "Hertzian waves" or "aetheric waves". Hertz's death in 1894 brought published reviews of his earlier discoveries, and a renewed interest on the part of Marconi. He was permitted to briefly study the subject under Augusto Righi, a University of

Bologna physicist and neighbour of Marconi who had done research on Hertz's work.

Marconi's first transmitter, consisting of a copper sheet capacitive antenna (top) connected to a Righi spark gap (left) powered by an induction coil (center) with a telegraph key (right) to switch it on and off to spell out text messages in Morse code.

Early experimental devices Marconi began to conduct experiments, building much of his own equipment in the attic of his home at the Villa Griffone in Pontecchio, Italy, with the help of his butler Mignani. His goal was to use radio waves to create a practical system of "wireless telegraphy"—i.e. the transmission of telegraph messages without connecting wires as used by the electric telegraph. This was

Marconi Rock, Salvan, Switzerland, site of 1895 experiments

not a new idea—numerous investigators had been exploring wireless telegraph technologies for over 50 years, but none had proven technically and commercially successful. Marconi's system had the following components:[16]

- A relatively simple oscillator, or spark-producing radio transmitter.

- A wire or metal sheet capacity area suspended at a height above the ground;

- A coherer receiver, which was a modification of Edouard Branly's original device, with refinements to increase sensitivity and reliability;

- A telegraph key to operate the transmitter to send short and long pulses, corresponding to the dots-and-dashes of Morse code; and

- A telegraph register, activated by the coherer, which recorded the received Morse code dots and dashes onto a roll of paper tape.

Similar configurations using spark-gap transmitters plus coherer-receivers had been tried by others, but many were unable to achieve transmission ranges of more than a few hundred metres.

Marconi, just twenty years old, began his first experiments working on his own with the help of his butler Mignani. In the summer of 1894, he built a storm alarm made up of a battery, a coherer, and an electric bell, which went off if there was lightning. Soon after he was able to make a bell ring on the other side of the room by pushing a telegraphic button on a bench.[17]

One night in December, Guglielmo woke his mother up and invited her into his secret workshop and showed her the experiment he had created. The next day he also showed his work to his father, who, when he was certain there were no wires, gave his son all of the money he had in his wallet so Guglielmo could buy more materials.

In the summer of 1895 Marconi moved his experimentation outdoors and continued to experiment on his father's estate in Bologna. He tried different arrangements and shapes of antenna. A breakthrough came when he found that much greater range could be achieved by attaching one side of the transmitter and receiver to a wire or metal sheet elevated above ground, and the other side to the earth.[18] This monopole antenna reduced the frequency of the waves compared to the dipole antennas used by Hertz, and radiated vertically polarized radio waves which could travel longer distances. Soon he was able to transmit signals over a hill, a distance of approximately 2.4 kilometres (1.5 mi).[19] By this point he concluded that with additional funding and research, a device could become capable of spanning greater distances and would prove valuable both commercially and militarily.

Marconi wrote to the Ministry of Post and Telegraphs, then under the direction of the honorable Pietro Lacava, explaining his wireless telegraph machine and asking for funding. He never received a response to his letter which was eventually dismissed by the Minister who wrote "to the Longara" on the document, referring to the insane asylum on Via della Lungara in Rome.[20]

In 1896, Marconi spoke with his family friend Carlo Gardini, Honorary Consul at the United States Consulate in Bologna, about leaving Italy to go to England. Gardini wrote a letter of introduction to the Ambassador of Italy in London, Annibale Ferrero, explaining who Marconi was and about these extraordinary discoveries. In his response, Ambassador Ferrero advised them not to reveal the results until after they had obtained the copyrights. He also encouraged him to come to England where he believed it would be easier to find the necessary funds to convert the findings from Marconi's experiment into a practical use. Finding little interest or appreciation for his work in Italy, Marconi travelled to London in early 1896 at the age of 21, accompanied by his mother, to seek support for his work; Marconi spoke fluent English in addition to Italian. Marconi arrived at Dover and at Customs the Customs officer

opened his case to find various contraptions and apparatus. The customs officer immediately contacted the Admiralty in London. While there, Marconi gained the interest and support of William Preece, the Chief Electrical Engineer of the British Post Office.

British Post Office engineers inspect Marconi's radio equipment during demonstration on Flat Holm Island, 13 May 1897. The transmitter is at center, the coherer receiver below it, the pole supporting the wire antenna is visible at top.

The British become interested   The apparatus that Marconi possessed at that time was similar to that of one in 1882 by A. E. Dolbear, of Tufts College, which used a spark coil generator and a carbon granular rectifier for reception.[21] A plaque[22] on the outside of BT Centre commemorates Marconi's first public transmission of wireless signals from that site.[23] A series of demonstrations for the British government followed—by March 1897, Marconi had transmitted Morse code signals over a distance of about 6 kilometres (3.7 mi) across Salisbury Plain. On 13 May 1897, Marconi sent the world's first ever wireless communication over open sea. The experiment, based in Wales, witnessed a message transversed over the Bristol Channel from Flat Holm Island to Lavernock Point in Penarth, a distance of 6 kilometres (3.7 mi). The message read "Are you ready".[24] The transmitting equipment was almost immediately relocated to Brean Down Fort on the Somerset coast, stretching the range to 16 kilometres (9.9 mi).

Impressed by these and other demonstrations, Preece introduced Marconi's ongoing work to the general public at two important London lectures: "Telegraphy without Wires", at the Toynbee Hall on 11 December 1896; and "Signaling through Space without Wires", given to the Royal Institution on 4 June 1897.

Numerous additional demonstrations followed, and Marconi began to receive international attention. In July 1897, he carried out a series of tests at La Spezia, in his home country, for the Italian government. A test for Lloyds

Plaque on the outside of BT Centre commemorates Marconi's first public transmission of wireless signals.

Marconi watching associates raising the kite (a "Levitor" by B.F.S. Baden-Powell[26]) used to lift the antenna at St. John's, Newfoundland, December 1901

between Ballycastle and Rathlin Island, Ireland, was conducted on 6 July 1898. The English channel was crossed on 27 March 1899, from Wimereux, France to South Foreland Lighthouse, England, and in the autumn of 1899, the first demonstrations in the United States took place, with the reporting of the America's Cup international yacht races at New York.

Marconi sailed to the United States at the invitation of the New York Herald newspaper to cover the America's Cup races off Sandy Hook, NJ. The transmission was done aboard the SS *Ponce*, a passenger ship of the *Porto Rico Line*.[25] Marconi left for England on 8 November 1899 on the American Line's SS *Saint Paul*, and he and his assistants installed wireless equipment aboard during the voyage. On 15 November *Saint Paul* became the first ocean liner to report her imminent return to Great Britain by wireless when Marconi's Royal Needles Hotel radio station contacted her sixty-six nautical miles off the English coast.

Transatlantic transmissions At the turn of the 20th century, Marconi began investigating the means to signal completely across the Atlantic, in order to compete with the transatlantic telegraph cables. Marconi established a wireless transmitting station at Marconi House, Rosslare Strand,

Co. Wexford in 1901 to act as a link between Poldhu in Cornwall, England and Clifden in Co. Galway, Ireland. He soon made the announcement that on 12 December 1901, using a 500-foot (150 m) kite-supported antenna for reception, the message was received at Signal Hill in St John's, Newfoundland (now part of Canada) signals transmitted by the company's new high-power station at Poldhu, Cornwall. The distance between the two points was about 2,200 miles (3,500 km). Heralded as a great scientific advance, there was—and continues to be—considerable skepticism about this claim. The exact wavelength used is not known, but it is fairly reliably determined to have been in the neighborhood of 350 meters. The tests took place at a time of day during which the entire transatlantic path was in daylight. We now know (although Marconi did not know then) that this was the worst possible choice. At this medium wavelength, long distance transmission in the daytime is not possible because of heavy absorption of the skywave in the ionosphere. It was not a blind test—Marconi knew in advance to listen for a repetitive signal of three clicks, signifying the Morse code letter *S*. The clicks were reported to have been heard faintly and sporadically. There was no independent confirmation of the reported reception, and the transmissions were difficult to distinguish from atmospheric noise. (A detailed technical review of Marconi's early transatlantic work appears in John S. Belrose's work of 1995.) The Poldhu transmitter was a two-stage circuit.[28][29]

Feeling challenged by skeptics, Marconi prepared a better organized and documented test. In February 1902, the SS *Philadelphia* sailed west from Great Britain with Marconi aboard, carefully recording signals sent daily from the Poldhu station. The test results produced coherer-tape reception up to 1,550 miles (2,490 km), and audio reception up to 2,100 miles (3,400 km). The maximum distances were achieved at night, and these tests were the first to show that for mediumwave and longwave transmissions, radio signals travel much farther at night than in the day. During the daytime, signals had only been received up to about 700 miles (1,100 km), less than half of the distance claimed

Marconi demonstrating apparatus similar to that used by him to transmit the first wireless signal across the Atlantic Ocean, 1901. The transmitter is at right, the receiver with paper tape recorder at left.

Marconi caricatured by Spy for Vanity Fair, 1905

tially limited to line-of-sight distances.

On 17 December 1902, a transmission from the Marconi station in Glace Bay, Nova Scotia, Canada, became the world's first radio message to cross the Atlantic from North America. In 1901, Marconi built a station near South Wellfleet, Massachusetts that on 18 January 1903 sent a message of greetings from Theodore Roosevelt, the President of the United States, to King Edward VII of the United Kingdom, marking the first transatlantic radio transmission originating in the United States. However, consistent transatlantic signalling was difficult to establish.

Marconi began to build high-powered stations on both sides of the Atlantic to communicate with ships at sea, in competition with other inventors. In 1904 a commercial service was established to transmit nightly news summaries to subscribing ships, which could incorporate them into their onboard newspapers. A regular transatlantic radio-telegraph service was finally begun on 17 October 1907[30][31] between Clifden Ireland and Glace Bay, but even after this the company struggled for many years to provide reliable communication to others.

Titanic The role played by Marconi Co. wireless in maritime rescues, particularly two famous ocean liner sinkings, the RMS *Titanic* on 15 October 1912 and the RMS *Lusitania* on 7 May 1915 raised public awareness of the value of radio and brought fame to Marconi.

The two radio operators aboard the RMS *Titanic*—Jack Phillips and Harold Bride—were not employed by the White Star Line, but by the Marconi International Marine Communication Company. After the sinking of the ocean liner on 15 April 1912, survivors were rescued by the RMS *Carpathia* of the Cunard Line.[32] Also employed by the Marconi Company was David Sarnoff, who would later head RCA. Wireless communications were reportedly maintained for 72 hours between *Carpathia* and Sarnoff,[33] but Sarnoff's involvement has been questioned by some modern historians. When *Carpathia* docked in New York, Marconi went aboard with a reporter from *The New York Times* to talk with Bride, the surviving operator.[32] On 18 June 1912, Marconi gave evidence to the Court of Inquiry into the loss of *Titanic* regarding the marine telegraphy's functions and the procedures for emergencies at sea.[34] Britain's postmaster-general summed up, referring to the *Titanic* disaster, "Those who have been saved, have been saved through one man, Mr. Marconi...and his marvelous invention."[35] Marconi was offered free passage on *Titanic* before she sank, but had taken *Lusitania* three days earlier. As his daughter Degna later explained, he had paperwork to do and preferred the public stenographer aboard that vessel.[36]

earlier at Newfoundland, where the transmissions had also taken place during the day. Because of this, Marconi had not fully confirmed the Newfoundland claims, although he did prove that radio signals could be sent for hundreds of kilometres, despite some scientists' belief they were essen-

Continuing work Over the years, the Marconi companies gained a reputation for being technically conservative, in particular by continuing to use inefficient spark-transmitter technology, which could only be used for radiotelegraph operations, long after it was apparent that the future of radio communication lay with continuous-wave transmissions, which were more efficient and could be used for audio transmissions. Somewhat belatedly, the company did begin significant work with continuous-wave equipment beginning in 1915, after the introduction of the oscillating vacuum tube (valve). In 1920, employing a vacuum tube transmitter, the New Street Works factory in Chelmsford was the location for the first entertainment radio broadcasts in the United Kingdom, featuring Dame Nellie Melba. In 1922 regular entertainment broadcasts commenced from the Marconi Research Centre at Great Baddow, forming the prelude to the BBC.

Later years

SIGNOR AND MRS. MARCONI.

Marconi with his wife c. 1910

In 1914, Marconi was made a Senator in the Italian Senate and appointed Honorary Knight Grand Cross of the Royal Victorian Order in the UK. During World War I, Italy joined the Allied side of the conflict, and Marconi was placed in charge of the Italian military's radio service. He attained the rank of lieutenant in the Italian Army and of commander in the Italian Navy. In 1929, he was made a

marquess by King Victor Emmanuel III.

Marconi joined the Italian Fascist party in 1923. In 1930, Italian dictator Benito Mussolini appointed him President of the Royal Academy of Italy, which made Marconi a member of the Fascist Grand Council.

Marconi died in Rome on 20 July 1937 at age 63, following a series of heart attacks, and Italy held a state funeral for him. As a tribute, shops on the street where he lived were "Closed for national mourning".[38] In addition, at 6 pm the next day, the time designated for the funeral, all BBC transmitters and wireless Post Office transmitters in the British Isles observed two minutes of silence in his honor. The British Post Office also sent a message requesting that all broadcasting ships honor Marconi with two minutes of broadcasting silence as well.[38] His remains are housed in the Villa Griffone at Sasso Marconi, Emilia-Romagna, which assumed that name in his honour in 1938.[39][40]

In 1943, the Supreme Court of the United States handed down a decision on Marconi's radio patents restoring some of the prior patents of Oliver Lodge, John Stone Stone, and Nikola Tesla.[41][42] The decision was not about Marconi's original radio patents[43] and the court declared that their decision had no bearing on Marconi's claim as the first to achieve radio transmission, just that since Marconi's claim to certain patents were questionable, he could not claim infringement on those same patents.[44] (There are claims the high court was trying to nullify a World War I claim against the U.S. government by the Marconi Company via simply restoring the non-Marconi prior patent.)[41]

2.22.2 Personal life

American electrical engineer Alfred Norton Goldsmith and Marconi on 26 June 1922.

Marconi had a brother, Alfonso, and a stepbrother, Luigi.

On 16 March 1905, Marconi married the Hon. Beatrice O'Brien (1882–1976), a daughter of Edward O'Brien, 14th Baron Inchiquin, having met her in Poole in 1904.[45] They had three daughters, Degna (1908–1998), Gioia (1916–1996), and Lucia (born and died 1906), and a son, Giulio, 2nd Marchese Marconi (1910–1971).

In 1913, the Marconis returned to Italy and became part of Rome society. Beatrice served as a lady-in-waiting to Queen Elena. The Marconis divorced in 1924, and, at Marconi's request, the marriage was annulled on 27 April 1927, so he could remarry.[46] Beatrice Marconi married her second husband, Liborio Marignoli, Marchese di Montecorona, on 3 March 1924 and had a daughter, Flaminia.[47]

On 12 June 1927 (religious 15 June), Marconi married Maria Cristina Bezzi-Scali (1900–1994), only daughter of Francesco, Count Bezzi-Scali. They had one daughter, Maria Elettra Elena Anna (born 1930), who married Prince Carlo Giovannelli (born 1942) in 1966; they later divorced. For unexplained reasons, Marconi left his entire fortune to his second wife and their only child, and nothing to the children of his first marriage.[48]

Later in life, Marconi was an active Italian Fascist[49] and an apologist for their ideology and actions such as the attack by Italian forces in Ethiopia.

Marconi wanted to personally introduce in 1931 the first radio broadcast of a Pope, Pius XI, and did announce at the microphone: "With the help of God, who places so many mysterious forces of nature at man's disposal, I have been able to prepare this instrument which will give to the faithful of the entire world the joy of listening to the voice of the Holy Father".[50]

2.22.3   Legacy and honours

Honours and awards

- In 1909, Marconi shared the Nobel Prize in Physics with Karl Braun for his contributions to radio communications.[3]

- In 1918, he was awarded the Franklin Institute's Franklin Medal.

- In 1929, he was made a marquess by King Victor Emmanuel III, thus becoming Marchese Marconi.

- In 1931, he was awarded with John Scott Medal by wireless telegraphy

- In 1934, he was awarded the Wilhelm Exner Medal.

- In 1977, Marconi was inducted into the National Broadcasters Hall of Fame.[51]

- In 1988, the Radio Hall of Fame (Museum of Broadcast Communications, Chicago) inducted Marconi as a Pioneer (soon after the inception of its awards).[52]

- In 1990, the Bank of Italy issued a 2000 lire banknote featuring his portrait on the front and on the back his accomplishments.[53]

- In 2001, Great Britain released a commemorative British two pound coin celebrating the 100th anniversary of Marconi's first wireless communication.

- Marconi's early experiments in wireless telegraphy were the subject of two IEEE Milestones; one in Switzerland in 2003[54] and most recently in Italy in 2011.[55]

- In 2009, Italy issued a commemorative silver €5 coin honouring the centennial of Marconi's Nobel Prize.

- In 2009, he was inducted into the New Jersey Hall of Fame.[56]

- The Dutch radio academy bestows the Marconi Awards annually for outstanding radio programmes, presenters and stations.

- The National Association of Broadcasters (US) bestows the annual NAB Marconi Radio Awards also for outstanding radio programs and stations.

Tributes

Guglielmo Marconi Memorial in Washington, D.C.

- A funerary monument to the effigy of Marconi can be seen in the Basilica of Santa Croce, Florence but his remains are in near the Mausoleum of Guglielmo Marconi in Pontecchio Marconi, near Bologna. His former villa, adjacent to the mausoleum is the Marconi Museum (Italy) with much of his equipment.

- A statue of Guglielmo Marconi stands in Church Square Park in Hoboken, NJ.

- A Guglielmo Marconi sculpture by Attilio Piccirilli stands in Washington, D.C.

- A large collection of Marconi artifacts was held by The General Electric Company, p.l.c. (GEC) of the United Kingdom which later renamed Marconi plc and Marconi Corporation plc. In December 2004 the extensive Marconi Collection, held at the former Marconi Research Centre at Great Baddow, Chelmsford, Essex UK was donated to the nation by the Company via the University of Oxford.[57] This consisted of the BAFTA award-winning MarconiCalling website, some 250+ physical artifacts and the massive ephemera collection of papers, books, patents and many other items. The artifacts are now held by The Museum of the History of Science and the ephemera Archives by the nearby Bodleian Library.[58] Following three years work at the Bodleian, an Online Catalogue to the Marconi Archives was released in November 2008.

- A granite obelisk stands on the clifftop near the site of Marconi's Marconi's Poldhu Wireless Station in Cornwall, commemorating the first transatlantic transmission.

Places and organizations named after Marconi

Europe

Italy

- Bologna Guglielmo Marconi Airport (IATA: BLQ – ICAO: LIPE), of Bologna, is named after Marconi, its native son.

- Marconi University in Rome, Italy (Università degli Studi "Guglielmo Marconi" di Roma)

- Ponte Guglielmo Marconi, bridge that connects Piazza Augusto Righi with Piazza Tommaso Edison, in Rome

Oceania

Australia

- Australian football (soccer) and social club Marconi Stallions

North America

Canada

- The 'Marconi's Wireless Telegraph Company of Canada' (now CMC Electronics, of Montreal, Canada, was created in 1903 by Guglielmo Marconi.[59] In 1925 the company was renamed to the 'Canadian Marconi Company', which was acquired by English Electric in 1953.[59] The company name changed again to CMC Electronics Inc. (French: CMC Électronique) in 2001.

- The Marconi National Historic Sites of Canada was created by Parks Canada as a tribute to Marconi's vision in the development of radio telecommunications. The first official wireless message was sent from this location by the Atlantic Ocean to England in 1902. The museum site is located in Glace Bay, Nova Scotia, at Table Head on Timmerman Street.

United States

California

- Marconi Conference Center and State Historic Park, site of the transoceanic Marshall Receiving Station, Marshall

Massachusetts

- Marconi Beach in Wellfleet, Massachusetts, part of the Cape Cod National Seashore, located near the site of his first transatlantic wireless signal from the U.S. to England. There are still remnants of the wireless tower at this beach and at Forest Road Beach in Chatham, Massachusetts.[60]

New Jersey

- New Brunswick Marconi Station, now the *Guglielmo Marconi Memorial Plaza* in Somerset, NJ. President Woodrow Wilson's Fourteen Points speech was transmitted from the site in 1918.

New York

- La Scuola d'Italia Guglielmo Marconi on New York City's Upper East Side

Pennsylvania

- Marconi Plaza, Philadelphia, Pennsylvania. Roman terrace-styled plaza originally designed by the architects Olmsted Brothers in 1914–1916, built as the grand entrance for the 1926 Sesquicentennial Exposition and renamed to honor Marconi.

2.22.4   Patents

British patents

- British patent No. 12,039 (1897) "*Improvements in Transmitting Electrical impulses and Signals, and in Apparatus therefor*". Date of Application 2 June 1896; Complete Specification Left, 2 March 1897; Accepted, 2 July 1897 (later claimed by Oliver Lodge to contain his own ideas which he failed to patent).

- British patent No. 7,777 (1900) "*Improvements in Apparatus for Wireless Telegraphy*". Date of Application 26 April 1900; Complete Specification Left, 25 February 1901; Accepted, 13 April 1901.

- British patent No. 10245 (1902)

- British patent No. 5113 (1904) "*Improvements in Transmitters suitable for Wireless Telegraphy*". Date of Application 1 March 1904; Complete Specification Left, 30 November 1904; Accepted, 19 January August 1905.

- British patent No. 21640 (1904) "*Improvements in Apparatus for Wireless Telegraphy*". Date of Application 8 October 1904; Complete Specification Left, 6 July 1905; Accepted, 10 August 1905.

- British patent No. 14788 (1904) "*Improvements in or relating to Wireless Telegraphy*". Date of Application 18 July 1905; Complete Specification Left, 23 January 1906; Accepted, 10 May 1906.

US patents

- U.S. Patent 586,193 "*Transmitting electrical signals*", (using Ruhmkorff coil and Morse code key) filed December 1896, patented July 1897

- U.S. Patent 624,516 "*Apparatus employed in wireless telegraphy*".

- U.S. Patent 627,650 "*Apparatus employed in wireless telegraphy*".

- U.S. Patent 647,007 "*Apparatus employed in wireless telegraphy*".

- U.S. Patent 647,008 "*Apparatus employed in wireless telegraphy*".

- U.S. Patent 647,009 "*Apparatus employed in wireless telegraphy*".

- U.S. Patent 650,109 "*Apparatus employed in wireless telegraphy*".

- U.S. Patent 650,110 "*Apparatus employed in wireless telegraphy*".

- U.S. Patent 668,315 "*Receiver for electrical oscillations*".

- U.S. Patent 760,463 "*Wireless signaling system*".

- U.S. Patent 792,528 "*Wireless telegraphy*". Filed 13 October 1903; Issued 13, 1905.

- U.S. Patent 676,332 "*Apparatus for wireless telegraphy*" (later practical version of system)

- U.S. Patent 757,559 "*Wireless telegraphy system*". Filed 19 November 1901; Issued 19 April 1904.

- U.S. Patent 760,463 "*Wireless signaling system*". Filed 10 September 1903; Issued 24 May 1904.

- U.S. Patent 763,772 "*Apparatus for wireless telegraphy*" (Four tuned system; this innovation was predated by N. Tesla, O. Lodge, and J. S. Stone)

- U.S. Patent 786,132 "*Wireless telegraphy*". Filed 13 October 1903

- U.S. Patent 792,528 "*Wireless telegraphy*". Filed 13 October 1903; Issued 13 June 1905.

- U.S. Patent 884,986 "*Wireless telegraphy*". Filed 28 November 1902; Issued 14 April 1908.

- U.S. Patent 884,987 "*Wireless telegraphy*".

- U.S. Patent 884,988 "*Detecting electrical oscillations*". Filed 2 February 1903; Issued 14 April 1908.

- U.S. Patent 884,989 "*Wireless telegraphy*".

- U.S. Patent 935,381 "*Transmitting apparatus for wireless telegraphy*". Filed 10 April 1908; Issued 28 September 1909.

- U.S. Patent 935,382 "*Apparatus for wireless telegraphy*".

- U.S. Patent 935,383 "*Apparatus for wireless telegraphy*". Filed 10 April 1908; Issued 28 September 1909.

- U.S. Patent 954,640 "*Apparatus for wireless telegraphy*". Filed 31 March 1909; Issued 12 April 1910.

- U.S. Patent 997,308 "*Transmitting apparatus for wireless telegraphy*". Filed 15 July 1910; Issued 11 July 1911.

- U.S. Patent 1,102,990 "*Means for generating alternating electric currents*". Filed 27 January 1914; Issued 7 July 1914.

- U.S. Patent 1,148,521 "*Transmitter for wireless telegraphy*". Filed 20 July 1908.

- U.S. Patent 1,226,099 "*Transmitting apparatus for use in wireless telegraphy and telephony*". Filed 31 December 1913; Issued 15 May 1917.

- U.S. Patent 1,271,190 "*Wireless telegraph transmitter*".

- U.S. Patent 1,377,722 "*Electric accumulator*". Filed 9 March 1918

- U.S. Patent 1,148,521 "*Transmitter for wireless telegraphy*". Filed 20 July 1908; Issued 3 August 1915.

- U.S. Patent 1,981,058 "*Thermionic valve*". Filed 14 October 1926; Issued 20 November 1934.

Reissued (US)

- U.S. Patent RE11,913 "*Transmitting electrical impulses and signals and in apparatus, there-for*". Filed 1 April 1901; Issued 4 June 1901.

2.22.5 See also

- History of radio

- Jagadish Chandra Bose

- List of people on stamps of Ireland

- List of covers of Time magazine during the 1920s – 6 December 1926

2.22.6 Notes

[1] Bondyopadhyay, Prebir K. (1995). "25th European Microwave Conference, 1995": 879. doi:10.1109/EUMA.1995.337090. |chapter= ignored (help)

[2] Sungook Hong, Wireless: From Marconi's Black-box to the Audion, MIT Press - 2001, page 1

[3] "Guglielmo Marconi: The Nobel Prize in Physics 1909"

[4] Bondyopadhyay, P.K. (1998). "Sir J.C. Bose diode detector received Marconi's first transatlantic wireless signal of December 1901 (the 'Italian Navy Coherer' Scandal Revisited)". *Proceedings of the IEEE* **86**: 259. doi:10.1109/5.658778.

[5] Roy, Amit (8 December 2008). "Cambridge 'pioneer' honour for Bose". *The Telegraph* (Kolkota). Retrieved 10 June 2010.

[6] *Icons of invention: the makers of the modern world from Gutenberg to Gates*. ABC-CLIO. Retrieved 7 August 2011.

[7] *Ingenious Ireland: A County-by-County Exploration of the Mysteries and Marvels of the Ingenious Irish*. Simon and Schuster. Retrieved 7 August 2011.

[8] *Atti della Accademia di scienze, lettere e arti di Palermo: Scienze*, Presso l'accademia, 1974, p. 11.

[9] Marconi: the Irish connection, Michael Sexton, Four Courts Press, 2005

[10] Alfonso, not Guglielmo, was a pupil at Bedford School; 'It is not generally known that the Marconi family at one time lived in Bedford, in the house on Bromham Road on the western corner of Ashburnham Road, and that the elder brother of the renowned Marchese Marconi attended this School for four years', *The Ousel* (June, 1936), p. 78 (Alfonso's obituary)

[11] *Bedfordshire Times* 23 July 1937, p. 9 (Guglielmo's obituary)

[12] "Guglielmo Marconi and Early Systems of Wireless Communication" (PDF). Retrieved 15 February 2014.

[13] McHenry, Robert, ed. (1993). "Guglielmo Marconi". *Encyclopædia Britannica*.

[14] Corradi, Giuliano, "Guglielmo Marconi," *Guglielmo Marconi. Tracce di un genio nel Tigullio*, 2009.

[15] Marconi, Maria Christina, *Marconi My Beloved. 2001. p. 19-24.*

[16] Marconi delineated his 1895 apparatus in his Nobel Award speech. See: Marconi, "Wireless Telegraphic Communication: Nobel Lecture, 11 December 1909." Nobel Lectures. Physics 1901–1921. Amsterdam: Elsevier Publishing Company, 1967: 196–222. p. 198.

[17] Guglielmo Marconi, padre della radio. Radiomarconi.com. Retrieved on 12 July 2012.

[18] Marconi, "Wireless Telegraphic Communication: Nobel Lecture, 11 December 1909." Nobel Lectures. Physics 1901–1921. Amsterdam: Elsevier Publishing Company, 1967: 196–222. p. 206.

[19] "The Nobel Prize in Physics 1909".

[20] Luigi Solari, *Guglielmo Marconi e la Marina Militare Italiana*, Rivista Marittima, febbraio 1948

[21] Alfred Thomas Story, *The Story of Wireless Telegraphy*. 1904. p. 58.

[22] Plaque #2389 on Open Plaques.

[23] "Flickr Photo".

[24] BBC Wales, Marconi's Waves at the Wayback Machine (archived January 20, 2007)

[25] Helgesen, Henry N. "Wireless Goes to Sea: Marconi's Radio and SS Ponce". *Sea History* (Spring 2008): 122.

[26] First Atlantic Ocean crossing by a wireless signal. Carnet-devol.org. Retrieved on 12 July 2012.

[27] Page, Walter Hines, and Arthur Wilson Page, *The World's Work*. Doubleday, Page & Company, 1908. p. 9625

[28] "Marconi and the History of Radio". *IEEE Antennas and Propagation Magazine* **46** (2): 130. 2004. doi:10.1109/MAP.2004.1305565.

[29] John S. Belrose, "Fessenden and Marconi: Their Differing Technologies and Transatlantic Experiments During the First Decade of this Century". International Conference on 100 Years of Radio – 5–7 September 1995.

[30] "The Clifden Station of the Marconi Wireless Telegraph System". *Scientific American*. 23 November 1907.

[31] Second Test of the Marconi Over-Ocean Wireless System Proved Entirely Successful. Sydney Daily Post. 24 October 1907.

[32] John P. Eaton & Charles A. Haas *Titanic – Triumph and Tragedy, A Chronicle in Words and Pictures*. 1994 ISBN 0857330241.

[33] Herron, Edward A. (1969). *Miracle of the Air Waves: A History of Radio*. Messner. ISBN 0-671-32079-3.

[34] Court of Inquiry *Loss of the S.S. Titanic* 1912

[35] "Titanic's Wireless Connection". Wireless History Foundation. April 2012. Retrieved 7 October 2013.

[36] Greg Daugherty (March 2012). "Seven Famous People Who Missed the Titanic". Smithsonian Magazine.

[37] William John Baker, *History Of The Marconi Company 1874–1965*. 1972. p. 296

[38] "Radio falls silent for death of Marconi"

[39] http://markpadfield.com/marconicalling/museum/html/places/places-i=13.html

[40] Guglielmo Marconi at *Find a Grave*

[41] *Jean-Michel Redouté, Michiel Steyaert, EMC of Analog Integrated Circuits, page 3*. Books.google.com. Retrieved 18 March 2013.

[42] *Charles T. Meadow, Making Connections: Communication through the Ages, Scarecrow Press 2002, page 193*. Books.google.com. Retrieved 15 February 2014.

[43] "Thomas H. White, Nikola Tesla: The Guy Who DIDN'T "Invent Radio", http://earlyradiohistory.us, 1 November 2012". Earlyradiohistory.us. Retrieved 15 February 2014.

[44] *Robert Sobot, Wireless Communication Electronics: Introduction to RF Circuits and Design Techniques, page 4*. Books.google.com. 18 February 2012. Retrieved 18 March 2013.

[45] Padfield, Mark. "Beatrice O'Brien". Marconi Calling.

[46] Degna Marconi, *My Father, Marconi* (Guernica Editions, 2001), pp. 218–227 ISBN 1550711512.

[47] *Kelly's Handbook to the Titled, Landed, and Official Classes* (Kelly's, 1969), p. 623

[48] Degna Marconi, *My Father, Marconi* (Guernica Editions, 2001), p. 232 ISBN 1550711512.

[49] Physicsworld.com, "*Guglielmo Marconi: radio star*", 2001 Archived September 17, 2013 at the Wayback Machine

[50] "80 Years of Vatican Radio, Pope Pius XI and Marconi. .. and Father Jozef Murgas?". Saint Benedict Center.

[51] National Broadcasters Hall of Fame. Accessed 10 February 2009

[52] "Pioneer: Guglielmo Marconi". *radiohof.org*. Retrieved 30 May 2012.

[53] Italy 2000 lira banknote (1990) Banknote Museum (banknote.ws). Retrieved on 2013-03-17.

[54] "Milestones:Marconi'{}s Early Wireless Experiments, 1895". *IEEE Global History Network*. IEEE. Retrieved 29 July 2011.

[55] "List of IEEE Milestones". *IEEE Global History Network*. IEEE. Retrieved 29 July 2011.

[56] New Jersey to Bon Jovi: You Give Us a Good Name. accesshollywood.com (2 February 2009).

[57] http://news.bbc.co.uk/1/hi/england/berkshire/4072929.stm

[58] http://www.ouls.ox.ac.uk/news/2008_nov_07

[59] "CMC Electronics' Profile". CMC Electronics Inc. Retrieved 12 January 2007.

[60] "Chatham Marconi Maritime Center". *www.arrl.org*. Retrieved 2015-11-09.

2.22.7 Further reading

Relatives and company publications

- Bussey, Gordon, *Marconi's Atlantic Leap*, Marconi Communications, 2000. ISBN 0-9538967-0-6

- Isted, G.A., *Guglielmo Marconi and the History of Radio – Part I*, General Electric Company, p.l.c., *GEC Review*, Volume 7, No. 1, p45, 1991, ISSN 0267-9337

- Isted, G.A., *Guglielmo Marconi and the History of Radio – Part II*, General Electric Company, p.l.c., *GEC Review*, Volume 7, No. 2, p110, 1991, ISSN 0267-9337

- Marconi, Degna, *My Father, Marconi*, James Lorimer & Co, 1982. ISBN 0-919511-14-7 (Italian version): *Marconi, mio padre*, Di Renzo Editore, 2008, ISBN 88-8323-206-2

- Marconi's Wireless Telegraph Company, *Year book of wireless telegraphy and telephony*, London: Published for the Marconi Press Agency Ltd., by the St. Catherine Press / Wireless Press. LCCN 14017875 sn 86035439

- Simons, R.W., *Guglielmo Marconi and Early Systems of Wireless Communication*, General Electric Company, p.l.c., *GEC Review*, Volume 11, No. 1, p37, 1996, ISSN 0267-9337

Other

- Ahern, Steve (ed), *Making Radio* (2nd Edition) Allen & Unwin, Sydney, 2006 ISBN 9781741149128.

- Aitken, Hugh G. J., *Syntony and Spark: The Origins of Radio*, New York: John Wiley & Sons, 1976. ISBN 0-471-01816-3

- Aitken, Hugh G. J., *The Continuous Wave: Technology and American Radio, 1900–1932*, Princeton, New Jersey: Princeton University Press, 1985. ISBN 0-691-08376-2.

- Anderson, Leland I., Priority in the Invention of Radio – Tesla vs. Marconi

- Baker, W. J., *A History of the Marconi Company*, 1970.

- Brodsky, Ira. "The History of Wireless: How Creative Minds Produced Technology for the Masses" (Telescope Books, 2008)

- Cheney, Margaret, "Tesla: Man Out of Time" Laurel Publishing, 1981. Chapter 7, esp pp 69, re: published lectures of Tesla in 1893, copied by Marconi.

- Clark, Paddy, "Marconi's Irish Connections Recalled," published in ";100 Years of Radio," IEE Conference Publication 411, 1995.

- Coe, Douglas and Kreigh Collins (ills), *Marconi, pioneer of radio*, New York, J. Messner, Inc., 1943. LCCN 43010048

- Garratt, G. R. M., *The early history of radio: from Faraday to Marconi*, London, Institution of Electrical Engineers in association with the Science Museum, History of technology series, 1994. ISBN 0-85296-845-0 LCCN gb 94011611

- Geddes, Keith, *Guglielmo Marconi, 1874–1937*, London : H.M.S.O., A Science Museum booklet, 1974. ISBN 0-11-290198-0 LCCN 75329825 (*ed.* Obtainable in the U.S.A. from Pendragon House Inc., Palo Alto, California.)

- Hancock, Harry Edgar, *Wireless at sea; the first fifty years: A history of the progress and development of marine wireless communications written to commemorate the jubilee of the Marconi International Marine Communication Company, Limited*, Chelmsford, Eng., Marconi International Marine Communication Co., 1950. LCCN 51040529 /L

- Hong, Sungook, *Wireless: From Marconi's Black-Box to the Audio*, Cambridge, Mass.: MIT Press, 2001. ISBN 0-262-08298-5.

- Hughes, Michael and Bosworth, Katherine, *Titanic Calling : Wireless Communications During the Great Disaster*, Oxford, The Bodleian Library, 2012, ISBN 978-1-85124-377-8

- Janniello, Maria Grace, Monteleone, Franco and Paoloni, Giovanni (eds) (1996), *One hundred years of radio: From Marconi to the future of the telecommunications*. Catalogue of the extension, Venice: Marsilio.

- Jolly, W. P., *Marconi*, 1972.

- Larson, Erik, *Thunderstruck*, New York: Crown Publishers, 2006. ISBN 1-4000-8066-5 A comparison of the lives of Hawley Harvey Crippen and Marconi. Crippen was a murderer whose Transatlantic escape was foiled by the new invention of shipboard radio.

- MacLeod, Mary K., *Marconi: The Canada Years – 1902–1946*, Halifax, Nova Scotia: Nimbus Publishing Limited, 1992, ISBN 1551093308

- Masini, Giancarlo, *Guglielmo Marconi*, Turin: Turinese typographical-publishing union, 1975. LCCN 77472455 (*ed.* Contains 32 tables outside of the text*)*

- Mason, H. B. (1908). *Encyclopaedia of ships and shipping*, Wireless Telegraphy. London: Shipping Encyclopaedia. 1908.

- Perry, Lawrence (1902). "Commercial Wireless Telegraphy". *The World's Work: A History of Our Time* **V**: 3194–3201. Retrieved 10 July 2009.

- Stone, Ellery W., *Elements of Radiotelegraphy*

- Weightman, Gavin, *Signor Marconi's magic box: the most remarkable invention of the 19th century & the amateur inventor whose genius sparked a revolution*, 1st Da Capo Press ed., Cambridge, MA : Da Capo Press, 2003. ISBN 0-306-81275-4

- Winkler, Jonathan Reed. *Nexus: Strategic Communications and American Security in World War I*. (Cambridge, MA: Harvard University Press, 2008). Account of rivalry between Marconi's firm and the U.S. government during World War I.

2.22.8 External links

Wikimedia

- "Marconi, Guglielmo". *Encyclopædia Britannica* (12th ed.). 1922.

General achievements

- Nobel Prize: Guglielmo Marconi biography

- Marconi il 5 marzo 1896, presenta a Londra la prima richiesta provvisoria di brevetto, col numero 5028 e col titolo "Miglioramenti nella telegrafia e relativi apparati" (Great Britain and France between 1896 and 1924)

- List of British and French patents (1896–1924) The first patent application number 5028 of 5 March 1896 (Provisional deprivation)

Foundations and academics

- University of Oxford Introduction to the Online Catalogue of the Marconi Collection

- University of Oxford Online Catalogue of the Marconi Archives

- Guglielmo Marconi Foundation, Pontecchio Marconi, Bologna, Italy

- Galileo Legacy Foundation: pictures of the Dedication of the Guglielmo Marconi Square, Johnston RI USA Dedication Photos

- History of Marconi House, Marconi House, Strand / Aldwych, London.

Multimedia and books

- MarconiCalling – The Life, Science and Achievements of Guglielmo Marconi, part of the Marconi Collection at the University of Oxford

- Canadian Heritage Minute featuring Marconi

- Guglielmo Marconi documentary, narrated by Walter Cronkite

- Review of *Signor Marconi's Magic Box*

Transatlantic "signals" and radio

- Robert (Bob) White, Guglielmo Marconi – Aerial Assistance with a Kite. Bridging the Atlantic By Wireless Signal – 12 December 1901. Kiting, The Journal of the American Kitefliers Association. Vol. 23, Issue 5 – Winter 2002. November 2001

- Faking the Waves, 1901

- Marconi and "wireless telegraphy" using kites

Keys and "signals"

- Sparks Telegraph Key Review An exhaustive listing of wireless telegraph key manufacturers including photos of most Marconi keys

- United States Senate Inquiry into the Titanic disaster – Testimony of Guglielmo Marconi

Priority of invention

vs Tesla

- PBS: Marconi and Tesla: Who invented radio?

- U.S. Supreme Court, "*Marconi Wireless Telegraph co. of America v. United States*". 320 U.S. 1. Nos. 369, 373. Argued 9–12 April 1943. Decided 21 June 1943.

- 21st Century Books: Priority in the Invention of Radio – Tesla vs. Marconi

Personal

- Information about Marconi and his yacht Elettra

- I diari di laboratorio di Guglielmo Marconi (The diaries of laboratory Guglielmo Marconi.)

- Comitato Guglielmo Marconi International, Bologna, Italy (Marconi's voice)

- August 1914 photo article on Marconi Belmar station in Wall, NJ, InfoAge. (See also, Marconi Period of Significance Historic Buildings.)

- Marconi, Guglielmo: Statue north of Meridian Hill Park in Washington, D.C. by Attilio Piccirilli

Other

- Guglielmo Marconi, 2000 Italian Lire (1990)

2.23 International Broadcasting Act

Signed in law in 1994 by U.S. President Bill Clinton, this act was meant to streamline the U.S. international broadcasting and provide a cost-effective way to continue Radio Free Europe/ Radio Liberty, Voice of America, and Radio Marti.[1] It placed control of the international broadcasting under the United States Information Agency.[2]

2.23.1 History

In 1958, President Eisenhower in an address to the United Nations proposed monitoring radio broadcasts:

> "I believe that this Assembly should [...] consider means for monitoring the radio broadcasts directed across national frontiers in the troubled Near East area. It should then examine complaints from these nations which consider their national security jeopardized by external propaganda."[3]

In the 1960s, President Kennedy to build an international broadcasting arm of the United States to as a way to promote foreign policy and overthrow socialism.[4] In 1976, President Gerald Ford signed the Voice of America charter that established it as the leading branch of US international broadcasting.

In 1993, the Clinton Administration proposed cutting the budget for Radio Free Europe and Radio Liberty in order to reduce budget expenditures.[1] However, after working with the Congress, the International Broadcasting Act was born.

2.23.2 Original law

This Act (Public Law 103-236) consolidated all nonmilitary, U.S. Government international broadcast services under a Broadcasting Board of Governors (BBG) and also created the International Broadcasting Bureau (IBB).[5] The BBG is an independent government agency created to replace the Board for International Broadcasting and consolidate Voice of America broadcasting.[6]

In this law, the President appoints one member of the board as the Chairman of the board. The Secretary of State also serves on the board.[7]

Besides combining current radio service, this Act also created the Radio Free Asia - a network aimed at Burma, China, Cambodia, Laos, North Korea, and Vietnam.[8]

2.23.3 Congressional updates

In September 2009, the 111th Congress amended the International Broadcasting Act to allow a one year extension of the operation of Radio Free Asia.[9]

In 2002, the Act was amended to include the Radio Free Afghanistan.[7]

In May 1994, the President announce the continuation of Radio Free Asia after 2009 was dependent on its increased international broadcasting and ability to reach its audience.[10]

2.23.4 References

[1] Raghavan, Sudarsan V., Stephen S. Johnson, Kristi K. Bahrenburg. Sending cross-border static: on the fate of Radio Free Europe and the influence of international broadcasting. Journal of International Affairs, Vol. 47, 1993.

[2] United States International Broadcasting Act, Pub. L. No. 103-236, title. III.

[3] Dept of State Bulletin 337-342 at 339. 1958 Statement to the UN, August 1958.

[4] Jon T. Powell, "Towards a Negotiable Definition of Propaganda for International Agreements Related to Direct Broadcast Satellites," Law & Contemporary Problems 45 (1982): 3, 25-26.

[5] IBB Fact Sheet. University of Illinois Chicago. Web

[6] Broadcasting Board of Governors FAQ

[7] U.S. Code. House of Representatives

[8] Price, Monroe. The Transformation of international broadcasting. Global Media and National Controls: Rethinking the Role of the State, MIT Press, 2002.

[9] Bill Text Versions for the 111th Congress, 2009 - 2010. The Library of Congress.

[10] Executive Order 12, 850, 3 C.F.R. 606, 607 § 1(b).

2.24 International   Broadcasting Convention

IBC logo

RAI Exhibition and Convention Centre

The **International Broadcasting Convention**, more commonly known by its acronym **IBC**, is an annual trade show for broadcasters, content creators/providers, equipment manufacturers, professional and technical associations, and other participants in the broadcasting industry. IBC is Europe's largest professional broadcast show and is held annually in September at the RAI Exhibition and Convention Centre in Amsterdam, the Netherlands.

In recent years IBC has de-highlighted its "Broadcasting" focus, and instead marketed its exhibition as the world's largest centered on media and entertainment. The name "International Broadcasting Convention" is no longer used on its website.

2.24.1 Dates

In 2009, the show was held on the following dates:

- Conference: 10–14 September 2009

- Exhibition: 11–15 September 2009

In 2010, the show was held on the following dates:

- Conference: 9–14 September 2010

- Exhibition: 10–14 September 2010

In 2011, the show was held on the following dates:

- Conference: 8-13 September 2011

- Exhibition: 9-13 September 2011

In 2012, the show was held on the following dates:

- Conference: 6-11 September 2012

- Exhibition: 7-11 September 2012

In 2013, the show took place at the Amsterdam RAI on the following dates:

- Conference: 12-17 September 2013

- Exhibition: 13-17 September 2013

In 2014, the show took place on the following dates:

- Conference: 11-15 September 2014

- Exhibition: 12-16 September 2014

In 2015, the show took place on the following dates:

- Conference: 10 - 14 September 2015

- Exhibition: 11 - 15 September 2015

2.24.2 References

2.24.3 External links

- www.ibc.org

2.25 Interval signal

An **interval signal**, or **tuning signal**, is a characteristic sound or musical phrase used in international broadcasting and by some domestic broadcasters, played before commencement or during breaks in transmission, but most commonly between programmes in different languages. It serves several purposes:

- It assists a listener to tune his or her radio to the correct frequency of the station. This is because most older and cheaper radio receivers do not have digital frequency readout.

- It informs other stations that the frequency is in use.

- It serves as a station identifier even if the language used in the subsequent broadcast is not one the listener understands.

The practice began in Europe in the 1920s and 1930s and was carried over into shortwave broadcasts. The use of interval signals has declined with the advent of digital tuning systems, but has not vanished. Interval signals were not required on commercial channels in the USA, where jingles were used as identification.

2.25.1 Broadcasting services and interval signals

- BBC World Service: *Bow Bells* (English programme), three notes tuned B-B-C (non-English programme, non-Europe), four notes tuned B-B-B-E (non-English programme, to Europe).

- China National Radio, China Radio International: Chime version of 义勇军进行曲 ("March of the Volunteers").

- Voice of the Strait News Radio: Bell version of 三大纪律八项注意 ("Three Rules of Discipline and Eight Points for Attention").

- Deutsche Welle: Piano version of *Es sucht der Bruder seine Brüder* from Fidelio by Ludwig van Beethoven.

- Radio Australia: Chimes version of *Waltzing Matilda* (chorus).

- Radio Belarus: *Radzima maja darahaja* ("My dear Motherland").

- Radio Canada International: First four notes of *O Canada*, played on piano or autoharp.

- Radio France Internationale: Electronic-disco, culminating in the last 8 measures of *La Marseillaise*.

- Radio Japan: *Kazoe-uta* (Japanese counting song), さくら さくら ("Sakura Sakura" - Cherry Blossoms).

- KBS World Radio: *Dawn*.

- Radio Habana Cuba: Melody of the *La Marcha del 26 de Julio* ("March of the 26th of July").

- Radio Netherlands: Chime version of the Eighty Years' War song *Merck toch hoe sterck*.

- Radio New Zealand International: The call of a New Zealand bellbird.

- Radio Republik Indonesia: *Rayuan Pulau Kelapa* ("Solace on Coconut Island"), composed by Ismail Marzuki.

- RTÉ Radio 1: Chime version of *O'Donnell Abú* ("O'Donnell Forever").[1]

- Radio Slovenia: Electronically generated cuckoo chirping.

- Radio Sweden: Chime version of *Ut i vida världen* ("Out in the Wide World"), composed by Ralph Lundsten.[2]

- Radio Ukraine International: *Reve ta stohne Dnipr shyrokyi*.

- Vatican Radio: *Christus Vincit*, played on flute.

- Voice of America: Brass band version of *Yankee Doodle*.

- Voice of Korea: Melody of 김일성장군의 노래 ("Song of General Kim Il-sung").

- Far East Broadcasting Company: *Lord Jesus to Save Sinners*.

- Ö1: Three notes tuned O-R-F, played on viola.

- RTL Radio: *Feierwon* by Michel Lentz, played on chimes.[3]

- DR P1: *Drømte mig en drøm i nat*, played on xylophone.[4]

- Polish Radio External Service: Excerpt from *Prząśniczka* by Stanisław Moniuszko, played on piano.[5]

- Voice of Mongolia: *Эх орон* ("Motherland").

- Trans World Radio: *What a Friend We Have in Jesus*.

- Radiodifusión Argentina al Exterior: First eight notes of *Mi Buenos Aires querido* by Carlos Gardel.

- Radio Nacional de Venezuela - Canal Internacional: Beginning of *Alma Llanera* by Pedro Elías Gutiérrez and Rafael Bolívar.

Formerly used

- Radio Austria International: Orchestral version of *An der schönen blauen Donau* ("Blue Danube Waltz") by Johann Strauss.[6]

- Radio Berlin International: Beginning of *Auferstanden aus Ruinen* ("Risen from Ruins"), played on chimes.

- Radio France Internationale: Trumpet version of a popular song *Nous n'irons plus au bois*.[6]

- Radio Peking (predecessor of China Radio International): Chimes version of 东方红 ("The East Is Red").

- Radio Moscow (former international service of the Soviet Union): *Песня о Родине* ("Wide Is My Motherland"); *Midnight in Moscow*, played by balalaika.

- Radio Norway International (former international service of NRK): Ancient folk tune from the Hallingdal region.[7]

- Radio RSA (former international service of Apartheid-era South African Broadcasting Corporation): Bokmakierie chirping and first bars of *Ver in die Wereld, Kittie*, played on guitar.[6][8]

- Radio Sweden: Opening notes of Carl Michael Bellman's *Storm och böljor tystna r'en*.[6]

- Radio Tirana: Trumpet version of *With Pickaxe and Rifle*.

- Radio Polonia: Piano version of *Etude No. 12* ("Revolutionary Etude") by Frédéric Chopin.

- Radio Prague: Trumpet version of *Kupředu levá* ("Forward, Left"); *Adagio – Allegro molto* from *Symphony No. 9* by Dvořák.

- Radio Yugoslavia, later International Radio of Serbia and Montenegro: *Jugoslavijo*.

- Rai Italia Radio: Mechanically generated canary chirping.

- Deutschlandfunk: Celesta version of *Dir, Land voll Lieb' und Leben* from "Ich hab mich ergeben" by Hans Ferdinand Maßmann.[9]

- Voice of Russia: "Majestic" chorus from the "Great Gate of Kiev" portion of *Pictures at an Exhibition* by Mussorgsky.

- Voice of Turkey: Makam, played on piano.

- Radio Serbia: *Bože pravde*.

- Rádio Nacional: *Luar do Sertão*.

- Radio Mayak: Vibraphone version of *Moscow Nights*.

- BBC World Service: Trumpet version of *Oranges and Lemons*, first four notes of *Symphony No. 5* by Beethoven, played on timpani; *Lillibullero* (signature tune, played on trumpet).[10]

- NPO: First seven notes of *Wilhelmus*, played on clarinet (Radio 1 and Radio 5), synthesizer (Radio 3), spinet (Radio 4) and carillon (Radio 2).[11]

- Berliner Rundfunk: Motif from the opera *Regina* by Albert Lortzing, played by trumpets.

- Kol Yisrael: Trumpet and drum version of *Hatikvah*.

- NRK P1: Motif from *Sigurd Josarfal* by Edward Grieg.[12]

- Radio Katowice: Sound of a hammer striking an anvil.

- Radio Beromünster: Zit isch daa, played on music box.

- RTBF International: Où peut-on être mieux qu'au sein de sa famille.

- Radio DDR: First few bars of Wann wir schreiten Seit' an Seit'.

Numbers stations interval signals

Numbers stations are often named after their interval signals, such as The Lincolnshire Poacher or Magnetic Fields after "Magnetic Fields Part 1" by Jean Michel Jarre.

2.25.2 References

[1]

[2] Radio Sweden interval signal Retrieved 2011-11-24.

[3] http://www.radioforen.de/index.php?threads/
pausenzeichen-und-ihre-musikalischen-quellen.21902/
page-3

[4] https://www.youtube.com/watch?v=lpDQYdoIQ0w

[5] https://www.youtube.com/watch?v=ztH0_2ueIYc

[6] Frost, J. M. *World Radio TV Handbook*. New York: Billboard Publications, 1983.

[7] Frost, Jens Mathiesen. *World Radio-TV Handbook*. London: Billboard Publications, 1974.

[8] DX LISTENING DIGEST 7-043

[9] http://www.kalter-krieg-im-radio.de/index.php?er=18#

[10] BBC World Service (Europe) interval signal Retrieved 2013-10-09.

[11] nl:Pauzeteken

[12] http://www.ontheshortwaves.com/Articles/The_Interval_
Signal.pdf

Frost, Jens Mathiesen (1974). *World Radio-TV Handbook 1974*. London: Billboard Publications. p. 408. ISBN 0823058980.

Sennitt, Andrew G.; David Bobbitt (December 2005). *World Radio and Television Handbook 2006*. Billboard Books. p. 608. ISBN 0-8230-7798-5.

Sennitt, Andrew G. *World Radio and Television Handbook 1997*. Billboard Books. p. 560. ISBN 0-8230-7797-7.

2.25.3 External links

- Interval Signals Online

- Nobuyuki Kawamura's Interval Signal Library

- TRS Consultants' Audio Bytes

- IntervalSignal DataBase (German) English version

- Uwe Volk's Sound Library (available both in English and in German)

2.26 Joe Adamov

Joe Adamov, Name in Russian: Иосиф Адамов (Yosif Adamov) [1](7 January 1920 - December 2005) was a journalist and presenter on Radio Moscow and its successor the Voice of Russia for over sixty years. Of Armenian descent, he was born in Batumi, Georgia . During most of his career he was a resident of Moscow, Russia. His death was reported on Voice of Russia's English-language news bulletins on December 12, 2005, although neither these bulletins nor the obituary on their website gave an exact date of death.

An expert English-speaker who spoke with a neutral American accent, Joe Adamov joined Radio Moscow as an announcer in 1942. Among English-speaking listeners he is best known as the presenter of the programme *Moscow Mailbag* which answered questions from listeners on all aspects of the USSR and Russia. During the Cold War the programme, like all other Radio Moscow output, was carefully studied by Western governments, a fact of which the Soviet authorities were well aware. Joe Adamov presented Moscow Mailbag from 1957 until shortly before his death when ill-health forced him to relinquish the role permanently. The programme remains on the air with different presenters.

In his capacities as a Radio Moscow journalist and presenter of Moscow Mailbag, Adamov conducted interviews with many western politicians and journalists, including Dwight D. Eisenhower, Eleanor Roosevelt, Walter Cronkite and Larry King.

Joe Adamov was also the official Soviet translator at the trial of Gary Powers, shot down in a U-2 spy plane in 1960.

The signature poem of Moscow Mailbag was:

> *You can't do better*
> *than send us that letter*
> *and in it tell Joe*
> *what you think of his show*

2.26.1 References

[1] pron. yos-eef adamof

2.26.2 External links

- Voice of Russia obituary

- Biography by Katherine Lawson

- Archive of Moscow Mailbag programmes presented by Joe Adamov

2.27 John Robles

John Robles

John Anthony Robles II (born 10 April 1966) is a journalist and presenter on the Voice of Russia[1][2] the Russian government's international radio broadcasting service. He is the presenter of the English Language program *Moscow Mailbag* which answers questions from listeners all over the world on all aspects of the USSR and Russia.

He was born a US Citizen in Rio Piedras, Puerto Rico, Puerto Rico being a US territory. Both he and Edward Snowden have been granted political asylum in Russia.

2.27.1 Education and employment

Robles attended the Pennsylvania State University where he took an undergraduate course in Soviet Studies, majoring in the Russian Language. He taught at the BKC-Ih School for Foreign Languages in Moscow, Russia, for thirteen years.[3]

2.27.2 Revocation of US passport

Living in Moscow in March 2007 John Robles applied to have his US passport renewed at the US Embassy. His passport was revoked however and the embassy refused to issue him with a new one. An accusation was made that Robles owed child support in Yolo County California. Robles disputed this as he had full custody of his 2 children and had raised them on his own. He claimed neither he nor his children had been in the US since 1995. He had been issued a new passport previously in 1998 after it had been stolen. It was after his passport was revoked that he was granted political asylum in Russia. Robles had been running a political website critical of the actions of President George W. Bush and hosted in Russia since July 2003.[4]

2.27.3 References

[1] "John Robles column at Voice of Russia". Retrieved 2013-08-31.

[2] "John Robles page at Voice of Russia". Retrieved 2012-02-01.

[3] "BKC-ih past teachers list". Retrieved 2009-11-01.

[4] "John Robles JAR2 website". Retrieved 2012-02-01.

2.28 Medium wave

Main article: Medium frequency

Medium wave (**MW**) is the part of the medium frequency (MF) radio band used mainly for AM radio broadcasting. For Europe the MW band ranges from 526.5 kHz to 1606.5 kHz,[1] using channels spaced every 9 kHz, and in North America an extended MW broadcast band goes from 535 kHz to 1705 kHz,[2] using 10 kHz spaced channels.

2.28.1 Propagation characteristics

Wavelengths in this band are long enough that radio waves are not blocked by buildings and hills and can propagate beyond the horizon following the curvature of the Earth; this is called the groundwave. Practical groundwave reception typically extends to 200–300 miles, with longer distances over terrain with higher ground conductivity, and greatest distances over salt water. Most broadcast stations use groundwave to cover their listening area.

Medium waves can also reflect off charged particle layers in the ionosphere and return to Earth at much greater distances; this is called the skywave. At night, especially in winter months and at times of low solar activity, the ionospheric D layer virtually disappears. When this happens, MF radio waves can easily be received many hundreds or even thousands of miles away as the signal will be reflected by the higher F layer. This can allow very long-distance

Typical mast radiator of a commercial medium wave AM broadcasting station, Chapel Hill, North Carolina, USA

broadcasting, but can also interfere with distant local stations. Due to the limited number of available channels in the MW broadcast band, the same frequencies are reallocated to different broadcasting stations several hundred miles apart. On nights of good skywave propagation, the skywave signals of distant station may interfere with the signals of local stations on the same frequency. In North America, the North American Radio Broadcasting Agreement (NARBA) sets aside certain channels for nighttime use over extended service areas via skywave by a few specially licensed AM broadcasting stations. These channels are called clear channels, and they are required to broadcast at higher powers of 10 to 50 kW.

2.28.2 Use in the Americas

See also: North American Radio Broadcasting Agreement

Initially broadcasting in the United States was restricted to two wavelengths: "entertainment" was broadcast at 360 meters (833 kHz), with stations required to switch to 485 me-

ters (619 kHz) when broadcasting weather forecasts, crop price reports and other government reports.[3] This arrangement had numerous practical difficulties. Early transmitters were technically crude and virtually impossible to set accurately on their intended frequency and if (as frequently happened) two (or more) stations in the same part of the country broadcast simultaneously the resultant interference meant that usually neither could be heard clearly. The Commerce Department rarely intervened in such cases but left it up to stations to enter into voluntary timesharing agreements amongst themselves. The addition of a third "entertainment" wavelength, 400 meters,[3] did little to solve this overcrowding.

In 1923, the Commerce Department realized that as more and more stations were applying for commercial licenses, it was not practical to have every station broadcast on the same three wavelengths. On 15 May 1923, Commerce Secretary Herbert Hoover announced a new bandplan which set aside 81 frequencies, in 10 kHz steps, from 550 kHz to 1350 kHz (extended to 1500, then 1600 and ultimately 1700 kHz in later years). Each station would be assigned one frequency (albeit usually shared with stations in other parts of the country and/or abroad), no longer having to broadcast weather and government reports on a different frequency than entertainment. Class A and B stations were segregated into sub-bands.[4]

Nowadays in most of the Americas, mediumwave broadcast stations are separated by 10 kHz and have two sidebands of up to ±5 kHz in theory, although in practice stations transmit audio of up to 10 kHz.[5] In the rest of the world, the separation is 9 kHz, with sidebands of ±4.5 kHz. Both provide adequate audio quality for voice, but are insufficient for high-fidelity broadcasting, which is common on the VHF FM bands. In the US and Canada the maximum transmitter power is restricted to 50 kilowatts, while in Europe there are medium wave stations with transmitter power up to 2 megawatts daytime.[6]

Most United States AM radio stations are required by the Federal Communications Commission (FCC) to shut down, reduce power, or employ a directional antenna array at night in order to avoid interference with each other due to nighttime only long-distance skywave propagation (sometimes loosely called 'skip'). Those stations which shut down completely at night are often known as "daytimers". Similar regulations are in force for Canadian stations, administered by Industry Canada; however, daytimers no longer exist in Canada, the last station having signed off in 2013, after migrating to the FM band.

2.28.3   Use in Europe

See also: Geneva Frequency Plan of 1975 and FM radio §
Adoption of FM broadcasting worldwide

In Europe, each country is allocated a number of frequencies on which high power (up to 2 MW) can be used; the maximum power is also subject to international agreement by the **International Telecommunication Union ITU** .[7] In most cases there are two power limits: a lower one for omnidirectional and a higher one for directional radiation with minima in certain directions. The power limit can also be depending on daytime and it is possible, that a station may not work at nighttime, because it would then produce too much interference. Other countries may only operate low-powered transmitters on the same frequency, again subject to agreement. For example, Russia operates a high-powered transmitter, located in its Kaliningrad exclave and used for external broadcasting, on 1386 kHz. The same frequency is also used by low-powered local radio stations in the United Kingdom, which has approximately 250 medium-wave transmitters of 1 kW and over;[8] other parts of the United Kingdom can still receive the Russian broadcast. International mediumwave broadcasting in Europe has decreased markedly with the end of the Cold War and the increased availability of satellite and Internet TV and radio, although the cross-border reception of neighbouring countries' broadcasts by expatriates and other interested listeners still takes place.

Due to the high demand for frequencies in Europe, many countries operate single frequency networks; in Britain, BBC Radio Five Live broadcasts from various transmitters on either 693 or 909 kHz. These transmitters are carefully synchronized to minimize interference from more distant transmitters on the same frequency.

Overcrowding on the Medium wave band is a serious problem in parts of Europe contributing to the early adoption of VHF FM broadcasting by many stations (particularly in Germany). However, in recent years several European countries (Including Ireland, Poland and, to a lesser extent Switzerland) have started moving away from Medium wave altogether with most/all services moving exclusively to other bands (usually VHF).

In Germany, almost all Medium wave public-radio broadcasts were discontinued between 2012 and 2015 to cut costs and save energy,[9] with the last such remaining programme (Deutschlandradio) being switched off on 2015-12-31.[10]

2.28.4   Stereo and digital transmissions

See also: AM stereo

Stereo transmission is possible and offered by some stations in the U.S., Canada, Mexico, the Dominican Republic, Paraguay, Australia, The Philippines, Japan, South Korea, South Africa, and France. However, there have been multiple standards for AM stereo. C-QUAM is the official standard in the United States as well as other countries, but receivers that implement the technology are no longer readily available to consumers. Used receivers with AM Stereo can be found. Names such as "FM/AM Stereo" or "AM & FM Stereo" can be misleading and usually do not signify that the radio will decode C-QUAM AM stereo, whereas a set labeled "FM Stereo/AM Stereo" or "AMAX Stereo" will support AM stereo.

In September 2002, the United States Federal Communications Commission approved the proprietary iBiquity in-band on-channel (IBOC) HD Radio system of digital audio broadcasting, which is meant to improve the audio quality of signals. The Digital Radio Mondiale (DRM) IBOC system has been approved by the ITU for use outside North America and U.S. territories. Some HD Radio receivers also support C-QUAM AM stereo, although this feature is usually not advertised by the manufacturer.

2.28.5   Antennas

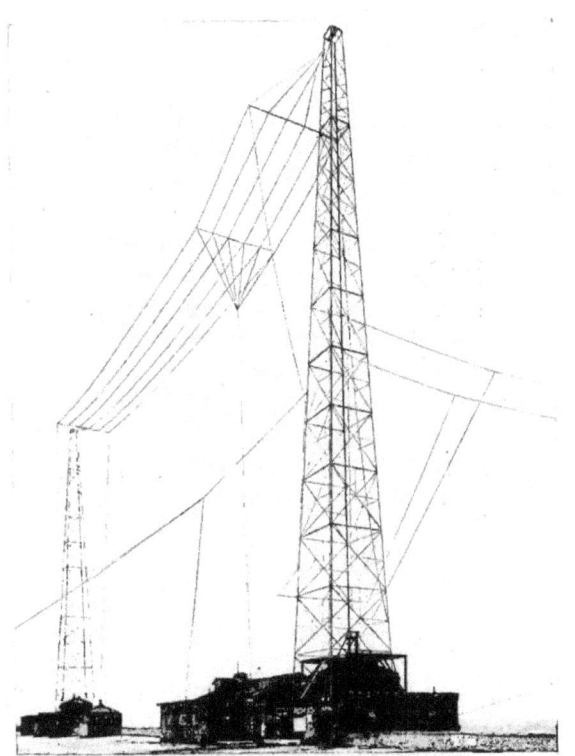

Multiwire T antenna of radio station WBZ, Massachusetts, USA, 1925. T antennas were the first antennas used for medium wave broadcasting, and are still used at lower power

For broadcasting, mast radiators are the most common type of antenna used, consisting of a steel lattice guyed mast in which the mast structure itself is used as the antenna. Stations broadcasting with low power can use masts with heights of a quarter-wavelength (about 310 millivolts per meter using one kilowatt at one kilometer) to 5/8 wavelength (225 electrical degrees; about 440 millivolts per meter using one kilowatt at one kilometer), while high power stations mostly use half-wavelength to 5/9 wavelength. The usage of masts taller than 5/9 wavelength (200 electrical degrees; about 410 millivolts per meter using one kilowatt at one kilometer) with high power gives a poor vertical radiation pattern, and 195 electrical degrees (about 400 millivolts per meter using one kilowatt at one kilometer) is generally considered ideal in these cases. Usually mast antennas are series-excited (base driven); the feedline is attached to the mast at the base, so the base of the antenna is at high electrical potential and must be supported on a ceramic insulator to insulate it from the ground. Shunt-excited masts, in which the base of the mast is at a node of the standing wave at ground potential and so does not need to be insulated from the ground have fallen into disuse, except in cases of exceptionally high power, 1 MW or more, where series excitement might be impractical. If grounded masts or towers are required, then cage aerials or long-wire aerials are used. Another possibility consists of feeding the mast or the tower by cables running from the tuning unit to the guys or crossbars in a certain height.

Directional aerials consist of multiple masts, which need not to be from the same height. It is also possible to realize directional aerials for mediumwave with cage aerials where some parts of the cage are fed with a certain phase difference.

For medium-wave (AM) broadcasting, quarter-wave masts are between 153 feet (47 m) and 463 feet (141 m) high, depending on the frequency. Because such tall masts can be costly and uneconomic, other types of antennas are often used, which employ capacitive top-loading (electrical lengthening) to achieve equivalent signal strength with vertical masts shorter than a quarter wavelength.[11] A "top hat" of radial wires is occasionally added to the top of mast radiators, to allow the mast to be made shorter. For local broadcast stations and amateur stations of under 5 kW, T- and L-antennas are often used, which consist of one or more horizontal wires suspended between two masts, attached to a vertical radiator wire. A popular choice for lower-powered stations is the umbrella antenna, which needs only one mast one tenth wavelength or less in height. This antenna uses a single mast insulated from ground and fed at the lower end against ground. At the top of the mast, radial top-load wires are connected (usually about six) which slope downwards at an angle of 40-45 degrees as far as about one-third of the total height, where they are terminated in insulators

and thence outwards to ground anchors. Thus the umbrella antenna uses the guy wires as the top-load part of the antenna. In all these antennas the smaller radiation resistance of the short radiator is increased by the capacitance added by the wires attached to the top of the antenna.

In some rare cases dipole antennas are used, which are slung between two masts or towers. Such antennas are intended to radiate a skywave. The medium-wave transmitter at Berlin-Britz for transmitting RIAS used a cross dipole mounted on five 30.5 metre high guyed masts to transmit the skywave to the ionosphere at nighttime.

Receiving antennas

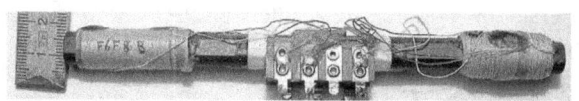

Typical ferrite rod antenna used in AM radio receivers

Because at these frequencies atmospheric noise is far above the receiver signal to noise ratio, inefficient antennas much smaller than a wavelength can be used for receiving. For reception at frequencies below 1.6 MHz, which includes long and medium waves, loop antennas are popular because of their ability to reject locally generated noise. By far the most common antenna for broadcast reception is the ferrite-rod antenna, also known as a loopstick antenna. The high permeability ferrite core allows it to be compact enough to be enclosed inside the radio's case and still have adequate sensitivity.

2.28.6   See also

- Medium frequency band

- AM radio

- Longwave

- MW DX

- Shortwave

- FM radio

- Satellite radio

- List of European medium wave transmitters

- Wave plan of Geneva

- DAB Radio

2.28.7 References

[1] "United Kingdom Frequency Allocation Table 2008" (PDF). Ofcom. p. 21. Retrieved 2010-01-26.

[2] "U.S. Frequency Allocation Chart" (PDF). National Telecommunications and Information Administration, U.S. Department of Commerce. October 2003. Retrieved 2009-08-11.

[3] "Building the Broadcast Band". Earlyradiohistory.us. Retrieved 2010-05-07.

[4] Christopher H. Sterling; John M. Kittross (2002). *Stay tuned: a history of American broadcasting.* Psychology Press. p. 95. ISBN 0-8058-2624-6.

[5] "Subpart A: AM Broadcast Stations, Sec. 73.44 AM transmission system emission limitations". *TITLE 47--TELECOMMUNICATION, CHAPTER I--FEDERAL COMMUNICATIONS COMMISSION (CONTINUED), PART 73_RADIO BROADCAST SERVICES.* U.S. Government Printing Office. Revised as of October 1, 2006. Retrieved 2009-04-24. Check date values in: |date= (help)

[6] "MWLIST quick and easy: Europe, Africa and Middle East". Retrieved 11 December 2015.

[7] "International Telecommunication Union". ITU. Retrieved 2009-04-24.

[8] "MW channels in the UK". Retrieved 11 December 2015.

[9] "Fast alle ARD-Radiosender stellen Mittelwelle ein". heise.de. 2015-01-06. Retrieved 2015-12-31.

[10] Heumann, Marcus (2015-12-17). "Abschied von der Mittelwelle. Der gefürchtete Wellensalat ist Geschichte". Deutschlandfunk.de. Retrieved 2015-12-31.

[11] Weeks, W.L 1968, *Antenna Engineering*, McGraw Hill Book Company, Section 2.6

2.28.8 External links

- "Building the Broadcast Band" the development of the 520-1700 kHz MW (AM) band

- M3 Map of Effective Ground Conductivity in the USA

- MWLIST worldwide database of MW and LW stations

- www.mwcircle.org The Medium Wave Circle. A UK-based club for Medium wave DX'ers and enthusiasts.

- - List of long- and mediumwave transmitters with GoogleMap-Links to transmission sites

2.29 Millennium Live

Millennium Live: Humanity's Broadcast is a title of a cancelled international television special, which was an unsuccessful attempt to broadcast an international celebration of the beginning of the Year 2000, or the so-called Millennium. Reports claimed that the show was to have involved broadcasters in up to 130 nations. *Millennium Live: Humanity's Broadcast* was going to compete against the *2000 Today* international broadcast that was supported by ABC in the United States, and led by BBC in the UK.

The programme was called off on 28 December 1999 when its organizers, the Millennium Television Network (MTN), announced that it had failed to obtain sufficient financing for the broadcast. MTN reportedly was not paying production and satellite companies for their services prior to the cancellation. The efforts to establish a global broadcast for the following New Year's Eve were shelved.

2.29.1 History

Millennium Television Network (MTN) was formed by Live Aid's American producer Hal Uplinger to prepare and conduct the broadcast.[1] Early planning among international representatives reportedly occurred in Cannes on October 1998.[2] *Millennium Live* was planned as a 24 or 25 hour broadcast from 11:00 UTC 31 December 1999. Pax TV (now known as ION Television) had the exclusive United States rights to broadcast the show which they billed as *Pax Millennium Live: A New World's Eve*, while the Associated Broadcasting Company and People's Television Network had the exclusive Philippines rights to broadcast the show, which they billed as *PTV Millennium Live: The Eve of A New World* and *Millennium Live on ABC-5: The Dawn of The New World*.[3]

The competitor of *Millennium Live: Humanity's Broadcast* would have been *2000 Today*, an international broadcast which was led by the BBC in the UK, and supported by PBS and ABC in the United States, RTL Television in Germany, and GMA Network in the Philippines.

Scheduled musical guests included Aerosmith, Bee Gees, Blondie, Chicago, Phil Collins, Destiny's Child, Ricky Martin, 'N Sync, The Pretenders, Sting, Santana and 10,000 Maniacs.[1][3][4][5][6] Bryan Adams, Simply Red and the Spice Girls were also sought as featured artists.[7]

New Year's Eve celebrations from various worldwide locations were to have been seen on the show.[5] Angelica Castro (Chile), Carmen Electra (US), Ramzi Malouki (France), Daniel Razon and Amelyn Veloso (Philippines), and Zam Nkosi (South Africa) were also scheduled as a program host.[1] The special was to have been hosted in Los Angeles on a set contained in a 90-foot geodesic dome at Manhattan Beach.[1]

The program was cancelled on 28 December 1999 with an announcement that MTN had failed to obtain sufficient financing for the broadcast.[6][8] MTN reportedly was not paying production and satellite companies for their services prior to the cancellation.[6] Pax aired a series of movies in *Millennium Live's* place, while some of its broadcasters moved to the *2000 Today* broadcast (PTV and ABC-5 were backed out).[9] Efforts to establish a global broadcast for the following New Year's Eve on the real Millennium (2001) were reported, but these plans never materialised.[10]

2.29.2 Broadcasters

The following broadcasters were reported as participants in *Millennium Live*: Note: RAI was the only broadcaster that moved to the *2000 Today* broadcast

- Australia: Nine Network[11]
- Brazil: Rede Bandeirantes[11]
- Canada: MuchMoreMusic[9]
- Chile: Chilevisión
- Estonia: TV1
- France: France 2[9]
- Germany: Sat.1[11]
- Hong Kong: STAR TV, ATV & TVB
- India: Zee TV[9][11]
- Italy: Rai 1, Rai 2, Rai 3, Italia1, Rete 4, Canale5
- Japan: Vibe TV (now MTV Japan)[11]
- Mexico: Televisa, & TV Azteca
- Philippines: People's Television Network & ABC-5
- South Africa: SABC[11]
- Spain: Telecinco[11]
- United States: Pax (now Ion Television)[11]
- Venezuela: Venevisión[11]

2.29.3 See also

- *2000 Today* - the successful international television special, which was broadcast in 78 countries (including broadcasters from the cancelled *Millennium Live* TV special)

2.29.4 References

[1] "'Millennium' adds talent". *Variety*. 10 November 1999. Retrieved 1 February 2014.

[2] "Pax TV To Broadcast 24-hour Millennium Show". Business Wire. 21 May 1999. Retrieved 1 February 2014.

[3] Carman, John (29 December 1999). "'Oh, the Humanity' -- If World Ends, TV's Got It". *San Francisco Chronicle*. Retrieved 1 February 2014.

[4] "TV rings in New Year". *Corpus Christi Caller-Times*. 26 December 1999. Retrieved 1 February 2014.

[5] Johnson, Allan (26 December 1999). "Ready To Pop". *Chicago Tribune*. Retrieved 1 February 2014.

[6] Schneider, Michael (28 December 1999). "Pax's 'Live' for Y2K DOA". Retrieved 1 February 2014.

[7] "Headliners Named For MTN Broadcast". PAX TV. 7 October 1999. Retrieved 1 February 2014.

[8] Carman, John (30 December 1999). "Producers Cancel Global Millennium Telecast". Archived from the original on 10 September 2006. Retrieved 1 February 2014.

[9] Schneider, Michael (29 December 1999). "Nixed show irks global b'casters". *Variety*. Retrieved 1 February 2014.

[10] "Satellite Television's Biggest Weekend: Live, Complex Millennium Networks". *Satellite Today*. 10 April 2000. Retrieved 1 February 2014.

[11] "Artists Confirmed For 'Millennium Live... Humanity's Broadcast'". PR Newswire. 7 October 1999. Retrieved 1 February 2014.

2.29.5 External links

- Everything 2000 - New Year's Eve Broadcast Plans 26 January 1999
- Everything 2000 - Millennium Live Worldwide Broadcast Cancelled 29 December 1999
- Kansas City Star - Y2K: The TV takeover 29 December 2001
- Kansas City Star - Funding glitch shuts down Pax's millennium special (via TV Barn) 30 December 1999
- Millennium Hell - Pax Backs Hacks. Lacks Facts. Backtracks. Sacks Millennium Quacks to the Max
- Egypt2000.com - The Twelve Dreams of the Sun - Jean-Michel Jarre performance

2.30 Modulation

For other uses, see Modulation (disambiguation).

In electronics and telecommunications, **modulation** is the process of varying one or more properties of a periodic waveform, called the *carrier signal*, with a modulating signal that typically contains information to be transmitted.

In telecommunications, modulation is the process of conveying a message signal, for example a digital bit stream or an analog audio signal, inside another signal that can be physically transmitted. Modulation of a sine waveform transforms a baseband message signal into a passband signal.

A **modulator** is a device that performs modulation. A **demodulator** (sometimes *detector* or *demod*) is a device that performs demodulation, the inverse of modulation. A modem (from **mo**dulator–**dem**odulator) can perform both operations.

The aim of **analog modulation** is to transfer an analog baseband (or lowpass) signal, for example an audio signal or TV signal, over an analog bandpass channel at a different frequency, for example over a limited radio frequency band or a cable TV network channel.

The aim of **digital modulation** is to transfer a digital bit stream over an analog bandpass channel, for example over the public switched telephone network (where a bandpass filter limits the frequency range to 300–3400 Hz) or over a limited radio frequency band.

Analog and digital modulation facilitate frequency division multiplexing (FDM), where several low pass information signals are transferred simultaneously over the same shared physical medium, using separate passband channels (several different carrier frequencies).

The aim of **digital baseband modulation** methods, also known as line coding, is to transfer a digital bit stream over a baseband channel, typically a non-filtered copper wire such as a serial bus or a wired local area network.

The aim of **pulse modulation** methods is to transfer a narrowband analog signal, for example a phone call over a wideband baseband channel or, in some of the schemes, as a bit stream over another digital transmission system.

In music synthesizers, modulation may be used to synthesise waveforms with an extensive overtone spectrum using a small number of oscillators. In this case the carrier frequency is typically in the same order or much lower than the modulating waveform (see frequency modulation synthesis or ring modulation synthesis).

2.30.1 Analog modulation methods

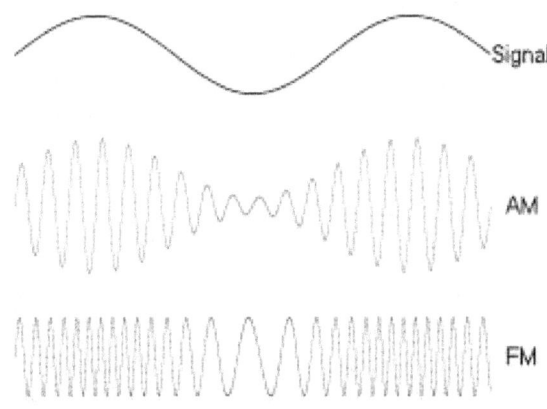

A low-frequency message signal (top) may be carried by an AM or FM radio wave.

In analog modulation, the modulation is applied continuously in response to the analog information signal.

List of common analog modulation techniques

Common analog modulation techniques are:

- Amplitude modulation (AM) (here the amplitude of the carrier signal is varied in accordance to the instantaneous amplitude of the modulating signal)

    - Double-sideband modulation (DSB)

        - Double-sideband modulation with carrier (DSB-WC) (used on the AM radio broadcasting band)
        - Double-sideband suppressed-carrier transmission (DSB-SC)
        - Double-sideband reduced carrier transmission (DSB-RC)

    - Single-sideband modulation (SSB, or SSB-AM)

        - Single-sideband modulation with carrier (SSB-WC)
        - Single-sideband modulation suppressed carrier modulation (SSB-SC)

    - Vestigial sideband modulation (VSB, or VSB-AM)

    - Quadrature amplitude modulation (QAM)

- Angle modulation, which is approximately constant envelope

- Frequency modulation (FM) (here the frequency of the carrier signal is varied in accordance to the instantaneous amplitude of the modulating signal)

- Phase modulation (PM) (here the phase shift of the carrier signal is varied in accordance with the instantaneous amplitude of the modulating signal)

2.30.2 Digital modulation methods

In digital modulation, an analog carrier signal is modulated by a discrete signal. Digital modulation methods can be considered as digital-to-analog conversion, and the corresponding demodulation or detection as analog-to-digital conversion. The changes in the carrier signal are chosen from a finite number of M alternative symbols (the *modulation alphabet*).

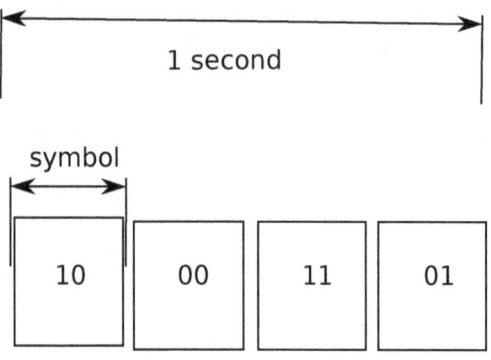

Schematic of 4 baud (8 bit/s) data link containing arbitrarily chosen values.

A simple example: A telephone line is designed for transferring audible sounds, for example tones, and not digital bits (zeros and ones). Computers may however communicate over a telephone line by means of modems, which are representing the digital bits by tones, called symbols. If there are four alternative symbols (corresponding to a musical instrument that can generate four different tones, one at a time), the first symbol may represent the bit sequence 00, the second 01, the third 10 and the fourth 11. If the modem plays a melody consisting of 1000 tones per second, the symbol rate is 1000 symbols/second, or baud. Since each tone (i.e., symbol) represents a message consisting of two digital bits in this example, the bit rate is twice the symbol rate, i.e. 2000 bits per second. This is similar to the technique used by dialup modems as opposed to DSL modems.

According to one definition of digital signal, the modulated signal is a digital signal. According to another definition, the modulation is a form of digital-to-analog conversion. Most textbooks would consider digital modulation schemes as a form of digital transmission, synonymous to data transmission; very few would consider it as analog transmission.

Fundamental digital modulation methods

The most fundamental digital modulation techniques are based on keying:

- PSK (phase-shift keying): a finite number of phases are used.

- FSK (frequency-shift keying): a finite number of frequencies are used.

- ASK (amplitude-shift keying): a finite number of amplitudes are used.

- QAM (quadrature amplitude modulation): a finite number of at least two phases and at least two amplitudes are used.

In QAM, an inphase signal (or I, with one example being a cosine waveform) and a quadrature phase signal (or Q, with an example being a sine wave) are amplitude modulated with a finite number of amplitudes, and then summed. It can be seen as a two-channel system, each channel using ASK. The resulting signal is equivalent to a combination of PSK and ASK.

In all of the above methods, each of these phases, frequencies or amplitudes are assigned a unique pattern of binary bits. Usually, each phase, frequency or amplitude encodes an equal number of bits. This number of bits comprises the *symbol* that is represented by the particular phase, frequency or amplitude.

If the alphabet consists of $M = 2^N$ alternative symbols, each symbol represents a message consisting of N bits. If the symbol rate (also known as the baud rate) is f_S symbols/second (or baud), the data rate is $N f_S$ bit/second.

For example, with an alphabet consisting of 16 alternative symbols, each symbol represents 4 bits. Thus, the data rate is four times the baud rate.

In the case of PSK, ASK or QAM, where the carrier frequency of the modulated signal is constant, the modulation alphabet is often conveniently represented on a constellation diagram, showing the amplitude of the I signal at the x-axis, and the amplitude of the Q signal at the y-axis, for each symbol.

Modulator and detector principles of operation

PSK and ASK, and sometimes also FSK, are often generated and detected using the principle of QAM. The I and Q signals can be combined into a complex-valued signal $I+jQ$ (where j is the imaginary unit). The resulting so called equivalent lowpass signal or equivalent baseband signal is a complex-valued representation of the real-valued modulated physical signal (the so-called passband signal or RF signal).

These are the general steps used by the modulator to transmit data:

1. Group the incoming data bits into codewords, one for each symbol that will be transmitted.

2. Map the codewords to attributes, for example amplitudes of the I and Q signals (the equivalent low pass signal), or frequency or phase values.

3. Adapt pulse shaping or some other filtering to limit the bandwidth and form the spectrum of the equivalent low pass signal, typically using digital signal processing.

4. Perform digital to analog conversion (DAC) of the I and Q signals (since today all of the above is normally achieved using digital signal processing, DSP).

5. Generate a high frequency sine carrier waveform, and perhaps also a cosine quadrature component. Carry out the modulation, for example by multiplying the sine and cosine waveform with the I and Q signals, resulting in the equivalent low pass signal being frequency shifted to the modulated passband signal or RF signal. Sometimes this is achieved using DSP technology, for example direct digital synthesis using a waveform table, instead of analog signal processing. In that case the above DAC step should be done after this step.

6. Amplification and analog bandpass filtering to avoid harmonic distortion and periodic spectrum.

At the receiver side, the demodulator typically performs:

1. Bandpass filtering.

2. Automatic gain control, AGC (to compensate for attenuation, for example fading).

3. Frequency shifting of the RF signal to the equivalent baseband I and Q signals, or to an intermediate frequency (IF) signal, by multiplying the RF signal with a local oscillator sinewave and cosine wave frequency (see the superheterodyne receiver principle).

4. Sampling and analog-to-digital conversion (ADC) (sometimes before or instead of the above point, for example by means of undersampling).

5. Equalization filtering, for example a matched filter, compensation for multipath propagation, time spreading, phase distortion and frequency selective fading, to avoid intersymbol interference and symbol distortion.

6. Detection of the amplitudes of the I and Q signals, or the frequency or phase of the IF signal.

7. Quantization of the amplitudes, frequencies or phases to the nearest allowed symbol values.

8. Mapping of the quantized amplitudes, frequencies or phases to codewords (bit groups).

9. Parallel-to-serial conversion of the codewords into a bit stream.

10. Pass the resultant bit stream on for further processing such as removal of any error-correcting codes.

As is common to all digital communication systems, the design of both the modulator and demodulator must be done simultaneously. Digital modulation schemes are possible because the transmitter-receiver pair have prior knowledge of how data is encoded and represented in the communications system. In all digital communication systems, both the modulator at the transmitter and the demodulator at the receiver are structured so that they perform inverse operations.

Non-coherent modulation methods do not require a receiver reference clock signal that is phase synchronized with the sender carrier signal. In this case, modulation symbols (rather than bits, characters, or data packets) are asynchronously transferred. The opposite is coherent modulation.

List of common digital modulation techniques

The most common digital modulation techniques are:

- Phase-shift keying (PSK)

  - Binary PSK (BPSK), using M=2 symbols
  - Quadrature PSK (QPSK), using M=4 symbols
  - 8PSK, using M=8 symbols

- 16PSK, using M=16 symbols
- Differential PSK (DPSK)
- Differential QPSK (DQPSK)
- Offset QPSK (OQPSK)
- π/4–QPSK

- Frequency-shift keying (FSK)

  - Audio frequency-shift keying (AFSK)
  - Multi-frequency shift keying (M-ary FSK or MFSK)
  - Dual-tone multi-frequency (DTMF)

- Amplitude-shift keying (ASK)

- On-off keying (OOK), the most common ASK form

  - M-ary vestigial sideband modulation, for example 8VSB

- Quadrature amplitude modulation (QAM), a combination of PSK and ASK

  - Polar modulation like QAM a combination of PSK and ASK

- Continuous phase modulation (CPM) methods

  - Minimum-shift keying (MSK)
  - Gaussian minimum-shift keying (GMSK)
  - Continuous-phase frequency-shift keying (CPFSK)

- Orthogonal frequency-division multiplexing (OFDM) modulation

  - Discrete multitone (DMT), including adaptive modulation and bit-loading

- Wavelet modulation

- Trellis coded modulation (TCM), also known as Trellis modulation

- Spread-spectrum techniques

  - Direct-sequence spread spectrum (DSSS)
  - Chirp spread spectrum (CSS) according to IEEE 802.15.4a CSS uses pseudo-stochastic coding
  - Frequency-hopping spread spectrum (FHSS) applies a special scheme for channel release
  - SIM31 (SIM) New digital Mode SIM31 SIM63 tks SWL Tunisian

MSK and GMSK are particular cases of continuous phase modulation. Indeed, MSK is a particular case of the sub-family of CPM known as continuous-phase frequency-shift keying (CPFSK) which is defined by a rectangular frequency pulse (i.e. a linearly increasing phase pulse) of one symbol-time duration (total response signaling).

OFDM is based on the idea of frequency-division multiplexing (FDM), but the multiplexed streams are all parts of a single original stream. The bit stream is split into several parallel data streams, each transferred over its own sub-carrier using some conventional digital modulation scheme. The modulated sub-carriers are summed to form an OFDM signal. This dividing and recombining helps with handling channel impairments. OFDM is considered as a modulation technique rather than a multiplex technique, since it transfers one bit stream over one communication channel using one sequence of so-called OFDM symbols. OFDM can be extended to multi-user channel access method in the orthogonal frequency-division multiple access (OFDMA) and multi-carrier code division multiple access (MC-CDMA) schemes, allowing several users to share the same physical medium by giving different sub-carriers or spreading codes to different users.

Of the two kinds of RF power amplifier, switching amplifiers (Class D amplifiers) cost less and use less battery power than linear amplifiers of the same output power. However, they only work with relatively constant-amplitude-modulation signals such as angle modulation (FSK or PSK) and CDMA, but not with QAM and OFDM. Nevertheless, even though switching amplifiers are completely unsuitable for normal QAM constellations, often the QAM modulation principle are used to drive switching amplifiers with these FM and other waveforms, and sometimes QAM demodulators are used to receive the signals put out by these switching amplifiers.

Automatic digital modulation recognition (ADMR)

Automatic digital modulation recognition in intelligent communication systems is one of the most important issues in software defined radio and cognitive radio. According to incremental expanse of intelligent receivers, automatic modulation recognition becomes a challenging topic in telecommunication systems and computer engineering. Such systems have many civil and military applications. Moreover, blind recognition of modulation type is an important problem in commercial systems, especially in software defined radio. Usually in such systems, there are some extra information for system configuration, but considering blind approaches in intelligent receivers, we can reduce information overload and increase transmission performance.[1] Obviously, with no knowledge of the transmitted data and many unknown parameters at the receiver, such as the signal power, carrier frequency and phase offsets, timing information, etc., blind identification of the modulation is a difficult task. This becomes even more challenging in real-world scenarios with multipath fading, frequency-selective and time-varying channels.[2]

There are two main approaches to automatic modulation recognition. The first approach uses likelihood-based methods to assign an input signal to a proper class. Another recent approach is based on feature extraction.

Digital baseband modulation or line coding

Main article: Line code

The term **digital baseband modulation** (or digital baseband transmission) is synonymous to line codes. These are methods to transfer a digital bit stream over an analog baseband channel (a.k.a. lowpass channel) using a pulse train, i.e. a discrete number of signal levels, by directly modulating the voltage or current on a cable. Common examples are unipolar, non-return-to-zero (NRZ), Manchester and alternate mark inversion (AMI) codings.[3]

2.30.3   Pulse modulation methods

Pulse modulation schemes aim at transferring a narrowband analog signal over an analog baseband channel as a two-level signal by modulating a pulse wave. Some pulse modulation schemes also allow the narrowband analog signal to be transferred as a digital signal (i.e., as a quantized discrete-time signal) with a fixed bit rate, which can be transferred over an underlying digital transmission system, for example, some line code. These are not modulation schemes in the conventional sense since they are not channel coding schemes, but should be considered as source coding schemes, and in some cases analog-to-digital conversion techniques.

Analog-over-analog methods

- Pulse-amplitude modulation (PAM)

- Pulse-width modulation (PWM) and Pulse-depth modulation (PDM)

- Pulse-position modulation (PPM)

Analog-over-digital methods

- Pulse-code modulation (PCM)

  - Differential PCM (DPCM)

- Adaptive DPCM (ADPCM)

- Delta modulation (DM or Δ-modulation)

- Delta-sigma modulation ($\sum\Delta$)

- Continuously variable slope delta modulation (CVSDM), also called Adaptive-delta modulation (ADM)

- Pulse-density modulation (PDM)

2.30.4   Miscellaneous modulation techniques

- The use of on-off keying to transmit Morse code at radio frequencies is known as continuous wave (CW) operation.

- Adaptive modulation

- Space modulation is a method whereby signals are modulated within airspace such as that used in instrument landing systems.

2.30.5   Further reading

- Multipliers vs. Modulators Analog Dialogue, June 2013

- The Easiest Way to Understand Modulation Blog Post and YouTube Video, 2014

2.30.6   See also

- Neuromodulation

- Demodulation

- Electrical resonance

- Modulation order

- Types of radio emissions

- Communications channel

- Channel access methods

- Channel coding

- Line code

- Telecommunication

- Modem

- RF modulator

- Codec

- Ring modulation

2.30.7 References

[1] M. Hadi Valipour, M. Mehdi Homayounpour and M. Amin Mehralian, Automatic digital modulation recognition in presence of noise using SVM and PSO, in Proceedings of 2012 Sixth International Symposium on Telecommunications (IST), pp 378-382, Nov 2012, Tehran, Iran.

[2] Dobre, Octavia A., Ali Abdi, Yeheskel Bar-Ness, and Wei Su. Communications, IET 1, no. 2 (2007): 137-156. (2007). "Survey of automatic modulation classification techniques: classical approaches and new trends" (PDF). *IET Communications*: 137–156.

[3] Ke-Lin Du and M. N. S. Swamy (2010). *Wireless Communication Systems: From RF Subsystems to 4G Enabling Technologies*. Cambridge University Press. p. 188. ISBN 978-0-521-11403-5.

2.30.8 External links

- Interactive presentation of soft-demapping for AWGN-channel in a web-demo Institute of Telecommunications, University of Stuttgart

- Modem(Modulation and Demodulation)

2.31 NEXUS International Broadcasting Association

NEXUS International Broadcasting Association (NEXUS-IBA) was officially founded in 1988 in Milan, Italy, as a non-profit association of broadcasters and program producers.

As the Latin word *nexus*, NEXUS-IBA is a link (or point of connection) between content producers and the association's technical facilities used to bring such content to the public.

2.31.1 Member services

NEXUS-IBA main focus is to provide broadcasts internationally by means of Shortwave radio, targeting Europe, Africa, Middle East and Asia with high power transmitters and directional HRS type antennas. Most programs on the air are produced by NEXUS-IBA members, and originate as multiple audio and video streams from NEXUS-IBA network operations center in Milano. From there they are transmitted live to each ground or satellite relay station. Internet audio and video channels are available to the public via the WorldDirector streaming network operated by Wornex International, and published either on the association's web sites or directly on its members'. To increase performance due to the international character of NEXUS-IBA audiences, live and on demand audio and video streams are made available to listeners via multiple servers or point of presence in Europe, Asia and the USA.

Services provided to NEXUS-IBA members also include: *ad hoc* research and development projects based on the Association's technical know-how, audio and video Streaming media, Shortwave and Satellite television broadcasting, internet hosting and consulting on international media.

2.31.2 NEXUS-IBA as an R&D organization

NEXUS International Broadcasting Association is also a research and development organization that developed the first known Content Delivery Network in 1995. As a non profit organization NEXUS-IBA acts as a research incubator with hi-tech professionals in the computing, Internet, telephone and media industries for a number of media related technologies. Since early 1995 NEXUS-IBA's Content delivery network technology - called WorldDirector and later incorporated by its spin off Wornex International - has been used for distribution of NEXUS-IBA members' audio & video streaming channels, as well as other archives and web services belonging to the Association's members.

NEXUS-IBA also participated in a three year research project under the European Union's 5th Framework Program (Information Society Technologies, Key Action I.4.2, On-line Support to Democratic Processes) for the development of DEMOS (Delphi Mediation Online System), an Internet based platform to favour debate among citizens and politicians, aimed at facilitating "online-democracy".

DEMOS has been completely re-designed and made available to government bodies and private enterprises by NEXUS-IBA's spin off Wornex International, with the continued support of NEXUS-IBA researchers.

2.31.3 See also

- Content delivery network

2.31.4 References

2.31.5 External links

- DEMOS at UNESCO

- DEMOS paper

- DEMOS at WORNEX International

- Conoscere DEMOS (white paper in Italian)
- padania.org a DEMOS-3 application
- Wornex International
- NEXUS-IBA home page
- NEXUS-IBA Shortwave schedules
- International Public Access Radio
- European Gospel Radio
- Undpi.org

2.32 NORMOB

NORMOB (Nordic Broadcasters Mobile Alliance) is an alliance between the Norwegian Broadcasting Corporation (NRK), Danish Radio (DR), Swedish TV (SVT), Swedish Radio (SR), the Swedish Educational Broadcasting Company (UR) and Finnish Broadcasting Corporation (YLE). NORMOB was established in 2005 in order to promote cooperation between the broadcasters on mobile services. The body meets 2-3 times a year.

2.33 Olympic Broadcasting Services

Olympic Broadcasting Services covering the men's 10 kilometre marathon swim at the 2012 Olympic Games

Olympic Broadcasting Services (**OBS**) is an agency of the International Olympic Committee, established in 2001 to be responsible for host broadcasting - the world feeds provided to all international broadcasters of the Olympic Games and Paralympic Games from 2008 on, and provide continuity in Olympic coverage from one Games to the next. Previously the host broadcaster role was delegated to the local organising committees or to third-party broadcasters.

2.33.1 History

Beijing 2008

Main article: Beijing Olympic Broadcasting

Its operations began with the 2008 Summer Olympics in Beijing, where Beijing Olympic Broadcasting, a joint venture between OBS and the Beijing Organizing Committee, acted as the host broadcasting consortium and the state television network, China Central Television which is one of the host nation broadcasters of the games.

Vancouver 2010

For the 2010 Winter Olympics in Vancouver, a wholly owned division, Olympic Broadcasting Services Vancouver, was set up. The executive director of OBSV is Nancy Lee, a former producer and executive for CBC Sports. The 2010 Olympics marked the first games where the host broadcasting facilities was provided solely by OBS.[1] The 2012 Olympics and Paralympics were broadcast by the OBS.[2]

London 2012

It is speculated that in the run-up to the London 2012 Olympic Games, an extra ring of security was put around the trailer of Danny Boyle (the ceremony director) following "friction" between his crew and Olympic Broadcasting Services. The home nation broadcaster was the BBC with Olympics specials. [3]

Rio 2016

It is speculated that at the Rio 2016 Olympic Games, Rede Globo, Rede Record and Rede Bandeirantes will be responsible for generating the international radio and TV signal.

2.33.2 References

[1] http://www.obsv.ca/obsintroduction.html[]

[2] http://www.obs.es/obslondon2012.html[]

[3] http://www.guardian.co.uk/sport/2012/jul/18/ olympic-opening-ceremony-drama?newsfeed=true

2.33.3 External links

- Official website

2.34 Our World (TV special)

For the American series of the same name, see Our World (TV series).

Our World was the first live, international, satellite television production, which was broadcast on 25 June 1967. Creative artists, including The Beatles, opera singer Maria Callas, and painter Pablo Picasso—representing nineteen nations—were invited to perform or appear in separate segments featuring their respective countries. The two-and-half-hour event had the largest television audience ever up to that date: an estimated 400 to 700 million people around the globe watched the broadcast. Today, it is most famous for the segment from the United Kingdom starring The Beatles. They performed their song "All You Need Is Love" for the first time to close the broadcast.

2.34.1 Planning

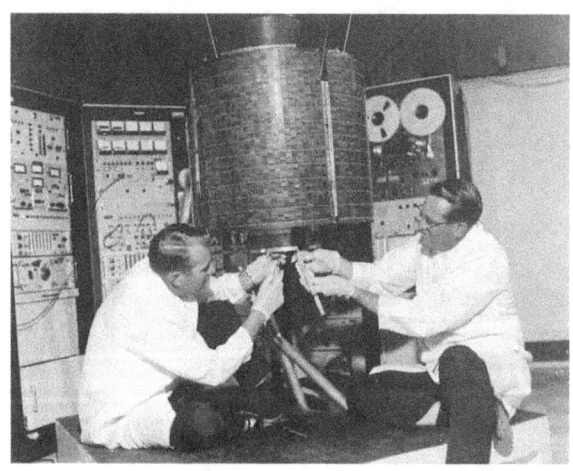

The Intelsat I nicknamed "Early Bird", one of the satellites used

The project was conceived by BBC producer Aubrey Singer. It was transferred to the European Broadcasting Union, but the master control room for the broadcast was still at the BBC in London. The satellites used were Intelsat I (known as "Early Bird"), Intelsat 2-2 ("Lani Bird"), Intelsat 2-3 ("Canary Bird"), and NASA's ATS-1.[2]

It took ten months to bring everything together. One hitch was the sudden pull-out of the Eastern Bloc countries headed by the Soviet Union in the week leading up to the broadcast. Apparently it was a protest at the Western nations' response to the Six-Day War.[1]

The ground rules included that no politicians or heads of state could participate in the broadcast. In addition, everything had to be 'live', so no use of videotape or film was permitted. Ten thousand technicians, producers, and interpreters took part in this gigantic broadcast. Each country would have its own announcers, due to language issues, and interpreters would voice-over the original sound when not in a country's native language. In the end 14 countries participated in the production that was transmitted to 31 countries with an estimated audience of between 400 and 700 million people.[1][3]

2.34.2 Participant countries

The participant countries were the following ones: [4][5]

Czechoslovakia, Poland, East Germany, the Soviet Union and Hungary withdrew prior to the broadcast, in protest for the Six-Day War.[6]

2.34.3 The broadcast

The opening credits were accompanied by the Our World theme sung in 22 different languages by the Vienna Boys' Choir.[7]

Canada's CBC Television had Marshall McLuhan being interviewed in a Toronto television control room. At 7:17 pm GMT, the show switched to the United States' segment about the Glassboro, New Jersey, conference between American president Lyndon Johnson and Soviet premier Alexei Kosygin; since *Our World* insisted that no politicians be shown, only the house where the conference was being held was televised. National Educational Television's (NET) Dick McCutcheon ended up talking about the impact of the new television technology on a global scale.[1]

The show switched back to Canada at 7:18 pm GMT. Segments that were beamed worldwide were from a Ghost Lake, Alberta ranch, showing a rancher, and his cutting horse, cutting out a herd of cattle. The last Canadian segment was from Kitsilano Beach, located in Vancouver's Point Grey district at 7:19 pm GMT.[1]

At 7:20 pm GMT, the program shifted continents to Asia, with Tokyo, Japan being the next segment. It was 4:20 a.m. local time and NHK showed the construction of the Tokyo subway system.[1]

The equator was crossed for the first time in the program when it switched to the Australian contribution, which was at 5:22 a.m. Australian Eastern Standard Time (AEST). This was the most technically complicated point in the broadcast, as both the Japanese and Australian satellite ground stations had to reverse their actions: Tokyo had to go from transmit mode to receive mode, while Melbourne had to switch from receive to transmit mode.[2] The segment dealt with Trams leaving the Hanna Street Depot in Melbourne with Australian Broadcasting Corporation's Brian King explaining that sunrise was many hours away as it was winter there.[1] A scientific segment, later on in the broadcast, was also included that dealt with the Parkes Observatory tracking a deep space object.[8]

2.34.4   The Beatles' sequence

The Beatles performing "All You Need is Love".

The broadcast took place at the height of the Vietnam War. The Beatles were asked to write a song with a positive message song and John Lennon was specifically asked to write it. [9] They gave a live performance, transmitted at 8:54 pm GMT, performing a new song, written primarily by John Lennon, entitled "All You Need Is Love", which according to Ringo Starr and George Martin was composed specially for the occasion.[9] The Beatles invited many of their friends to the event to create a festive atmosphere and to join in on the song's chorus. Among the friends were members of The Rolling Stones, Eric Clapton, Marianne Faithfull, Keith Moon and Graham Nash.[9] The performance was preceded by just a single rehearsal.[9][10]

Although the entire program was originally transmitted in black-and-white (and thus the videotape recording was also in black and white), for its usage in the 1995 TV special *The Beatles Anthology*, The Beatles' performance on the 1967 programme was colourized—using colour photographs taken at the event as a reference.[11] The sequence opens in its original monochromatic format and rapidly morphs into full colour conveying the brightly coloured "flower power" and "psychedelic"-style clothing worn by The Beatles and their guests that was popular in 1967 during what was subsequently dubbed the "Summer of Love".[11]

2.34.5   In Popular Culture

In the novel *The Light of Other Days* by Arthur C. Clarke and Stephen Baxter the global media empire run by Hiram

Patterson is called OurWorld, the name chosen after the character saw the program as a child and was inspired to change the world.[12]

2.34.6   References and notes

[1] Burke, Stanley (25 June 1967). "Our World – Five continents linked via satellite". *CBC Archives* (Toronto: Canadian Broadcasting Corporation). Archived from the original on 7 January 2016. Retrieved 4 June 2007.

[2] Huntington, Tom. "The Whole World's Watching". *Air and Space Magazine* (Smithsonian Institution) **10** (April/May 1996). Archived from the original on 25 February 1999. Retrieved 25 June 2014.

[3] Harrington, Richard (24 November 2002). "His Musical Notes Have Become TV Landmarks". *The Washington Post*. p. Y06. Archived from the original on 21 April 2004. Retrieved 4 June 2007.

[4] "The Beatles on Our World: All You Need Is Love". *The Beatles Bible*. Cardiff, Wales, UK. Archived from the original on 7 January 2016. Retrieved 1 January 2015.

[5] "1967 – Our World – the first live, international, satellite television production". *Internet History Library*. 4 April 2012. Archived from the original on 7 January 2016. Retrieved 1 January 2015.

[6] McKellar, Colin. "Cooby Creek – Our World". *A Tribute to Honeysuckle Creek Tracking Station*. Australia. Archived from the original on 7 January 2016. Retrieved 1 January 2015.

[7] Flowers, Brian (4 July 2007). "The Technical History of Eurovision" (PDF). *EBU Technical Review* (European Broadcasting Union). Archived from the original on 25 June 2013. Retrieved 28 November 2007.

[8] Rowsthorn, Peter (4 May 2007). "Moment in Time Episode 12: First Satellite Broadcast". *Can We Help* (Sydney: Australian Broadcasting Corporation). Archived from the original on 25 June 2013. Retrieved 4 June 2007.

[9] Sheppard, John (3 June 1987). "It was 20 Years Ago Today". Granada TV. Retrieved 4 June 2007. Granada TV documentary shows The Beatles' *Our World* broadcast segment.

[10] Hastings, Chris (22 July 2007). "Beatles never told of protests at satellite show". *The Daily Telegraph* (London). Archived from the original on 25 June 2013. Retrieved 25 June 2013.

[11] Sella, Tom (1996). "Anthology Home Video". Beatles Reference Library. Retrieved 27 June 2010. Laserdisc 7, Side 1, Chapter 1

[12] Clarke, Arthur C.; Baxter, Stephen. *The Light of Other Days*. Tor Books. p. 15. ISBN 0-312-87199-6.

2.34.7 External links

- *Our World* at the Internet Movie Database

2.35 Pangea Day

Pangea Day was an international multimedia event conducted on May 10, 2008. Cairo, Kigali, London, Los Angeles, Mumbai and Rio de Janeiro were linked to produce a 4-hour program of films, music and speakers. The program was broadcast live across the globe from 1800 to 2200 UTC, culminating in a global drum circle, symbolizing the common heartbeat of the world. According to the festival organizers, "Pangea Day plans to use the power of film to bring the world a little closer together." [1] [2]

Pangea Day originated in 2006 when documentary filmmaker Jehane Noujaim won the TED Prize. Jehane wished to use film to bring the world together.[3]

Pangea refers to the supercontinent from which all current continents eventually separated. It serves as a reminder of the "connectedness" or unitary nature of all people on Planet Earth.

2.35.1 Goals

- Bring together millions of people from all over the world in a unique shared experience.

- Use the power of film to create a better understanding of one another.

- Form a global community striving for a better future.

[2]

2.35.2 Live broadcast locations

Pangea Day was broadcast live from seven cities:

- Cairo - The Pyramids

- Kigali - Jali Gardens

- London - Somerset House

- Los Angeles - Sony Pictures Studios

- Mumbai - National Centre for the Performing Arts

- Rio de Janeiro - Morro da Urca

- Buenos Aires - [KONEX Theater]

In the United States, Current TV was the exclusive, English-language broadcaster.
[4]

2.35.3 Featured Films

- *A Thousand Words* directed by Ted Chung

- *More* directed by Mark Osborne

- *L'Homme Sans Tete* directed by Juan Diego Solanas

2.35.4 Global partner

Nokia was Pangea Day's premier global partner. In addition to providing financial support, Nokia sent video enabled devices to film schools and programs in disadvantaged areas and conflict zones, and to UNHCR refugee camps. Some of the films made in these locations were included in the Pangea Day broadcast. [5] [6]

2.35.5 Key participants

Hosts

- June Arunga

- Lisa Ling

- Max Lugavere

- Jason Silva

Advisory board

- J. J. Abrams

- Lawrence Bender

- Nancy Buirski

- Alan Cumming

- Richard Curtis

- Ami Dar

- Cameron Diaz

- Matthew Freud

- Bob Geldof

- Goldie Hawn

- Jim Hornthal

- Judy McGrath
- Pat Mitchell
- Vik Muniz
- Clare Munn
- Mira Nair
- Dr. Tero Ojanperä
- Eboo Patel
- Alexander Payne
- Richard Rogers
- Meg Ryan
- Deborah Scranton
- Paul Simon
- Jeffrey Skoll
- Sir Martin Sorrell
- Philippe Starck
- Dave Stewart
- Yossi Vardi
- Kevin Wall
- Forest Whitaker
- will.i.am
- Paul Zilk

[7]

Speakers

- Christiane Amanpour
- Bassam Aramin
- Karen Armstrong
- June Arunga
- Ali Abu Awwad
- Ishmael Beah
- Donald Brown
- Assaad Chaftari
- Muhieddine Chehab

- Robi Damelin
- Jonathan Harris
- Robert Kurzban
- Lisa Ling
- Max Lugavere
- Khaled Aboul Naga
- Queen Noor of Jordan
- Eboo Patel
- Carolyn Porco
- Jean-Paul Samputu
- Yonathan Shapira
- Jason Silva

[8]

Musicians

- Dave Stewart
- Gilberto Gil
- Hypernova
- Rokia Traoré

[9]

2.35.6   Notes and references

[1] http://www.hollywoodreporter.com/hr/content_display/
film/news/e3i12490cf38336fdda945e6048e35c6f19
retrieved January 12/08

[2] Pangea Day

[3] http://www.iht.com/articles/ap/2007/11/07/arts/
NA-A-E-CEL-US-Goldie-Hawn.php?WT.mc_id=
rssap_news retrieved January 12/08

[4] Pangea Day

[5] Pangea Day

[6] Pangea Day

[7] Pangea Day: Advisory Board

[8] Pangea Day: Speakers

[9] Pangea Day: Musical Performers

2.35.7 External links

- PangeaDay.org - The Official Pangea Day Website

- "Can Your Film Change the World?" - Pangea Day promo video at YouTube

- Press Release: Pangea Day Partners with Avenue A | Razorfish

2.36 Pirate radio

For other uses, see Pirate Radio (disambiguation).

Pirate radio is illegal or unregulated radio transmission for entertainment or political purposes. While *pirate* just refers to the illegal nature of the broadcasts, there have also been notable pirate offshore radio transmissions.

Pirate radio generally describes the unlicensed broadcast of FM radio, AM radio, or short wave signals over a wide range. In some cases radio stations are deemed legal where the signal is transmitted, but illegal where the signals are received—especially when the signals cross a national boundary. In other cases, a broadcast may be considered "pirate" due to the nature of its content, its transmission format (especially a failure to transmit a station identification according to regulations), or the transmit power (wattage) of the station, even if the transmission is not technically illegal (such as a web cast or an amateur radio transmission). Pirate radio stations are sometimes called **bootleg stations** (a term especially associated with two-way radio), **clandestine stations** (associated with heavily politically motivated operations) or **free radio** stations.

2.36.1 Pirate-radio history and examples

Radio "piracy" began with the advent of regulations of the public airwaves in the United States at the dawn of the age of radio. Initially, radio, or wireless as it was more commonly called, was an open field of hobbyists and early inventors and experimenters. The United States Navy began using radio for time signals and weather reports on the east coast of the United States in the 1890s. Before the advent of valve (vacuum tube) technology, early radio enthusiasts used noisy spark-gap transmitters, such as the first spark-gap modulation technology pioneered by the first real audio (rather than telegraph code) radio broadcaster, Charles D. Herrold, in San Jose, California, or the Ruhmkorff coil used by almost all early experimenters. The navy soon began complaining to a sympathetic press that amateurs were disrupting naval transmissions. The May 25, 1907, edition of *Electrical World* in an article called "Wireless and Lawless"[1] reported authorities were unable to prevent an amateur from interfering with the operation of a government station at the Washington, D.C. Navy Yard using legal means.

In the run-up to the London Radiotelegraph Convention in 1912 (essentially an international gentlemen's agreement on use of the radio band, non-binding and, on the high seas, completely null), and amid concerns about the safety of marine radio following the sinking of the *RMS* Titanic on April 15 of that year (although there were never allegations of radio interference in that event), the *New York Herald* of April 17, 1912, headlined President William Howard Taft's initiative to regulate the public airwaves in an article titled "President Moves to Stop Mob Rule of Wireless."

When the "Act to Regulate Radio Communication" was passed on August 13, 1912, amateurs and experimenters were not banned from broadcasting; rather, amateurs were assigned their own frequency spectrum, and licensing and call-signs were introduced. By regulating the public airwaves, President Taft thus created the legal space for illicit broadcasts to take place. An entire federal agency, the Federal Radio Commission, was formed in 1927 and succeeded in 1934 by the Federal Communications Commission. These agencies would enforce rules on call-signs, assigned frequencies, licensing and acceptable content for broadcast.

The Radio Act of 1912 gave the president legal permission to shut down radio stations "in time of war", and during the first two and a half years of World War One, before US entry, President Wilson tasked the US Navy with monitoring US radio stations, nominally to "ensure neutrality." The navy used this authority to shut down amateur radio in the western part of the US (the US was divided into two civilian radio "districts" with corresponding call-signs, beginning with "K" in the west and "W" in the east, in the regulatory measures; the navy was assigned call-signs beginning with "N"). When Wilson declared war on Germany on April 6, 1917, he also issued an executive order closing most radio stations not needed by the US government. The navy took it a step further and declared it was illegal to listen to radio or possess a receiver or transmitter in the US, but there were doubts they had the authority to issue such an order even in war time. The ban on radio was lifted in the US in late 1919.[2]

In 1924, New York City station WHN was accused of being an "outlaw station" by the American Telephone and Telegraph Company (AT&T) for violating trade licenses which permitted only AT&T stations to sell airtime on their transmitters. As a result of the AT&T interpretation a landmark case was heard in court, which even prompted comments from Secretary of Commerce Herbert Hoover when he took a public stand in the station's defense. Although AT&T won its case, the furor created was such that those restrictive provisions of the transmitter license were never enforced.

In 1948, the United Nations brought into being the Universal Declaration of Human Rights, of which Article 19 states: "Everyone has the right to freedom of opinion and expression; this right includes the freedom to hold opinions without interference and to seek, receive and impart information and ideas through any media and regardless of frontiers."

In Europe, Denmark had the first known radio station in the world to broadcast commercial radio from a vessel in international waters without permission from the authorities in the country that it broadcast to (Denmark in this case). The station was named Radio Mercur and began transmission on August 2, 1958. In the Danish newspapers it was soon called a "pirate radio".

In the 1960s in the UK, the term referred to not only a perceived unauthorized use of the state-run spectrum by the unlicensed broadcasters but also the risk-taking nature of offshore radio stations that actually operated on anchored ships or marine platforms. The term had been used previously in Britain and the US to describe unlicensed land-based broadcasters and even border blasters (for example, a 1940 British comedy about an unauthorized TV broadcaster, *Band Waggon*, uses the phrase "pirate station" several times). A good example of this kind of activity was Radio Luxembourg located in the Grand Duchy of Luxembourg. The English language evening broadcasts from Radio Luxembourg were beamed by Luxembourg-licensed transmitters. The audience in the United Kingdom originally listened to their radio sets by permission of a wireless license issued by the British General Post Office (GPO). However, under terms of that wireless licence, it was an offence under the Wireless Telegraphy Act to listen to *unauthorised* broadcasts, which possibly included those transmitted by Radio Luxembourg. Therefore, as far as the British authorities were concerned, Radio Luxembourg was a "pirate radio station" and British listeners to the station were breaking the law (although as the term 'unauthorised' was never properly defined it was somewhat of a legal grey area). This did not stop British newspapers from printing programme schedules for the station, or a British weekly magazine aimed at teenage girls, *Fab 208*, from promoting the DJs and their lifestyle (Radio Luxembourg's wavelength was 208 metres (1439, then 1440 kHz)).

Radio Luxembourg was later joined by other well-known pirate stations received in the UK in violation of UK licensing, including Radio Caroline and Radio Atlanta (subsequently Radio Carolines North and South respectively, following their merger and the original ship's relocation) and Radio London, all of which broadcast from vessels anchored outside of territorial limits and were therefore legitimate. Radio Jackie, for instance (although transmitting illegally), was registered for VAT and even had its address and telephone number in local telephone directories.

Where actual seafaring vessels are not involved, the term *pirate radio* is a political term of convenience as the word "pirate" suggests an illegal venture, regardless of the broadcast's actual legal status. The radio station XERF located at Ciudad Acuña, Coahuila, Mexico, just across the Rio Grande from Del Rio, Texas, US, is an example.

While Mexico issued radio station XERF with a license to broadcast, the power of its 250 kW transmitter was far greater than the maximum of 50 kW authorized for commercial use by the government of the United States of America. Consequently, XERF and many other radio stations in Mexico, which sold their broadcasting time to sponsors of English-language commercial and religious programs, were labelled as "border blasters", but not "pirate radio stations", even though the content of many of their programs could not have been aired by a US-regulated broadcaster. Predecessors to XERF, for instance, had originally broadcast in Kansas, advocating "goat-gland surgery" for improved masculinity, but moved to Mexico to evade US laws about advertising medical treatments, particularly unproven ones.

Free radio

Another variation on the term *pirate radio* came about during the "Summer of Love" in San Francisco during the 1960s. These were "**Free radio**", which usually referred to clandestine and unlicensed land-based transmissions. These were also tagged as being pirate radio transmissions. Free Radio was only ever used to refer to radio transmissions that were beyond government control, as was offshore radio in the UK and Europe.

The term *free radio* was adopted by the Free Radio Association of listeners who defended the rights of the offshore "radio stations" broadcasting from ships and marine structures off the coastline of the United Kingdom.

Félix Guattari points out:

In Europe, in addition to adopting the term *free radio*, supportive listeners of what had been called pirate radio adopted the term **offshore radio**, which was usually the term used by the owners of the marine broadcasting stations.

More recently the term "free radio" implied that the broadcasts were commercial-free and the station was only there for the output, be it a type of music or spoken opinion. In this context, 'pirate' radio thus refers to stations that do advertise and plug various gigs and raves.

2.36.2 Pirate radio by geographical area

See also: Pirate radio in Asia, Pirate radio in Australasia, Pirate radio in Central America and Caribbean Sea, Pirate radio in Europe, Pirate radio in the Middle East and Pirate radio in North America

Since this subject covers national territories, international waters and international airspace, the only effective way to treat this subject is on a country by country, international waters and international airspace basis. Because the laws vary, the interpretation of the term *pirate radio* also varies considerably.

2.36.3 Propaganda broadcasting

Propaganda broadcasting may be authorized by the government at the transmitting site, but may be considered unwanted or illegal by the government of the intended reception area. Propaganda broadcasting conducted by national governments against the interests of other national governments has created radio jamming stations transmitting noises on the same frequency to prevent reception of the incoming signal. While the United States transmitted its programs towards the Soviet Union, which attempted to jam them, in 1970 the government of the United Kingdom decided to employ a jamming transmitter to drown out the incoming transmissions from the commercial station Radio North Sea International, which was based aboard the motor vessel (MV) Mebo II anchored off southeast England in the North Sea. Other examples of this type of unusual broadcasting include the USCGC *Courier* (WAGR-410), a United States Coast Guard cutter which both originated and relayed broadcasts of the *Voice of America* from an anchorage at the Greek island of Rhodes to Soviet bloc countries. Balloons have been flown above Key West, Florida, to support the TV transmissions of TV Martí, which are directed at Cuba (the Cuban government jams the signals). Military broadcasting aircraft have been flown over Vietnam, Iraq, and many other nations by the United States Air Force.

2.36.4 Piracy in amateur and two-way radio

Illegal use of licensed radio spectrum (also known as bootlegging in CB circles) is fairly common and takes several forms.

- Unlicensed operation—Particularly associated with amateur radio and licensed personal communication services such as GMRS, this refers to use of radio equipment on a section of spectrum for which the

equipment is designed but on which the user is not licensed to operate (most such operators are informally known as "bubble pack pirates" from the sealed plastic retail packaging common to such walkie-talkies). While piracy on the US GMRS band, for example, is widespread (some estimates have the number of total GMRS users outstripping the number of licensed users by several orders of magnitude), such use is generally disciplined only in cases where the pirate's activity interferes with a licensee. A notable case is that of former United States amateur operator Jack Gerritsen operating under the revoked call sign KG6IRO[4] who was successfully prosecuted by the FCC for unlicensed operation and malicious interference.[5] A subcategory of this is free banding, the use of allocations nearby a legal allocation, most typically the 27 MHz Citizen's Band using modified or purpose-built gear.

- Inadvertent interference—Common when personal communications gear is brought into countries where it is not certified to operate. Such interference results from clashing frequency allocations, and occasionally requires wholesale reallocation of an existing band due to an insurmountable interference problem; for example, the 2004 approval in Canada of the unlicensed use of the United States General Mobile Radio Service frequencies due to interference from users of FRS/GMRS radios from the United States, where Industry Canada had to transfer a number of licensed users on the GMRS frequencies to unoccupied channels to accommodate the expanded service.

- Deliberate or malicious interference—refers to the use of two-way radio to harass or jam other users of a channel. Such behaviour is widely prosecuted, especially when it interferes with mission-critical services such as aviation radio or marine VHF radio.

- Illegal equipment—This refers to the use of illegally modified equipment or equipment not certified for a particular band. Such equipment includes illegal linear amplifiers for CB radio, antenna or circuit modifications on walkie-talkies, the use of "export" radios for free banding, or the use of amateur radios on unlicensed bands that amateur gear is not certified for. The use of marine VHF radio gear for inland mobile radio operations is common in some countries, with enforcement difficult since marine VHF is generally the province of maritime authorities.

2.36.5 Examples of known pirate radio stations

- Beat Radio, Minneapolis, Minnesota

- Booger Brothers Broadcasting System, Queens, New York (Operated in the 1930s by Les Paul)[6]

- Britain Radio, United Kingdom

- Citizens Radio 102.8 FM, Hong Kong

- Dread Broadcasting Corporation, London's first black music radio station

- Free Radio Santa Cruz California, USA

- Laser 558, North Sea

- Portland Radio Authority, Portland, Oregon

- Radio 270, United Kingdom

- Radio 390, United Kingdom

- Radio Caroline, United Kingdom

- Radio City, United Kingdom

- Radio Essex (later BBMS "Britain's Better Music Station"), United Kingdom (Operated by Paddy Roy Bates)

- Radio First Termer, Saigon, Vietnam 1971

- Radio Hauraki, New Zealand (ship, Tiri 1 and Tiri 2)

- Radio Jackie, United Kingdom (now licensed and legal)

- Radio Mercur, Denmark

- Radio Milinda, Dublin

- Radio North Sea International

- Radio Newyork International, Jones Beach, New York, USA (1987 & 1988 pirate ship)

- Radio Scotland, United Kingdom (1960s pirate ship)

- Radio Veronica, Netherlands

- Rinse FM, London (gained a licence in 2010)

- Swinging Radio England, United Kingdom

- Thameside Radio London

- TSF, Lisbon, Portugal (Now licensed)

- Voice of Peace, Israel (pirate ship)

- Wonderful Radio London, United Kingdom

- BMIR, Black Rock City, Nevada

2.36.6   In popular culture

The films *The Boat That Rocked* (titled Pirate Radio overseas), and *On the Air Live with Captain Midnight*, as well as the TV series *People Just Do Nothing* are set in the world of pirate radio, while *Born in Flames* features pirate radio stations as being part of an underground political movement.

2.36.7   See also

- Offshore radio

- Community radio

- Open spectrum

- Pirate radio in Europe

- Pirate radio in the United Kingdom

- Pirate television

- Software piracy

- KDIC railroad track antenna incident

2.36.8   References

[1] *Electrical world*. McGraw-Hill. 1907. pp. 1023–. Retrieved 6 March 2012.

[2] "Thomas H. White. "United States Early Radio History"". Earlyradiohistory.us. Retrieved 2011-06-16.

[3] Félix Guattari. "Plan for the Planet". In *Molecular Revolution. Psychiatry and Politics*. London: Penguin Books, 1984. p. 269.

[4] "W5YI Report". W5yi.org. Retrieved 2011-06-16.

[5] "Apologetic Radio Jammer Jack Gerritsen Gets Seven Years, Fines". *ARRL Web*. Sep 19, 2006. Archived from the original on Apr 2, 2007.

[6] Glasscapsule.com: Les Paul's Pirate Radio Station in Queens

2.36.9   External links

- Dave Rabbit: Radio First Termer, Saigon, Vietnam 1970-71

- Thomas H. White "United States Early Radio History"

- Harvey J. Levin: Pioneering the Economics of the Airwaves

- *The Invisible Resource: Use and Regulation of the Radio Spectrum*

- The Pirate Radio Hall of Fame

- Audio interview to the editorial staff of an italian pirate radio

- Solidarity Underground Radio Association, Poland

- Information and discussion on current and historic pirate radio in the United Kingdom

- http://www.radiokaleidoscope.com/

- http://www.radiocaroline.co.uk The official Radio Caroline website

2.37 Radio frequency

This article is about the generic oscillation. For the radiation, see Radio wave. For the electronics, see Radio frequency engineering.
"RF" redirects here. For other uses, see RF (disambiguation).

Radio frequency (RF) is any of the electromagnetic wave frequencies that lie in the range extending from around 3 kHz to 300 GHz, which include those frequencies used for communications or radar signals.[1] RF usually refers to electrical rather than mechanical oscillations. However, mechanical RF systems do exist (see mechanical filter and RF MEMS).

Although radio *frequency* is a rate of oscillation, the term "radio frequency" or its abbreviation "RF" are used as a synonym for radio – i.e., to describe the use of wireless communication, as opposed to communication via electric wires. Examples include:

- Radio-frequency identification

- ISO/IEC 14443−2 *Radio frequency power and signal interface*[2]

2.37.1 Special properties of RF current

Electric currents that oscillate at radio frequencies have special properties not shared by direct current or alternating current of lower frequencies.

- The energy in an RF current can radiate off a conductor into space as electromagnetic waves (radio waves); this is the basis of radio technology.

- RF current does not penetrate deeply into electrical conductors but tends to flow along their surfaces; this is known as the skin effect. For this reason, when the human body comes in contact with high power RF currents it can cause superficial but serious burns called *RF burns (Radiation burns)*.

- RF currents applied to the body often do not cause the painful sensation of electric shock as do lower frequency currents.[3][4] This is because the current changes direction too quickly to trigger depolarization of nerve membranes.

- RF current can easily ionize air, creating a conductive path through it. This property is exploited by "high frequency" units used in electric arc welding, which use currents at higher frequencies than power distribution uses.

- Another property is the ability to appear to flow through paths that contain insulating material, like the dielectric insulator of a capacitor.

- When conducted by an ordinary electric cable, RF current has a tendency to reflect from discontinuities in the cable such as connectors and travel back down the cable toward the source, causing a condition called standing waves. Therefore, RF current must be carried by specialized types of cable called transmission line.

2.37.2 Radio communication

To receive radio signals an antenna must be used. However, since the antenna will pick up thousands of radio signals at a time, a radio tuner is necessary to *tune into* a particular frequency (or frequency range).[5] This is typically done via a resonator – in its simplest form, a circuit with a capacitor and an inductor form a tuned circuit. The resonator amplifies oscillations within a particular frequency band, while reducing oscillations at other frequencies outside the band. Another method to isolate a particular radio frequency is by oversampling (which gets a wide range of frequencies) and picking out the frequencies of interest, as done in software defined radio.

The distance over which radio communications is useful depends significantly on things other than wavelength, such as transmitter power, receiver quality, type, size, and height of antenna, mode of transmission, noise, and interfering signals. Ground waves, tropospheric scatter and skywaves can all achieve greater ranges than line-of-sight propagation. The study of radio propagation allows estimates of useful range to be made.

2.37.3   Frequency bands

Main article: Radio spectrum

2.37.4   In medicine

Radio frequency (RF) energy, in the form of radiating waves or electrical currents, has been used in medical treatments for over 75 years,[7] generally for minimally invasive surgeries, using radiofrequency ablation and cryoablation, including the treatment of sleep apnea.[8] Magnetic resonance imaging (MRI) uses radio frequency waves to generate images of the human body.

Radio frequencies at non-ablation energy levels are sometimes used as a form of cosmetic treatment that can tighten skin, reduce fat (lipolysis), or promote healing.[9]

RF diathermy is a medical treatment that uses RF induced heat as a form of physical or occupational therapy and in surgical procedures. It is commonly used for muscle relaxation. It is also a method of heating tissue electromagnetically for therapeutic purposes in medicine. Diathermy is used in physical therapy and occupational therapy to deliver moderate heat directly to pathologic lesions in the deeper tissues of the body. Surgically, the extreme heat that can be produced by diathermy may be used to destroy neoplasms, warts, and infected tissues, and to cauterize blood vessels to prevent excessive bleeding. The technique is particularly valuable in neurosurgery and surgery of the eye. Diathermy equipment typically operates in the short-wave radio frequency (range 1–100 MHz) or microwave energy (range 434–915 MHz).

Pulsed electromagnetic field therapy (PEMF) is a medical treatment that purportedly helps to heal bone tissue reported in a recent NASA study. This method usually employs electromagnetic radiation of different frequencies - ranging from static magnetic fields, through extremely low frequencies (ELF) to higher radio frequencies (RF) administered in pulses.

2.37.5   Effects on the human body

Extremely low frequency RF

High-power extremely low frequency RF with electric field levels in the low kV/m range are known to induce perceivable currents within the human body that create an annoying tingling sensation. These currents will typically flow to ground through a body contact surface such as the feet, or arc to ground where the body is well insulated.[10][11]

Microwaves

Main article: Microwave burn

Microwave exposure at low-power levels below the Specific absorption rate set by government regulatory bodies are considered harmless non-ionizing radiation and have no effect on the human body. However, levels above the Specific absorption rate set by the U.S. Federal Communications Commission are considered potentially harmful (see Mobile phone radiation and health).

Long-term human exposure to high-levels of microwaves is recognized to cause cataracts according to experimental animal studies and epidemiological studies. The mechanism is unclear but may include changes in heat sensitive enzymes that normally protect cell proteins in the lens. Another mechanism that has been advanced is direct damage to the lens from pressure waves induced in the aqueous humor.

High-power exposure to microwave RF is known to create a range of effects from lower to higher power levels, ranging from unpleasant burning sensation on the skin and microwave auditory effect, to extreme pain at the midrange, to physical burning and blistering of skin and internals at high power levels (see microwave burn).

General RF exposure

The 1999 revision of Canadian Safety Code 6 recommended electric field limits of 100 kV/m for pulsed EMF to prevent air breakdown and spark discharges, mentioning rationale related to auditory effect and energy-induced unconsciousness in rats.[12] The pulsed EMF limit was removed in later revisions, however.[13]

For health effects see electromagnetic radiation and health.

For high-power RF exposure see radiation burn.

For low-power RF exposure see radiation-induced cancer.

2.37.6   As a weapon

See also: Directed energy weapons § Microwave weapons

A heat ray is an RF harassment device that makes use of microwave radio frequencies to create an unpleasant heating effect in the upper layer of the skin. A publicly known heat ray weapon called the Active Denial System was developed by the US military as an experimental weapon to deny the enemy access to an area. A death ray is a weapon that delivers heat ray electromagnetic energy at levels that injure human tissue. The inventor of the death ray, Harry

Grindell Matthews, claims to have lost sight in his left eye while developing his death ray weapon based on a primitive microwave magnetron from the 1920s (note that a typical microwave oven induces a tissue damaging cooking effect inside the oven at about 2 kV/m.)

2.37.7 Measurement

Since radio frequency radiation has both an electric and a magnetic component, it is often convenient to express intensity of radiation field in terms of units specific to each component. The unit *volts per meter* (V/m) is used for the electric component, and the unit *amperes per meter* (A/m) is used for the magnetic component. One can speak of an electromagnetic field, and these units are used to provide information about the levels of electric and magnetic field strength at a measurement location.

Another commonly used unit for characterizing an RF electromagnetic field is *power density*. Power density is most accurately used when the point of measurement is far enough away from the RF emitter to be located in what is referred to as the far field zone of the radiation pattern. In closer proximity to the transmitter, i.e., in the "near field" zone, the physical relationships between the electric and magnetic components of the field can be complex, and it is best to use the field strength units discussed above. Power density is measured in terms of power per unit area, for example, milliwatts per square centimeter (mW/cm^2). When speaking of frequencies in the microwave range and higher, power density is usually used to express intensity since exposures that might occur would likely be in the far field zone.

2.37.8 See also

- Amplitude modulation
- Electromagnetic Interference
- Electromagnetic radiation
- Electromagnetic spectrum
- EMF measurement
- Frequency allocation
- Frequency bandwidth
- Frequency modulation
- Plastic welding
- Pulsed electromagnetic field therapy
- Spectrum management

2.37.9 References

[1] "Definition of RADIO FREQUENCY". *Merriam-Webster*. Encyclopædia Britannica. n.d. Retrieved 6 August 2015.

[2] "ISO/IEC 14443-2:2001 Identification cards — Contactless integrated circuit(s) cards — Proximity cards — Part 2: Radio frequency power and signal interface". Iso.org. 2010-08-19. Retrieved 2011-11-08.

[3] Curtis, Thomas Stanley (1916). *High Frequency Apparatus: Its Construction and Practical Application*. USA: Everyday Mechanics Company. p. 6.

[4] Mieny, C. J. (2003). *Principles of Surgical Patient Care* (2nd ed.). New Africa Books. p. 136. ISBN 9781869280055.

[5] Brain, Marshall (2000-12-07). "How Radio Works". HowStuffWorks.com. Retrieved 2009-09-11.

[6] Jeffrey S. Beasley; Gary M. Miller (2008). *Modern Electronic Communication* (9th ed.). pp. 4–5. ISBN 978-0132251136.

[7] Ruey J. Sung and Michael R. Lauer (2000). *Fundamental approaches to the management of cardiac arrhythmias*. Springer. p. 153. ISBN 978-0-7923-6559-4.

[8] Melvin A. Shiffman, Sid J. Mirrafati, Samuel M. Lam and Chelso G. Cueteaux (2007). *Simplified Facial Rejuvenation*. Springer. p. 157. ISBN 978-3-540-71096-7.

[9] Noninvasive Radio Frequency for Skin Tightening and Body Contouring, Frontline Medical Communications, 2013

[10] Limits of Human Exposure to Radiofrequency Electromagnetic Fields in the Frequency Range from 3 kHz to 300 GHz, Canada Safety Code 6, page 63

[11] Extremely Low Frequency Fields Environmental Health Criteria Monograph No.238, chapter 5, page 121, WHO

[12] Limits of Human Exposure to Radiofrequency Electromagnetic Fields in the Frequency Range from 3 kHz to 300 GHz, Canada Safety Code 6,page 62

[13] http://www.hc-sc.gc.ca/ewh-semt/pubs/radiation/radio_guide-lignes_direct/index-eng.php Safety Code 6: Health Canada's Radiofrequency Exposure Guidelines - Environmental and Workplace Health - Health Canada

2.37.10 External links

- Definition of frequency bands (VLF, ELF ... etc.) IK1QFK Home Page (vlf.it)
- Radio, light, and sound waves, conversion between wavelength and frequency
- RF Terms Glossary

2.38 Radio propagation

Radio propagation is the behavior of radio waves when they are transmitted, or propagated from one point on the Earth to another, or into various parts of the atmosphere.[1] As a form of electromagnetic radiation, like light waves, radio waves are affected by the phenomena of reflection, refraction, diffraction, absorption, polarization, and scattering.[2]

Radio propagation is affected by the daily changes of water vapor in the troposphere and ionization in the upper atmosphere, due to the Sun. Understanding the effects of varying conditions on radio propagation has many practical applications, from choosing frequencies for international shortwave broadcasters, to designing reliable mobile telephone systems, to radio navigation, to operation of radar systems.

Radio propagation is also affected by several other factors determined by its path from point to point. This path can be a direct line of sight path or an over-the-horizon path aided by refraction in the ionosphere, which is a region between approximately 60 and 600 km.[3] Factors influencing ionospheric radio signal propagation can include sporadic-E, spread-F, solar flares, geomagnetic storms, ionospheric layer tilts, and solar proton events.

Radio waves at different frequencies propagate in different ways. At extremely low frequencies (ELF) and very low frequencies the wavelength is much larger than the separation between the earth's surface and the D layer of the ionosphere, so electromagnetic waves may propagate in this region as a waveguide. Indeed, for frequencies below 20 kHz, the wave propagates as a single waveguide mode with a horizontal magnetic field and vertical electric field.[4] The interaction of radio waves with the ionized regions of the atmosphere makes radio propagation more complex to predict and analyze than in free space. Ionospheric radio propagation has a strong connection to space weather. A sudden ionospheric disturbance or shortwave fadeout is observed when the x-rays associated with a solar flare ionize the ionospheric D-region. Enhanced ionization in that region increases the absorption of radio signals passing through it. During the strongest solar x-ray flares, complete absorption of virtually all ionospherically propagated radio signals in the sunlit hemisphere can occur. These solar flares can disrupt HF radio propagation and affect GPS accuracy.

Since radio propagation is not fully predictable, such services as emergency locator transmitters, in-flight communication with ocean-crossing aircraft, and some television broadcasting have been moved to communications satellites. A satellite link, though expensive, can offer highly predictable and stable line of sight coverage of a given area.

2.38.1 Free space propagation

In free space, all electromagnetic waves (radio, light, X-rays, etc.) obey the inverse-square law which states that the power density of an electromagnetic wave is proportional to the inverse of the square of the distance from a point source[5] or:

$$\rho_P \propto \frac{1}{r^2}.$$

Doubling the distance from a transmitter means that the power density of the radiated wave at that new location is reduced to one-quarter of its previous value.

The power density per surface unit is proportional to the product of the electric and magnetic field strengths. Thus, doubling the propagation path distance from the transmitter reduces each of their received field strengths over a free-space path by one-half.

2.38.2 Modes

Surface modes (groundwave)

Main article: Surface wave

Lower frequencies (between 30 and 3,000 kHz) have the property of following the curvature of the earth via groundwave propagation in the majority of occurrences.

In this mode the radio wave propagates by interacting with the semi-conductive surface of the earth. The wave "clings" to the surface and thus follows the curvature of the earth. Vertical polarization is used to alleviate short circuiting the electric field through the conductivity of the ground. Since the ground is not a perfect electrical conductor, ground waves are attenuated rapidly as they follow the earth's surface. Attenuation is proportional to the frequency making this mode mainly useful for LF and VLF frequencies (see also Earth-ionosphere waveguide).

Today LF and VLF are mostly used for time signals, and for military communications, especially one-way transmissions to ships and submarines, although radio amateurs have an allocation at 137 kHz in some parts of the world. Radio broadcasting using surface wave propagation uses the higher portion of the LF range in Europe, Africa and the Middle East.

Early commercial and professional radio services relied exclusively on long wave, low frequencies and ground-wave propagation. To prevent interference with these services, amateur and experimental transmitters were restricted to the higher (HF) frequencies, felt to be useless since their ground-wave range was limited. Upon discovery of the other propagation modes possible at medium wave and short wave frequencies, the advantages of HF for commercial and military purposes became apparent. Amateur experimentation was then confined only to authorized frequency segments in that range.[6]

Direct modes (line-of-sight)

Line-of-sight is the direct propagation of radio waves between antennas that are visible to each other. This is probably the most common of the radio propagation modes at VHF and higher frequencies. Because radio signals can travel through many non-metallic objects, radio can be picked up through walls. This is still line-of-sight propagation. Examples would include propagation between a satellite and a ground antenna or reception of television signals from a local TV transmitter.

Ground plane reflection effects are an important factor in VHF line of sight propagation. The interference between the direct beam line-of-sight and the ground reflected beam often leads to an effective inverse-fourth-power ($1/distance^4$) law for ground-plane limited radiation. [Need reference to inverse-fourth-power law + ground plane. Drawings may clarify]

Ionospheric modes (skywave)

Main article: Skywave

Skywave propagation, also referred to as skip, is any of the modes that rely on refraction of radio waves in the ionosphere, which is made up of one or more ionized layers in the upper atmosphere. F2-layer is the most important ionospheric layer for long-distance, multiple-hop HF propagation, though F1, E, and D-layers also play significant roles. The D-layer, when present during sunlight periods, causes significant amount of signal loss, as does the E-layer whose maximum usable frequency can rise to 4 MHz and above and thus block higher frequency signals from reaching the F2-layer. The layers, or more appropriately "regions", are directly affected by the sun on a daily diurnal cycle, a seasonal cycle and the 11-year sunspot cycle and determine the utility of these modes. During solar maxima, or sunspot highs and peaks, the whole HF range up to 30 MHz can be used usually around the clock and F2 propagation up to 50 MHz is observed frequently depending upon daily solar flux 10.7cm radiation values. During solar minima, or minimum sunspot counts down to zero, propagation of frequencies above 15 MHz is generally unavailable.

Although the claim is commonly made that two-way HF propagation along a given path is reciprocal, that is, if

the signal from location A reaches location B at a good strength, the signal from location B will be similar at station A because the same path is traversed in both directions. However, the ionosphere is far too complex and constantly changing to support the reciprocity theorem. The path is never exactly the same in both directions.[7] In brief, conditions at the two terminii of a path generally cause dissimilar polarization shifts, dissimilar splits into ordinary rays and extraordinary or *Pedersen rays* which are erratic and impossibly identical or similar due to variations in ionization density, shifting zenith angles, effects of the earth's magnetic DIPOLE contours, antenna radiation patterns, ground conditions and other variables.

Forecasting of skywave modes is of considerable interest to amateur radio operators and commercial marine and aircraft communications, and also to shortwave broadcasters. Real-time propagation can be assessed by listening for transmissions from specific beacon transmitters.

Meteor scattering Meteor scattering relies on reflecting radio waves off the intensely ionized columns of air generated by meteors. While this mode is very short duration, often only from a fraction of second to couple of seconds per event, digital Meteor burst communications allows remote stations to communicate to a station that may be hundreds of miles up to over 1,000 miles (1,600 km) away, without the expense required for a satellite link. This mode is most generally useful on VHF frequencies between 30 and 250 MHz.

Auroral backscatter Intense columns of Auroral ionization at 100 km altitudes within the auroral oval backscatter radio waves, perhaps most notably on HF and VHF. Backscatter is angle-sensitive—incident ray vs. magnetic field line of the column must be very close to right-angle. Random motions of electrons spiraling around the field lines create a Doppler-spread that broadens the spectra of the emission to more or less noise-like—depending on how high radio frequency is used. The radio-auroras are observed mostly at high latitudes and rarely extend down to middle latitudes. The occurrence of radio-auroras depends on solar activity (flares, coronal holes, CMEs) and annually the events are more numerous during solar cycle maxima. Radio aurora includes the so-called afternoon radio aurora which produces stronger but more distorted signals and after the Harang-minima, the late-night radio aurora (sub-storming phase) returns with variable signal strength and lesser doppler spread. The propagation range for this predominantly back-scatter mode extends up to about 2000 km in east-west plane, but strongest signals are observed most frequently from the north at nearby sites on same latitudes.

Rarely, a strong radio-aurora is followed by Auroral-E, which resembles both propagation types in some ways.

Sporadic-E propagation Main article: Sporadic E propagation

Sporadic E (Es) propagation can be observed on HF and VHF bands.[8] It must not be confused with ordinary HF E-layer propagation. Sporadic-E at mid-latitudes occurs mostly during summer season, from May to August in the northern hemisphere and from November to February in the southern hemisphere. There is no single cause for this mysterious propagation mode. The reflection takes place in a thin sheet of ionisation around 90 km height. The ionisation patches drift westwards at speeds of few hundred km per hour. There is a weak periodicity noted during the season and typically Es is observed on 1 to 3 successive days and remains absent for a few days to reoccur again. Es do not occur during small hours; the events usually begin at dawn, and there is a peak in the afternoon and a second peak in the evening.[9] Es propagation is usually gone by local midnight.

Observation of radio propagation beacons operating around 28.2 MHz, 50 MHz and 70 MHz, indicates that maximum observed frequency (MOF) for Es is found to be lurking around 30 MHz on most days during the summer season, but sometimes MOF may shoot up to 100 MHz or even more in ten minutes to decline slowly during the next few hours. The peak-phase includes oscillation of MOF with periodicity of approximately 5...10 minutes. The propagation range for Es single-hop is typically 1000 to 2000 km, but with multi-hop, double range is observed. The signals are very strong but also with slow deep fading.

Tropospheric modes

Tropospheric scattering Main article: Tropospheric scattering

At VHF and higher frequencies, small variations (turbulence) in the density of the atmosphere at a height of around 6 miles (10 km) can scatter some of the normally line-of-sight beam of radio frequency energy back toward the ground, allowing over-the-horizon communication between stations as far as 500 miles (800 km) apart. The military developed the White Alice Communications System covering all of Alaska, using this tropospheric scattering principle.

Tropospheric ducting Main article: Tropospheric ducting

Sudden changes in the atmosphere's vertical moisture content and temperature profiles can on random occasions make microwave and UHF & VHF signals propagate hundreds of kilometers up to about 2,000 kilometers (1,300 mi)—and for ducting mode even farther—beyond the normal radio-horizon. The inversion layer is mostly observed over high pressure regions, but there are several tropospheric weather conditions which create these randomly occurring propagation modes. Inversion layer's altitude for non-ducting is typically found between 100 meters (300 ft) to about 1 kilometer (3,000 ft) and for ducting about 500 meters to 3 kilometers (1,600 to 10,000 ft), and the duration of the events are typically from several hours up to several days. Higher frequencies experience the most dramatic increase of signal strengths, while on low-VHF and HF the effect is negligible. Propagation path attenuation may be below free-space loss. Some of the lesser inversion types related to warm ground and cooler air moisture content occur regularly at certain times of the year and time of day. A typical example could be the late summer, early morning tropospheric enhancements that bring in signals from distances up to few hundred kilometers for a couple of hours, until undone by the Sun's warming effect.

Tropospheric delay This is a source of error in radio ranging techniques, such as the Global Positioning System (GPS).[10] See also the page of GPS meteorology.

Rain scattering Rain scattering is purely a microwave propagation mode and is best observed around 10 GHz, but extends down to a few gigahertz—the limit being the size of the scattering particle size vs. wavelength. This mode scatters signals mostly forwards and backwards when using horizontal polarization and side-scattering with vertical polarization. Forward-scattering typically yields propagation ranges of 800 km. Scattering from snowflakes and ice pellets also occurs, but scattering from ice without watery surface is less effective. The most common application for this phenomenon is microwave rain radar, but rain scatter propagation can be a nuisance causing unwanted signals to intermittently propagate where they are not anticipated or desired. Similar reflections may also occur from insects though at lower altitudes and shorter range. Rain also causes attenuation of point-to-point and satellite microwave links. Attenuation values up to 30 dB have been observed on 30 GHz during heavy tropical rain.

Airplane scattering Airplane scattering (or most often reflection) is observed on VHF through microwaves and, besides back-scattering, yields momentary propagation up to 500 km even in mountainous terrain. The most common back-scatter applications are air-traffic radar, bistatic forward-scatter guided-missile and airplane-detecting trip-wire radar, and the US space radar.

Lightning scattering Lightning scattering has sometimes been observed on VHF and UHF over distances of about 500 km. The hot lightning channel scatters radiowaves for a fraction of a second. The RF noise burst from the lightning makes the initial part of the open channel unusable and the ionization disappears quickly because of recombination at low altitude and high atmospheric pressure. Although the hot lightning channel is briefly observable with microwave radar, no practical use for this mode has been found in communications.

Other effects

Diffraction Knife-Edge diffraction is the propagation mode where radio waves are bent around sharp edges. For example, this mode is used to send radio signals over a mountain range when a line-of-sight path is not available. However, the angle cannot be too sharp or the signal will not diffract. The diffraction mode requires increased signal strength, so higher power or better antennas will be needed than for an equivalent line-of-sight path.

Diffraction depends on the relationship between the wavelength and the size of the obstacle. In other words, the size of the obstacle in wavelengths. Lower frequencies diffract around large smooth obstacles such as hills more easily. For example, in many cases where VHF (or higher frequency) communication is not possible due to shadowing by a hill, it is still possible to communicate using the upper part of the HF band where the surface wave is of little use.

Diffraction phenomena by small obstacles are also important at high frequencies. Signals for urban cellular telephony tend to be dominated by ground-plane effects as they travel over the rooftops of the urban environment. They then diffract over roof edges into the street, where multipath propagation, absorption and diffraction phenomena dominate.

Absorption Low-frequency radio waves travel easily through brick and stone and VLF even penetrates sea-water. As the frequency rises, absorption effects become more important. At microwave or higher frequencies, absorption by molecular resonances in the atmosphere (mostly from water, H_2O and oxygen, O_2) is a major factor in radio propagation. For example, in the 58–60 GHz band, there is a major absorption peak which makes this band useless for long-distance use. This phenomenon was first discovered during radar research in World War II. Above about 400 GHz, the Earth's atmosphere blocks most of the spectrum while still passing some - up to UV light, which is blocked by ozone - but visible light and some of the near-infrared is transmitted. Heavy rain and falling snow also affect microwave absorption.

2.38.3 Measuring HF propagation

HF propagation conditions can be simulated using radio propagation models, such as the Voice of America Coverage Analysis Program, and realtime measurements can be done using chirp transmitters. For radio amateurs the WSPR mode provides maps with real time propagation conditions between a network of transmitters and receivers.[11] Even without special beacons the realtime propagation conditions can be measured: a worldwide network of receivers decodes morse code signals on amateur radio frequencies in realtime and provides sophisticated search functions and propagation maps for every station received.[12]

2.38.4 Practical effects

The average person can notice the effects of changes in radio propagation in several ways.

In AM broadcasting, the dramatic ionospheric changes that occur overnight in the mediumwave band drive a unique broadcast license scheme, with entirely different transmitter power output levels and directional antenna patterns to cope with skywave propagation at night. Very few stations are allowed to run without modifications during dark hours, typically only those on clear channels in North America. Many stations have no authorization to run at all outside of daylight hours. Otherwise, there would be nothing but interference on the entire broadcast band from dusk until dawn without these modifications.

For FM broadcasting (and the few remaining low-band TV stations), weather is the primary cause for changes in VHF propagation, along with some diurnal changes when the sky is mostly without cloud cover. These changes are most obvious during temperature inversions, such as in the late-night and early-morning hours when it is clear, allowing the ground and the air near it to cool more rapidly. This not only causes dew, frost, or fog, but also causes a slight "drag" on the bottom of the radio waves, bending the signals down such that they can follow the Earth's curvature over the normal radio horizon. The result is typically several stations being heard from another media market — usually a neighboring one, but sometimes ones from a few hundred kilometers away. Ice storms are also the result of inversions, but these normally cause more scattered omnidirection propagation, resulting mainly in interference, often among weather radio stations. In late spring and early summer, a combination of other atmospheric factors can occasionally cause skips that duct high-power signals to places well over 1000km away.

Non-broadcast signals are also affected. Mobile phone signals are in the UHF band, ranging from 700 to over 2600 Megahertz, a range which makes them even more prone to weather-induced propagation changes. In urban (and to some extent suburban) areas with a high population density, this is partly offset by the use of smaller cells, which use lower effective radiated power and beam tilt to reduce interference, and therefore increase frequency reuse and user capacity. However, since this would not be very cost-effective in more rural areas, these cells are larger and so more likely to cause interference over longer distances when propagation conditions allow.

While this is generally transparent to the user thanks to the way that cellular networks handle cell-to-cell handoffs, when cross-border signals are involved, unexpected charges for international roaming may occur despite not having left the country at all. This often occurs between southern San Diego and northern Tijuana at the western end of the U.S./Mexico border, and between eastern Detroit and western Windsor along the U.S./Canada border. Since signals can travel unobstructed over a body of water far larger than the Detroit River, and cool water temperatures also cause inversions in surface air, this "fringe roaming" sometimes occurs across the Great Lakes, and between islands in the Caribbean. Signals can skip from the Dominican Republic to a mountainside in Puerto Rico and vice versa, or between the U.S. and British Virgin Islands, among others. While unintended cross-border roaming is often automatically removed by mobile phone company billing systems, inter-island roaming is typically not.

2.38.5   See also

Main article: List of radio propagation terms

- Diversity scheme
- Earth bulge
- Earth-ionosphere waveguide
- Electromagnetic radiation
- Fading
- Fresnel zone
- Free space
- Inversion (meteorology)
- Kennelly–Heaviside layer

- Near and far field
- Radio atmospherics
- Radio frequency
- Radio horizon
- Radio propagation model
- Rayleigh fading
- Ray tracing (physics)
- Schumann resonance
- Skip (radio)
- Skip zone
- Skywave
- Tropospheric propagation
- TV and FM DX
- Upfade
- VOACAP - Free professional HF propagation prediction software

2.38.6   References

[1] H. P. Westman et al., (ed), *Reference Data for Radio Engineers, Fifth Edition*, 1968, Howard W. Sams and Co., no ISBN, Library of Congress Card No. 43-14665 page 26-1

[2] Demetrius T Paris and F. Kenneth Hurd, *Basic Electromagnetic Theory*, McGraw Hill, New York 1969 ISBN 0-07-048470-8, Chapter 8

[3] Radiowave propagation, edited by M.Hall and L.Barclay, page 2, published by Peter Peregrinus Ltd., (1989), ISBN 0-86341-156-8

[4] Radiowave propagation, edited by M.Hall and L.Barclay, published by Peter Peregrinus Ltd., page 3, (1989), ISBN 0-86341-156-8

[5] Westman *Reference data* page 26-19

[6] Clinton B. DeSoto (1936). *200 meters & Down - The Story of Amateur Radio*. W. Hartford, CT: The American Radio Relay League. pp. 132–146. ISBN 0-87259-001-1.

[7] G.W. Hull, "Nonreciprocal characteristics of a 1500km HF Ionospheric Path," *Proceedings of the IEEE*, 55, March 1967, pp. 426-427; "Origin of non-reciprocity on high-frequency ionospheric paths," *Nature*, pp. 483-484, and cited references.

[8] Davies, Kenneth (1990). *Ionospheric Radio*. IEE Electromagnetic Waves Series #31. London, UK: Peter Peregrinus Ltd/The Institution of Electrical Engineers. pp. 184–186. ISBN 0-86341-186-X.

[9] George Jacobs and Theodore J. Cohen, *Shortwave Propagation Handbook*. Hicksville, New York: CQ Publishing (1982), pp. 130-135. ISBN 978-0-943016-00-9

[10] Frank Kleijer (2004), Troposphere Modeling and Filtering for Precise GPS Leveling. Ph. D. thesis, Department of Mathematical Geodesy and Positioning, Delft University of Technology

[11] WSPR Propagation Conditions Map: http://wsprnet.org/drupal/wsprnet/map

[12] Network of CW Signal Decoders for Realtime Analysis: http://www.reversebeacon.net/

2.38.7 Further reading

- Lucien Boithais: *Radio Wave Propagation*. McGraw-Hill Book Company, New York. 1987. ISBN 0-07-006433-4

- Karl Rawer:*Wave Propagatiom im the Ionosphere*.Kluwer Acad.Publ.,Dordrecht 1993. ISBN 0-7923-0775-5

- H. Ward Silver and Mark J. Wilson, (eds), "Propagation of Radio Signals" (Ch. 19, by Emil Pocock), in *The ARRL Handbook for Radio Communications (88th edition, 2010), ARRL, Newington CT USA ISBN 0-87259-095-X*

- Yuri Blanarovich, VE3BMV, K3BU: "Electromagnetic Wave Propagation by Conduction" CQ Magazine June 1980, p. 44, http://k3bu.us/propagation.htm

2.38.8 External links

- Solar widget Propagation widget based on NOAA data. Also available as WordPress plugin.

- ARRL Propagation Page The American Radio Relay League page on radio propagation.

- HF Radio and Ionospheric Prediction Service - Australia

- NASA Space Weather Action Center

- HF Propagation Tutorial by the late NM7M

- Space Weather and Radio Propagation Resource Center Live data and images of space weather and radio propagation.

- Solar Terrestrial Dispatch

- Online Propagation Tools, HF Solar Data, and HF Propagation Tutorials

- DXing.info - Propagation links

- HF Radio Propagation Software for Firefox - Propfire Firefox plug-in for monitoring propagation, website utility to display HF propagation status, and article on understanding HF radio propagation forecasting

- The Basics of Radio Wave Propagation A resource by Edwin C. Jones (AE4TM), MD, PhD, Department of Physics and Astronomy, University of Tennessee.

- Dynamic Radio Propagation Data Constantly updated radio propagation data pulled from various sources.

- Solar Cycle 24 prediction and MF/HF/6M radiowave propagation forecast webpage (www.solarcycle24.org)

- 160 Meter (Medium Frequency) Radiowave Propagation Theory Notes webpage (www.wcflunatall.com/nz4o5.htm)

- Unusual HF Propagation Phenomena. 13 Apr 2009 Includes useful recordings each type. Retrieved 9 Oct 2009.

- Overview of radio propagation modes

- Propagation: Es & Thunderstorms by Thomas F. Giella, NZ4O, ex KN4LF.

- Radio propagation tutorial

The following external references provide practical examples of radio propagation concepts as demonstrated using software built on the VOACAP model.

- Online MOF/LOF HF Propagation Prediction Tool

- High Frequency radio propagation de-mystified.

- Is High Frequency radio propagation reciprocal?

- How does noise affect radio signals?

The following external link is designed for use by cell phones and mobile devices that can display content using Wireless Markup Language and the Wireless Application Protocol:

- WAP/WML Space Weather and Radio Propagation Resources Space weather and radio propagation resources.

2.39 Radio spectrum

The **radio spectrum** is the part of the electromagnetic spectrum from 3 Hz to 3000 GHz (3 THz) allocated to some 40 Radiocommunication services in line to the Radio Regulations (RR) of the International Telecommunication Union (ITU).[1] The transmission, emission and/or reception of radio waves for specific telecommunication purposes of radio waves is strictly regulated by the national administration.[2]

Different parts of the radio spectrum are allocated for different radio transmission technologies and applications. In some cases, parts of the radio spectrum is sold or licensed to operators of private radio transmission services (for example, cellular telephone operators or broadcast television stations). Ranges of allocated frequencies are often referred to by their provisioned use (for example, cellular spectrum or television spectrum).[3]

2.39.1 By frequency

A **band** is a small section of the spectrum of radio communication frequencies, in which channels are usually used or set aside for the same purpose.

Above 300 GHz, the absorption of electromagnetic radiation by Earth's atmosphere is so great that the atmosphere is effectively opaque, until it becomes transparent again in the near-infrared and optical window frequency ranges.

To prevent interference and allow for efficient use of the radio spectrum, similar services are allocated in bands. For example, broadcasting, mobile radio, or navigation devices, will be allocated in non-overlapping ranges of frequencies.

Each of these bands has a basic bandplan which dictates how it is to be used and shared, to avoid interference and to set protocol for the compatibility of transmitters and receivers. See detail of bands:http://www.ntia.doc.gov/files/ntia/Spectrum_Use_Summary_Master-06212010.pdf

As a matter of convention, bands are divided at wavelengths of 10^n metres, or frequencies of 3×10^n hertz. For example, 30 MHz or 10 m divides shortwave (lower and longer) from VHF (shorter and higher). These are the parts of the radio spectrum, and not its frequency allocation.

ITU

The **ITU radio bands** are designations defined in the ITU Radio Regulations. Article 2, provision No. 2.1 states that "the radio spectrum shall be subdivided into nine frequency bands, which shall be designated by progressive whole numbers in accordance with the following table[4]".

The table originated with a recommendation of the IVth CCIR meeting, held in Bucharest in 1937, and was approved by the International Radio Conference held at Atlantic City in 1947. The idea to give each band a number, in which the number is the logarithm of the approximate geometric mean of the upper and lower band limits in Hz, originated with B.C. Fleming-Williams, who suggested it in a letter to the editor of *Wireless Engineer* in 1942. (For example, the approximate geometric mean of Band 7 is 10 MHz, or 10^7 Hz.)[5]

† This column does not form part of the table in Provision No. 2.1 of the Radio Regulations

2.40 Reception report

A **reception report** is a means by which radio stations (usually short- and medium-wave broadcasters) receive detailed feedback from their listeners about the quality and content of their broadcasts. A reception report consists of several pieces of information which help the station verify that the report confirms coverage of their transmission, and usually include the following information:

- Date, time and frequency (in kHz) of the transmission

- Station name

- Description of the interval signal, if heard

- Programme details

- Name of announcers or programme host, if heard

- Details of the overall signal quality (normally using the SINPO code)

The listener's location relative to the station is also useful; this indicates how well the station's transmitter is performing and in which direction(s) its antenna is beaming the signal. The station also evaluates a reception report in light of the listener's receiver and antenna. Upon receipt of a correct report, a broadcaster sometimes issues a letter or postcard (known as a QSL card) to the sender, thanking them and confirming that the details are correct. "QSL" is part of the amateur radio Q code, meaning "I acknowledge receipt". One aspect of DXing is collecting QSL cards and letters from stations heard (similar to a birdwatcher's "life list"). QSLing a radio station involves writing an accurate reception report, mailing it to the station and awaiting a reply. Since QSLing is a voluntary act on the station's part, several techniques are used to improve a listener's success rate.

2.40.1 Useful reception reports

Station engineers and other personnel are primarily interested in whether or not their station is heard, and how well; therefore, a complete and accurate reception report (whether by postal or e-mail) is generally appreciated. To begin, report the frequency, date and time the station was heard. For medium-wave (AM) stations, the time should be that of the time zone in which the station is located. Thus, if a listener hears stations from the Eastern time zone of North America (such as Ontario and New York), EST or EDT should be given (depending on the time of year). Stations in the Central time zone (such as Chicago) use CST or CDT, which is one hour earlier than Eastern time. If a listener is uncertain of the time zone, they should clearly indicate their local time or even better use UTC

It is helpful to jot down programming as it is heard. Useful details include:

- Station identification
- Program name
- Names of station announcers (such as a talk show host)
- Commercials (good indicators, because the station keeps a logbook of commercials
- Names of network programs, such as "CNN News" or "TSN Sports"

The purpose of providing this information is to prove to the station that you heard their program and not that of another station. The more detail a listener can provide, the better the chance of eliciting a QSL. Include information, also, about how well the signal was received. For medium-wave reports, an indication of signal strength and any interference (co- or adjacent-channel interference, with identity of interfering station if possible) is usually sufficient.[1] International shortwave broadcasters are familiar with the SINPO code:

- **S**ignal strength
- **I**nterference (includes human-generated noise—for example, power-line hum)
- **N**oise ("white noise", or thunderstorm static)
- **P**ropagation disturbance (fading)
- **O**verall reception quality

Each letter receives a value between 5 and 1, where 5 is the best and 1 the worst. Many shortwave listeners (known as "program listeners") desire nothing more than music and

news from a broadcaster; however, for DXers a QSL collection is tangible proof of what they have heard. Some listeners use pre-printed forms if they are unfamiliar with the language spoken by station personnel.[2] While major international broadcasters have not required return postage for a QSL, the growing popularity of e-mailed reception reports and e-QSLs has largely eliminated the necessity for international reply coupons (IRCs), mint stamps from the verifying station's country or "green stamps" (US dollar bills).

2.40.2 References

[1] Quick Guide to Reception Reports Archived May 31, 2008 at the Wayback Machine

[2] Writing Useful Reception Reports http://www.dxinginfo.com/dx-reception-reports.html Archived May 31, 2008 at the Wayback Machine

2.41 Satellite radio

*This article is about **Satellite radio** as Broadcasting-satellite service (in line to article 1.39, ITU RR) and should not be confused with a Satellite radio system in the meaning of a Radiocommunication service in line to article 1.111 (ITU RR).*

Satellite radio is – according to *article 1.39* of the International Telecommunication Union´s (ITU) ITU Radio Regulations (RR) – a *Broadcasting-satellite service.*[1]

The satellite's signals are broadcast nationwide, across a much wider geographical area than terrestrial radio stations, and the service is primarily intended for the occupants of motor vehicles.[2][3] It is available by subscription, mostly commercial free, and offers subscribers more stations and a wider variety of programming options than terrestrial radio.[4]

Satellite radio technology was inducted into the Space Foundation Space Technology Hall of Fame in 2002.[5] Satellite radio uses the 2.3 GHz S band in North America for nationwide digital audio broadcasting (DAB).[6] In other parts of the world, satellite radio uses the 1.4 GHz L band allocated for DAB.[7]

Satellite radio subscribers purchase a receiver and pay a monthly subscription fee to listen to programming. They can listen through built-in or portable receivers in automobiles; in the home and office with a portable or tabletop receiver equipped to connect the receiver to a stereo system; or on the Internet.[8]

Ground stations transmit signals to the satellites, which are orbiting over 22,000 miles above the surface of the Earth. The satellites send the signals back down to radio receivers in cars and homes. This signal contains scrambled broadcasts, along with meta data about each specific broadcast. The signals are unscrambled by the radio receiver modules, which display the broadcast information. In urban areas, ground repeaters enable signals to be available even if the satellite signal is blocked. The technology allows for nationwide broadcasting, so that, for instance US listeners can hear the same stations anywhere in the country.[7][9]

2.41.1 History

United States

Main article: Sirius XM Radio

Sirius Satellite Radio was founded by Martine Rothblatt, David Margolese and Robert Briskman.[10][11] In June 1990, Rothblatt's shell company, Satellite CD Radio, Inc., petitioned the Federal Communications Commission (FCC) to assign new frequencies for satellites to broadcast digital sound to homes and cars.[2] The company identified and argued in favor of the use of the S-band frequencies that the FCC subsequently decided to allocate to digital audio broadcasting. The National Association of Broadcasters contended that satellite radio would harm local radio stations.[3]

In April 1992, Rothblatt resigned as CEO of Satellite CD Radio[10] and former NASA engineer Robert Briskman, who designed the company's satellite technology, was then appointed chairman and CEO.[12][13] Six months later, Rogers Wireless co-founder David Margolese, who had provided financial backing for the venture, acquired control of the company and succeeded Briskman. Margolese renamed the company CD Radio, and spent the next five years lobbying the FCC to allow satellite radio to be deployed, and the following five years raising $1.6 billion, which was used to build and launch three satellites into elliptical orbit from Kazakhstan in July 2000.[13][14][15][16] In 1997, after Margolese had obtained regulatory clearance and "effectively created the industry," the FCC also sold a license to the American Mobile Radio Corporation,[17] which changed its name to XM Satellite Radio in October 1998.[18] XM was founded by Lon Levin and Gary Parsons, who served as chairman until November 2009.[19][20]

CD Radio purchased their license for $83.3 million, and American Mobile Radio Corporation bought theirs for $89.9 million. Digital Satellite Broadcasting Corporation and Primosphere were unsuccessful in their bids for licenses.[21] Sky Highway Radio Corporation had also expressed interest in creating a satellite radio network, before being bought out by CD Radio in 1993 for $2 million.[22] In November 1999, Margolese changed the name of CD Radio to Sirius Satellite Radio.[11] In November 2001, Margolese stepped down as CEO, remaining as chairman until November 2003, with Sirius issuing a statement thanking him "for his great vision, leadership and dedication in creating both Sirius and the satellite radio industry."[23]

XM's first satellite was launched on March 18, 2001 and its second on May 8, 2001.[7] Its first broadcast occurred on September 25, 2001, nearly four months before Sirius.[24] Sirius launched the initial phase of its service in four cities on February 14, 2002,[25] expanding to the rest of the contiguous United States on July 1, 2002.[24] The two companies spent over $3 billion combined to develop satellite radio technology, build and launch the satellites, and for various other business expenses.[5] Stating that it was the only way satellite radio could survive, Sirius and XM announced their merger on February 19, 2007, becoming Sirius XM Radio.[26][27] The FCC approved the merger on July 25, 2008, concluding that it was not a monopoly, primarily due to Internet audio-streaming competition.[28]

Canada

Main article: Sirius XM Canada

XM satellite radio was launched in Canada on November 29, 2005. Sirius followed two days later on December 1, 2005. Sirius Canada and XM Radio Canada announced their merger into Sirius XM Canada on November 24, 2010.[29] It was approved by the Canadian Radio-television and Telecommunications Commission on April 12, 2011.[30]

Africa and Eurasia

Main article: 1worldspace

WorldSpace was founded by Ethiopia-born lawyer Noah Samara in Washington, D.C., in 1990,[31] with the goal of making satellite radio programming available to the developing world.[32] On June 22, 1991, the FCC gave WorldSpace permission to launch a satellite to provide digital programming to Africa and the Middle East.[2] WorldSpace first began broadcasting satellite radio on October 1, 1999, in Africa.[33] India would ultimately account for over 90% of WorldSpace's subscriber base.[34] In 2008, WorldSpace announced plans to enter Europe, but those plans were set aside when the company filed for Chapter 11 bankruptcy in November 2008.[35] In March 2010, the company announced it would be de-commissioning its two satellites (one served Asia, the other served Africa). Liberty Media, which owns 50% of Sirius XM Radio, had considered purchasing WorldSpace's assets, but talks between the companies collapsed.[32][36]

Japan

Main article: MobaHo!

MobaHo! was a mobile satellite digital audio/video broadcasting service in Japan, whose services began on October 20, 2004, and ended on March 31, 2009.[37]

2.41.2 System design

Satellite radio uses the 2.3 GHz S band in North America for nationwide digital audio broadcasting (DAB).[6] MobaHO! operated at 2.6 GHz. In other parts of the world, satellite radio uses the 1.4 GHz L band allocated for DAB.[7]

Satellite radio subscribers purchase a receiver and pay a monthly subscription fee to listen to programming. They can listen through built-in or portable receivers in automobiles; in the home and office with a portable or tabletop receiver equipped to connect the receiver to a stereo system; or on the Internet.[8]

Ground stations transmit signals to the satellites, which are orbiting over 22,000 miles above the surface of the Earth. The satellites send the signals back down to radio receivers in cars and homes. This signal contains scrambled broadcasts, along with meta data about each specific broadcast. The signals are unscrambled by the radio receiver modules, which display the broadcast information. In urban areas, ground repeaters enable signals to be available even if the satellite signal is blocked. The technology allows for nationwide broadcasting, so that, for instance US listeners can hear the same stations anywhere in the country.[7][9]

2.41.3 Content, availability and market penetration

Satellite radio in the US offers commercial-free music stations, as well as talk, news and sports, some of which include commercials.[38] In 2004, satellite radio companies in the United States began providing background music to hotels, retail chains, restaurants, airlines and other businesses.[39][40] Sirius XM Music for Business is a cheaper alternative to the costlier Muzak, and it doesn't come with a long-term commitment.[40][41] On April 30, 2013, Sirius XM CEO Jim Meyer stated that the company

would be pursuing opportunities over the next few years to provide in-car services through their existing satellites, including telematics (automated security and safety, such as stolen vehicle tracking and roadside assistance) and entertainment (such as weather and gas prices).[42]

As of April 2013, Sirius XM had 24.4 million subscribers.[43] This was primarily due to the company's partnerships with automakers and car dealers. Roughly 60% of new cars sold come equipped with Sirius XM, and just under half of those units gain paid subscriptions. The company has long-term deals with General Motors, Ford, Toyota, Kia, Bentley, BMW, Volkswagen, Nissan, Hyundai and Mitsubishi.[44] The presence of Howard Stern, whose show attracts over 12 million listeners per week, has also been a factor in the company's steady growth.[44][45] As of 2013, the main competition to satellite radio is streaming Internet services, such as Pandora and Spotify, as well as FM and AM Radio.[42]

2.41.4 Satellite radio vs. other formats

Satellite radio differs from AM or FM radio and digital television radio (or DTR) in the following ways. The table applies primarily to the United States.

[2] The sound quality with both satellite radio providers and DTR providers varies with each channel. Some channels have near CD-quality audio, and others use low-bandwidth audio suitable only for speech. Since only a certain amount of bandwidth is available within the licenses available, adding more channels means that the quality on some channels must be reduced. Both the frequency response and the dynamic range of satellite channels can be superior to most, but not all AM or FM radio stations, as most AM and FM stations clip the audio peaks to sound louder; even the worst channels are still superior to most AM radios, but a very few AM tuners are equal to or better than the best FM or satellite broadcasts when tuned to a local station, even if not capable of stereo. The use of HD Radio technology can allow AM and FM broadcasts to exceed the quality of satellite. AM does not suffer from multipath distortion or flutter in a moving vehicle like FM, nor does it become silent as you go behind a big hill like satellite radio.

[3] Some satellite radio services and DTR services act as *in situ* repeaters for local AM/FM stations and thus feature a high frequency of interruption.

[4] Nonprofit stations and public radio networks such as CBC/Radio-Canada, NPR, and PRI-affiliated stations and the BBC are commercial-free. In the US, all stations are required to have periodic station identifications and public service announcements.

[5] In the United States, the FCC regulates technical broad-

cast spectrum only. Program content is unregulated. However, the FCC has tried in the past to expand its reach to regulate content to satellite radio and cable television, and its options are still open to attempt such in the future. The FCC does issue licenses to both satellite radio providers (XM and Sirius) and controls who holds these licenses to broadcast.

[6] Degree of content regulation varies by country; however, the majority of industrialized nations have regulations regarding obscene and/or objectionable content.

2.41.5 See also

- Digital Multimedia Broadcasting

- List of United States radio networks

- Ripping music from satellite radio broadcasts

- Sirius XM Radio

- MobaHo!

- Satellite subcarrier audio

2.41.6 References

[1] ITU Radio Regulations, Section IV. Radio Stations and Systems – Article 1.39, definition: *Broadcasting-satellite service*

[2] Edmund L. Andrews, "F.C.C. Plan For Radio By Satellite," *New York Times*, October 8, 1992.

[3] Laurent Belsie, "Digital Audio Broadcasting Plays to Global Audience," *Christian Science Monitor*, March 9, 1992.

[4] Anita Jain, "Sirius Satellite Moves," *New York Sun*, p. 11, October 29, 2002.

[5] "Satellite Radio Technology," spacefoundation.org, 2002. Accessed May 1, 2013.

[6] "Satellite S Band Radio Frequency Table," CSG Network, updated August 15, 20011. Accessed April 23, 2013.

[7] Kevin Bonsor, "How Satellite Radio Works," HowStuffWorks. Accessed May 1, 2013.

[8] http://shop.siriusxm.com/

[9] Kathleen Kingsbury, "Satellite radio captures ears of millions," CNN, August 4, 2004.

[10] Matthew Herper, "From Satellites to Pharmaceuticals," *Forbes*, April 22, 2010.

[11] Steve Warren, *Radio: The Book*, Focal Press, 2004, p. 166.

[12] "Robert Briskman appointed chairman and CEO," *Satellite News*, June 1, 1992.

[13] Bethany McLean, "Satellite Killed The Radio Star," *Fortune*, January 22, 2001, pp. 94-100.

[14] Nancy Dillon, "Beaming Radio Into High-Tech Fast Lane," *New York Daily News*, June 5, 2000.

[15] Christopher H. Sterling, *Encyclopedia of Radio, Volume 1*, Taylor & Francis, 2003, p. 750.

[16] Simon Romero, "XM Satellite Radio Completes Its Financing," *New York Times*, July 10, 2000.

[17] Simon Houpt, "Radio Flyer," *Report on Business*, September 2001, pp. 14-16.

[18] "AMRC changes name to XM Satellite Radio," XM Satellite Radio press release, New York, BBC Archive, November 16, 1998.

[19] Vince Beiser, "Hotel Biz Zillionaire's Next Venture? Inflatable Space Pods," *Wired*, October 23, 2007.

[20] Kathy Shwiff, "Parsons Resigns as Chairman of Sirius XM Radio," *Wall Street Journal*, November 12, 2009.

[21] "Revolutions in Radio," PBS, May 4, 2005.

[22] "Sirius Satellite Radio, Inc. History," fundinguniverse.com. Accessed May 7, 2013.

[23] "David Margolese Steps Down as Sirius CEO," PRNewswire, October 16, 2001.

[24] Steve Parker, "XM plus Sirius = Satellite Radio Monopoly," *Huffington Post*, July 24, 2008.

[25] "Sirius Begins Satellite Service," *Radio*, February 14, 2002.

[26] Kim Hart, "Satellite Radio Merger Approved," *Washington Post*, July 26, 2008.

[27] Richard Siklos and Andrew Ross Sorkin, "Merger Would End Satellite Radio's Rivalry," *New York Times*, February 20, 2007.

[28] Olga Kharif, "The FCC Approves the XM-Sirius Merger," *Bloomberg Businessweek*, July 25, 2008.

[29] Emil Protalinski, "XM and Sirius to finally merge in Canada," techspot.com, November 25, 2010.

[30] "CRTC Approves Sirius XM Merger In Canada," All Access, April 12, 2011.

[31] Alex Benady, "Clockwork meets satellite in a revolution for Third World radio," *The Independent*, June 1, 1998.

[32] David S. Hilzenrath, "WorldSpace announces potential decommissioning of satellites," *Washington Post*, March 18, 2010.

[33] Denise Caruso, "Digital Commerce," *New York Times*, October 11, 1999.

[34] Dilip Maitra, "WorldSpace India to shut shop on December 31," *Deccan Herald*, December 24, 2009.

[35] Eric Pfanner, "As AM signal fades, Europe moves hesitantly to digital radio," *New York Times*, January 11, 2009.

[36] Roger Collis, "The Frequent Traveler: Keeping in touch on the road through satellite radio," *New York Times*, December 20, 2002.

[37] Tim Conneally, "Toshiba to shut down mobile broadcast TV service," betanews.com, July 30, 2008.

[38] http://www.siriusxm.com/channellineup

[39] Nick Bunkley, "Satellite radio scores with exclusive programming, in-car deals," *USA Today*, January 5, 2005.

[40] Barnaby J. Feder, "Tuning In to Music That People Tune Out," *New York Times*, February 16, 2004.

[41] http://www.siriusxm.com/siriusxmforbusiness

[42] Liana B. Baker, "New CEO to expand Sirius beyond satellite radio in cars," *Reuters*, April 30, 2013.

[43] Georg Szalai, "Sirius XM Names Jim Meyer Permanent CEO, Boosts Subs, Profit in First Quarter," *Hollywood Reporter*, April 30, 2013.

[44] Trefis Team, "Can Sirius XM Tune In Big Subscriber Growth This Year?" *Forbes*, April 12, 2013.

[45] Jeff Bercovici, "Sirius XM's Mel Karmazin: 'I'm One of the Most Underpaid Executives in the History of Executive Payment'," *Forbes*, April 3, 2012.

2.41.7   Further reading

- Navis, Chad, and Mary Ann Glynn. "How new market categories emerge: Temporal dynamics of legitimacy, identity, and entrepreneurship in satellite radio, 1990–2005." *Administrative Science Quarterly* (2010) 55#3 pp: 439-471.

2.42   Satellite television

Satellite television is – according to *article 1.39* of the International Telecommunication Union´s (ITU) ITU Radio Regulations (RR) – a *Broadcasting-satellite service*.[1]

It is delivering television programming using signals relayed from space radio stations (e.g. DVB satellites). The signals are received via an outdoor parabolic reflector antenna usually referred to as a satellite dish and a low-noise block downconverter (LNB). A satellite receiver then decodes the desired television programme for viewing on a television set. Receivers can be external set-top boxes, or a built-in television tuner. Satellite television provides a wide range of channels and services, especially to geographic areas without terrestrial television or cable television.

The most common method of reception is direct-broadcast satellite television (DBSTV), also known as "direct to home" (DTH).[2] In DBSTV systems, signals are relayed from a direct broadcast satellite on the K_u wavelength and are completely digital.[3] Satellite TV systems formerly used systems known as television receive-only. These systems received analog signals transmitted in the C-band spectrum from FSS type satellites, and required the use of large dishes. Consequently, these systems were nicknamed "big dish" systems, and were more expensive and less popular.[4]

The direct-broadcast satellite television signals were earlier analog signals and later digital signals, both of which require a compatible receiver. Digital signals may include high-definition television (HDTV). Some transmissions and channels are unencrypted and therefore free-to-air or free-to-view, while many other channels are transmitted with encryption (pay television), requiring a subscription.[5]

2.42.1   Technology

Satellites used for television signals are generally in either naturally highly elliptical (with inclination of +/–63.4 degrees and orbital period of about twelve hours, also known as Molniya orbit) or geostationary orbit 37,000 km (23,000 mi) above the earth's equator.[6]

Satellite television, like other communications relayed by satellite, starts with a transmitting antenna located at an uplink facility.[7] Uplink satellite dishes are very large, as much as 9 to 12 meters (30 to 40 feet) in diameter.[7] The increased diameter results in more accurate aiming and increased signal strength at the satellite.[7] The uplink dish is pointed toward a specific satellite and the uplinked signals are transmitted within a specific frequency range, so as to be received by one of the transponders tuned to that frequency range aboard that satellite.[8] The transponder 'retransmits' the signals back to Earth but at a different frequency band (a process known as translation, used to avoid interference with the uplink signal), typically in the C-band (4–8 GHz) or K_u-band (12–18 GHz) or both.[7] The leg of the signal path from the satellite to the receiving Earth station is called the downlink.[9]

A typical satellite has up to 32 transponders for Ku-band and up to 24 for a C-band only satellite, or more for hybrid satellites.[10] Typical transponders each have a bandwidth between 27 and 50 MHz.[10] Each geostationary C-band satellite needs to be spaced 2° from the next satellite to avoid interference; for K_u the spacing can be 1°.[11] This means that there is an upper limit of 360/2 = 180 geostation-

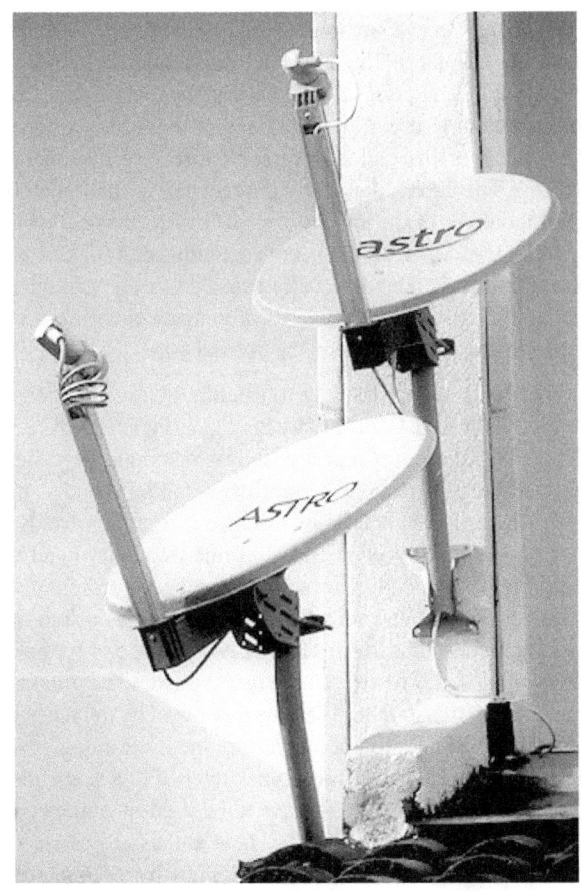

Satellite television dishes in Malaysia.

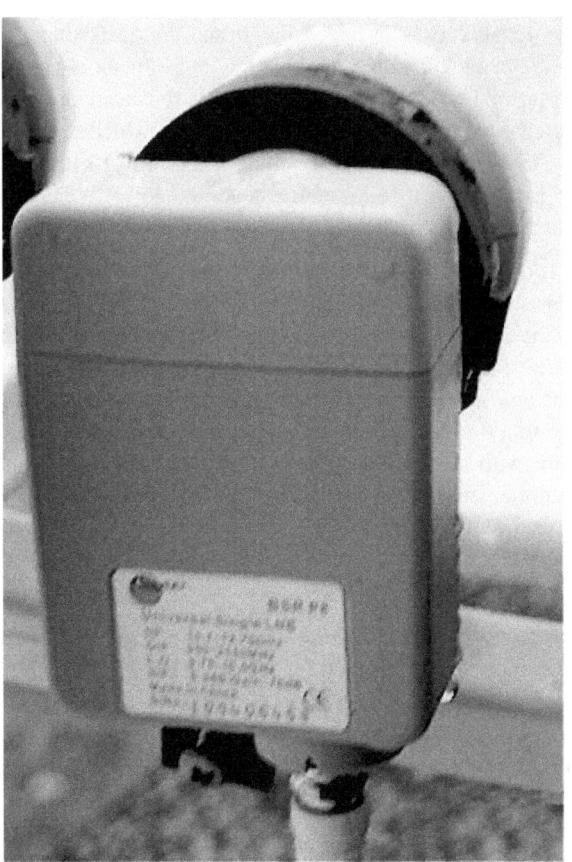

Back view of a linear polarised LNB.

An Inview Neelix set-top box.

Corrugated feedhorn and LNB on a Hughes DirecWay satellite dish.

ary C-band satellites or 360/1 = 360 geostationary K_u-band satellites.[11] C-band transmission is susceptible to terrestrial interference while K_u-band transmission is affected by rain (as water is an excellent absorber of microwaves at this particular frequency).[12] The latter is even more adversely affected by ice crystals in thunder clouds.[12]

On occasion, sun outage will occur when the sun lines up directly behind the geostationary satellite the reception an-

tenna is pointing to.[13] The downlinked satellite signal, quite weak after traveling the great distance (see inverse-square law), is collected with a parabolic receiving dish, which reflects the weak signal to the dish's focal point.[14] Mounted on brackets at the dish's focal point is a device called a feedhorn or collector.[15] The feedhorn is essentially the flared front-end of a section of waveguide that gathers the signals at or near the focal point and 'conducts' them to a probe or pickup connected to a low-noise block downconverter or LNB.[16] The LNB amplifies the relatively weak signals, filters the block of frequencies in which the satellite television signals are transmitted, and converts the block of frequencies to a lower frequency range in the L-band range.[16]

The original C-Band satellite television systems used a low-noise amplifier connected to the feedhorn at the focal point of the dish.[17] The amplified signal was then fed via very expensive and sometimes 50 ohm impedance gas filled hardline coaxial cable to an indoor receiver or, in other designs, fed to a downconverter (a mixer and a voltage tuned oscillator with some filter circuitry) for downconversion to an intermediate frequency.[17] The channel selection was controlled, typically by a voltage tuned oscillator with the tuning voltage being fed via a separate cable to the headend, but this design evolved.[17]

Designs for microstrip based converters for amateur radio frequencies were adapted for the 4 GHz C-Band.[18] Central to these designs was concept of block downconversion of a range of frequencies to a lower, and technologically more easily handled block of frequencies (intermediate frequency).[18]

The advantages of using an LNB are that cheaper cable could be used to connect the indoor receiver with the satellite television dish and LNB, and that the technology for handling the signal at L-Band and UHF was far cheaper than that for handling the signal at C-Band frequencies.[19] The shift to cheaper technology from the 50 Ohm impedance cable and N-Connectors of the early C-Band systems to the cheaper 75 Ohm technology and F-Connectors allowed the early satellite television receivers to use, what were in reality, modified UHF television tuners which selected the satellite television channel for down conversion to another lower intermediate frequency centered on 70 MHz where it was demodulated.[19] This shift allowed the satellite television DTH industry to change from being a largely hobbyist one where receivers were built in low numbers and complete systems were expensive (costing thousands of dollars) to a far more commercial one of mass production.[19] Direct broadcast satellite dishes are fitted with an LNBF, which integrates the feedhorn with the LNB.[19]

In the United States, service providers use the intermediate frequency ranges of 950-2150 MHz to carry the signal to the receiver. This allows for transmission of UHF band signals along the same span of coaxial wire at the same time. In some applications (DirecTV AU9-S and AT-9), ranges the lower B-Band and upper 2250-3000 MHz, are used. Newer LNBFs in use by DirecTV referred to as SWM (Single Wire Multiswitch), See also Single Cable Distribution, use a less limited frequency range of 2-2150 MHz.

The satellite receiver or set-top box demodulates and converts the signals to the desired form (outputs for television, audio, data, etc.).[20] Sometimes, the receiver includes the capability to unscramble or decrypt the received signal; the receiver is then called an integrated receiver/decoder or IRD.[21] The cable connecting the receiver to the LNBF or LNB should be of the low loss type RG-6, quad shield RG-6 or RG-11, etc.[16] RG-59 is not recommended for this application as it is not technically designed to carry frequencies above 950 MHz, but will work in many circumstances, depending on the quality of the coaxial wire.[16]

A practical problem relating to satellite home reception is that basically an LNB can only handle a single receiver.[22] This is due to the fact that the LNB is mapping two different circular polarizations – right hand and left hand – and in the case of the K-band two different reception bands – lower and upper – to one and the same frequency band on the cable.[22] Depending on which frequency a transponder is transmitting at and on what polarization it is using, the satellite receiver has to switch the LNB into one of four different modes in order to receive a specific desired program on a specific transponder.[22] This is handled by the receiver using the DiSEqC protocol to control the LNB mode.[22] If several satellite receivers are to be attached to a single dish a so-called multiswitch will have to be used in conjunction with a special type of LNB.[22] There are also LNBs available with a multiswitch already integrated.[22] This problem becomes more complicated when several receivers are to use several dishes (or several LNBs mounted in a single dish) pointing to different satellites.[22]

A common solution for consumers wanting to access multiple satellites is to deploy a single dish with a single LNB and to rotate the dish using an electric motor. The axis of rotation has to be set up in the north-south direction and, depending on the geographical location of the dish, have a specific vertical tilt. Set up properly the motorized dish when turned will sweep across all possible positions for satellites lined up along the geostationary orbit directly above the equator. The disk will then be capable of receiving any geostationary satellite that is visible at the specific location, i.e. that is above the horizon. The DiSEqC protocol has been extended to encompass commands for steering dish rotors.

2.42.2 Standards

Analog television which was distributed via satellite was usually sent scrambled or unscrambled in NTSC, PAL, or SECAM television broadcast standards. The analog signal is frequency modulated and is converted from an FM signal to what is referred to as baseband. This baseband comprises the video signal and the audio subcarrier(s). The audio subcarrier is further demodulated to provide a raw audio signal.

Later signals were digitized television signal or multiplex of signals, typically QPSK. In general, digital television, including that transmitted via satellites, is based on open standards such as MPEG and DVB-S/DVB-S2 or ISDB-S.

The conditional access encryption/scrambling methods include NDS, BISS, Conax, Digicipher, Irdeto, Cryptoworks, DG Crypt, Beta digital, SECA Mediaguard, Logiways, Nagravision, PowerVu, Viaccess, Videocipher, and VideoGuard. Many conditional access systems have been compromised.

2.42.3　Categories of usage

There are three primary types of satellite television usage: reception direct by the viewer, reception by local television affiliates, or reception by headends for distribution across terrestrial cable systems.

Direct to the viewer reception includes direct broadcast satellite (or DBS) and television receive-only (or TVRO), both used for homes and businesses including hotels, among other properties.

Direct broadcast via satellite

Main article: Direct-broadcast satellite television
Direct broadcast satellite, (DBS) also known as "Direct-

DBS satellite dishes installed on an apartment complex.

To-Home" can either refer to the communications satellites themselves that deliver DBS service or the actual television service.[2] Most satellite television customers in developed television markets get their programming through a direct broadcast satellite provider.[2] Signals are transmitted using K_u band and are completely digital which means it has high picture and stereo sound quality.[3]

Programming for satellite television channels comes from

A Sky Digital "minidish".

multiple sources and may include live studio feeds.[23] The broadcast centre assembles and packages programming into channels for transmission and, where necessary, encrypts the channels. The signal is then sent to the uplink [24] where it is transmitted to the satellite. With some broadcast centres, the studios, administration and uplink are all part of the same campus.[25] The satellite then translates and broadcasts the channels.[26]

Most of the DBS systems use the DVB-S standard for transmission.[2] With pay television services, the datastream is encrypted and requires proprietary reception equipment. While the underlying reception technology is similar, the pay television technology is proprietary, often consisting of a conditional-access module and smart card. This measure assures satellite television providers that only authorised, paying subscribers have access to pay television content but at the same time can allow free-to-air (FTA) channels to be viewed even by the people with standard equipment (DBS receivers without the conditional-access modules) available in the market.

Television receive-only

Main article: Television receive-only
The term Television receive-only, or TVRO, arose during the early days of satellite television reception to differentiate it from commercial satellite television uplink and downlink operations (transmit and receive). This was the primary method of satellite television transmissions before the satellite television industry shifted, with the launch of higher

A C-band satellite dish used by TVRO systems.

powered DBS satellites in the early 1990s which transmitted their signals on the K_u band frequencies.[4][27] Satellite television channels at that time were intended to be used by cable television networks rather than received by home viewers.[28] Early satellite television receiver systems were largely constructed by hobbyists and engineers. These early TVRO systems operated mainly on the C-band frequencies and the dishes required were large; typically over 3 meters (10 ft) in diameter.[29] Consequently, TVRO is often referred to as "big dish" or "Big Ugly Dish" (BUD) satellite television.

TVRO systems were designed to receive analog and digital satellite feeds of both television or audio from both C-band and K_u-band transponders on FSS-type satellites.[30][31] The higher frequency K_u-band systems tend to resemble DBS systems and can use a smaller dish antenna because of the higher power transmissions and greater antenna gain. TVRO systems tend to use larger rather than smaller satellite dish antennas, since it is more likely that the owner of a TVRO system would have a C-band-only setup rather than a K_u band-only setup. Additional receiver boxes allow for different types of digital satellite signal reception, such as DVB/MPEG-2 and 4DTV.

The narrow beam width of a normal parabolic satellite antenna means it can only receive signals from a single satellite at a time.[32] Simulsat or the Vertex-RSI TORUS, is a quasi-parabolic satellite earthstation antenna that is capable of receiving satellite transmissions from 35 or more C- and K_u-band satellites simultaneously.[33]

2.42.4 History

Early milestones

In 1945 British science fiction writer Arthur C. Clarke proposed a world-wide communications system which would function by means of three satellites equally spaced apart in earth orbit.[34][35] This was published in the October 1945 issue of the Wireless World magazine and won him the Franklin Institute's Stuart Ballantine Medal in 1963.[36][37]

The first public satellite television signals from Europe to North America were relayed via the Telstar satellite over the Atlantic ocean on 23 July 1962, although a test broadcast had taken place almost two weeks earlier on 11 July.[38] The signals were received and broadcast in North American and European countries and watched by over 100 million.[38] Launched in 1962, the *Relay 1* satellite was the first satellite to transmit television signals from the US to Japan.[39] The first geosynchronous communication satellite, Syncom 2, was launched on 26 July 1963.[40]

The world's first commercial communications satellite, called Intelsat I and nicknamed "Early Bird", was launched into geosynchronous orbit on April 6, 1965.[41] The first national network of television satellites, called Orbita, was created by the Soviet Union in October 1967, and was based on the principle of using the highly elliptical Molniya satellite for rebroadcasting and delivering of television signals to ground downlink stations.[42] The first commercial North American satellite to carry television transmissions was Canada's geostationary Anik 1, which was launched on 9 November 1972.[43] ATS-6, the world's first experimental educational and Direct Broadcast Satellite (DBS), was launched on 30 May 1974.[44] It transmitted at 860 MHz using wideband FM modulation and had two sound channels. The transmissions were focused on the Indian subcontinent but experimenters were able to receive the signal in Western Europe using home constructed equipment that drew on UHF television design techniques already in use.[45]

The first in a series of Soviet geostationary satellites to carry Direct-To-Home television, Ekran 1, was launched on 26 October 1976.[46] It used a 714 MHz UHF downlink frequency so that the transmissions could be received with existing UHF television technology rather than microwave technology.[47]

Beginning of the satellite TV industry, 1976–1980

The satellite television industry developed first in the US from the cable television industry as communication satellites were being used to distribute television programming to remote cable television headends. Home Box Office (HBO), Turner Broadcasting System (TBS), and Christian Broadcasting Network (CBN, later The Family Channel) were among the first to use satellite television to deliver programming. Taylor Howard of San Andreas, California became the first person to receive C-band satellite signals with his home-built system in 1976.[48]

In the US, PBS, a non-profit public broadcasting service, began to distribute its television programming by satellite in 1978.[49]

In 1979 Soviet engineers developed the Moskva (or Moscow) system of broadcasting and delivering of TV signals via satellites. They launched the Gorizont communication satellites later that same year. These satellites used geostationary orbits.[50] They were equipped with powerful on-board transponders, so the size of receiving parabolic antennas of downlink stations was reduced to 4 and 2.5 metres.[50] On October 18, 1979, the Federal Communications Commission (FCC) began allowing people to have home satellite earth stations without a federal government license.[51] The front cover of the 1979 Neiman-Marcus Christmas catalogue featured the first home satellite TV stations on sale for $36,500.[52] The dishes were nearly 20 feet (6.1 m) in diameter[53] and were remote controlled.[54] The price went down by half soon after that, but there were only eight more channels.[55] The Society for Private and Commercial Earth Stations (SPACE), an organisation which represented consumers and satellite TV system owners, was established in 1980.[56]

Early satellite television systems were not very popular due to their expense and large dish size.[57] The satellite television dishes of the systems in the late 1970s and early 1980s were 10 to 16 feet (3.0 to 4.9 m) in diameter,[58] made of fibreglass or solid aluminum or steel,[59] and in the United States cost more than $5,000, sometimes as much as $10,000.[60] Programming sent from ground stations was relayed from eighteen satellites in geostationary orbit located 22,300 miles (35,900 km) above the Earth.[61][62]

TVRO/C-band satellite era, 1980–1986

Further information: Television receive-only

By 1980, satellite television was well established in the USA and Europe. On 26 April 1982, the first satellite channel in the UK, Satellite Television Ltd. (later Sky1), was launched.[63] Its signals were transmitted from the ESA's Orbital Test Satellites.[63] Between 1981 to 1985, TVRO systems' sales rates increased as prices fell. Advances in receiver technology and the use of gallium arsenide FET technology enabled the use of smaller dishes. Five hundred thousand systems, some costing as little as $2000, were sold in the US in 1984.[60][64] Dishes pointing to one satellite were even cheaper.[65] People in areas without local broadcast stations or cable television service could obtain good-quality reception with no monthly fees.[60][62] The

large dishes were a subject of much consternation, as many people considered them eyesores, and in the US most condominiums, neighborhoods, and other homeowner associations tightly restricted their use, except in areas where such restrictions were illegal.[4] These restrictions were altered in 1986 when the Federal Communications Commission ruled all of them illegal.[57] A municipality could require a property owner to relocate the dish if it violated other zoning restrictions, such as a setback requirement, but could not outlaw their use.[57] The necessity of these restrictions would slowly decline as the dishes got smaller.[57]

Originally, all channels were broadcast in the clear (ITC) because the equipment necessary to receive the programming was too expensive for consumers. With the growing number of TVRO systems, the program providers and broadcasters had to scramble their signal and develop subscription systems.

In October 1984, the U.S. Congress passed the Cable Communications Policy Act of 1984, which gave those using TVRO systems the right to receive signals for free unless they were scrambled, and required those who did scramble to make their signals available for a reasonable fee.[62][66] Since cable channels could prevent reception by big dishes, other companies had an incentive to offer competition.[67] In January 1986, HBO began using the now-obsolete VideoCipher II system to encrypt their channels.[58] Other channels used less secure television encryption systems. The scrambling of HBO was met with much protest from owners of big-dish systems, most of which had no other option at the time for receiving such channels, claiming that clear signals from cable channels would be difficult to receive.[68] Eventually HBO allowed dish owners to subscribe directly to their service for $12.95 per month, a price equal to or higher than what cable subscribers were paying, and required a descrambler to be purchased for $395.[68] This led to the attack on HBO's transponder Galaxy 1 by John R. MacDougall in April 1986.[68] One by one, all commercial channels followed HBO's lead and began scrambling their channels.[69] The Satellite Broadcasting and Communications Association (SBCA) was founded on December 2, 1986 as the result of a merger between SPACE and the Direct Broadcast Satellite Association (DBSA).[64]

Videocipher II used analog scrambling on its video signal and Data Encryption Standard–based encryption on its audio signal. VideoCipher II was defeated, and there was a black market for descrambler devices which were initially sold as "test" devices.[69]

Late 1980s and 1990s to present

By 1987, nine channels were scrambled, but 99 others were available free-to-air.[66] While HBO initially charged a monthly fee of $19.95, soon it became possible to unscramble all channels for $200 a year.[66] Dish sales went down from 600,000 in 1985 to 350,000 in 1986, but pay television services were seeing dishes as something positive since some people would never have cable service, and the industry was starting to recover as a result.[66] Scrambling also led to the development of pay-per-view events.[66] On November 1, 1988, NBC began scrambling its C-band signal but left its K_u band signal unencrypted in order for affiliates to not lose viewers who could not see their advertising.[70] Most of the two million satellite dish users in the United States still used C-band.[70] ABC and CBS were considering scrambling, though CBS was reluctant due to the number of people unable to receive local network affiliates.[70] The piracy on satellite television networks in the US led to the introduction of the Cable Television Consumer Protection and Competition Act of 1992. This legislation enabled anyone caught engaging in signal theft to be fined up to $50,000 and to be sentenced to a maximum of two years in prison.[71] A repeat offender can be fined up to $100,000 and be imprisoned for up to five years.[71]

Satellite television had also developed in Europe but it initially used low power communication satellites and it required dish sizes of over 1.7 metres. On 11 December 1988 Luxembourg launched Astra 1A, the first satellite to provide medium power satellite coverage to Western Europe.[72] This was one of the first medium-powered satellites, transmitting signals in K_u band and allowing reception with small dishes (90 cm).[72] The launch of Astra beat the winner of the UK's state Direct Broadcast Satellite licence holder, British Satellite Broadcasting, to the market.

In the US in the early 1990s, four large cable companies launched PrimeStar, a direct broadcasting company using medium power satellites. The relatively strong transmissions allowed the use of smaller (90 cm) dishes. Its popularity declined with the 1994 launch of the Hughes DirecTV and Dish Network satellite television systems.

On March 4, 1996 EchoStar introduced Digital Sky Highway (Dish Network) using the EchoStar 1 satellite.[73] EchoStar launched a second satellite in September 1996 to increase the number of channels available on Dish Network to 170.[73] These systems provided better pictures and stereo sound on 150–200 video and audio channels, and allowed small dishes to be used. This greatly reduced the popularity of TVRO systems. In the mid-1990s, channels began moving their broadcasts to digital television transmission using the DigiCipher conditional access system.[74]

In addition to encryption, the widespread availability, in the US, of DBS services such as PrimeStar and DirecTV had been reducing the popularity of TVRO systems since the early 1990s. Signals from DBS satellites (operating in the more recent K_u band) are higher in both frequency and power (due to improvements in the solar panels and energy efficiency of modern satellites) and therefore require much smaller dishes than C-band, and the digital modulation methods now used require less signal strength at the receiver than analog modulation methods.[75] Each satellite also can carry up to 32 transponders in the K_u band, but only 24 in the C band, and several digital subchannels can be multiplexed (MCPC) or carried separately (SCPC) on a single transponder.[76] Advances in noise reduction due to improved microwave technology and semiconductor materials have also had an effect.[76] However, one consequence of the higher frequencies used for DBS services is rain fade where viewers lose signal during a heavy downpour. C-band satellite television signals are less prone to rain fade.[77]

2.42.5 See also

- Dish Home (HD Panorama)
- Commercialization of space
- Free-to-air
- FTA Receiver
- Microwave antenna
- Molniya orbit
- Satellite dish
- Satellite subcarrier audio
- Satellite television by region
- Smart TV: provides television via internet connection
- SMATV
- Television antenna

2.42.6 References

Notes

[1] ITU Radio Regulations, Section IV. Radio Stations and Systems – Article 1.39, definition: *Broadcasting-satellite service*

[2] Antipolis, Sophia (September 1997). Digital Video Broadcasting (DVB); Implementation of Binary Phase Shift Keying (BPSK) modulation in DVB satellite transmission systems (PDF) (Report). European Telecommunications Standards Institute. p. 1-7. TR 101 198. Retrieved 20 July 2014.

[3] "Frequency letter bands". *Microwaves101.com*. 25 April 2008.

[4] "Installing Consumer-Owned Antennas and Satellite Dishes". FCC. Retrieved 2008-11-21.

[5] Campbell, Dennis; Cotter, Susan (1998). *Copyright Infringement*. Kluwer Law International. ISBN 90-247-3002-3. Retrieved 18 September 2014.

[6] Pattan 1993, p. 65.

[7] Pattan 1993, p. 207.

[8] Pattan 1993, p. 330.

[9] Pattan 1993, p. 327.

[10] Mott, Sheldon 2000, p. 253.

[11] Mott, Sheldon 2000, p. 268.

[12] Mott, Sheldon 2000, p. 115.

[13] Tirro 1993, p. 279.

[14] Minoli 2009, p. 60.

[15] Minoli 2009, p. 27.

[16] Minoli 2009, p. 194.

[17] "Europe's Best Kept Secret". *Electronics World + Wireless World* (Reed Business Publishing) **95**: 60–62. 1985. Retrieved 28 July 2014.

[18] "Microstrip Impedance Program". *Ham Radio Magazine* (Communications Technology, Incorporated) **17**: 84. 1984. Retrieved 28 July 2014.

[19] "Microwave Journal International". *Microwave Journal International* (Horizon House) **43** (10-12): 26–28. 2000. Retrieved 28 July 2014.

[20] Dodd 2002, p. 308.

[21] Dodd 2002, p. 72.

[22] Fox, Barry (1995). "Leaky dishes drown out terrestrial TV". *New Scientist* (Reed Business Information) **145**: 19–22. Retrieved 28 July 2014.

[23] "JEDI Innovation report".

[24] Bruce R. Elbert (2008). "9 Earth Stations and Network Technology". *Introduction To Satellite Communications*. Artech House. ISBN 9781596932111.

[25] "Space TV". *Popular Mechanics* (Hearst Magazines) **171** (8): 57–60. August 1994. ISSN 0032-4558.

[26] "Intelsat New Media Brochure" (PDF).

[27] James, Meg. NBC tacks on Telemundo oversight to Gaspin's tasks. Los Angeles Times, July 26, 2007. Retrieved on May 14, 2010.

[28] "Satellite Communications Training from NRI!". *Popular Science* (Bonnier Corporation) **228**. February 1986. Retrieved 16 December 2014.

[29] Prentiss 1989, p. 274.

[30] Prentiss 1989, p. 246.

[31] Prentiss 1989, p. 1.

[32] Prentiss 1989, p. 293.

[33] "Sensing SATCOM Success Is New Simulsat From ATCi". *Satnews*. 1 November 2009. Retrieved 16 December 2014.

[34] The Arthur C. Clarke Foundation at the Wayback Machine (archived July 25, 2011)

[35] Campbell, Richard; Martin, Christopher R.; Fabos, Bettina (23 February 2011). *Media and Culture: An Introduction to Mass Communication*. London, UK: Macmillan Publishers. p. 152. ISBN 978-1457628313. Retrieved 15 August 2014.

[36] The 1945 Proposal by Arthur C. Clarke for Geostationary Satellite Communications

[37] *Wireless technologies and the national information infrastructure*. DIANE Publishing. September 1995. p. 138. ISBN 0160481805. Retrieved 15 August 2014.

[38] Klein, Christopher (23 July 2012). "The Birth of Satellite TV, 50 Years Ago". *History.com*. History Channel. Retrieved 5 June 2014.

[39] "Relay 1". *NASA.gov*. NASA.

[40] Darcey, RJ (16 August 2013). "Syncom 2". *NASA.gov*. NASA. Retrieved 5 June 2014.

[41] "Encyclopedia Astronautica - Intelsat I". Retrieved 5 April 2010.

[42] "Soviet-bloc Research in Geophysics, Astronomy, and Space" (Press release). Springfield Virginia: U.S. Joint Publications Research Service. 1970. p. 60. Retrieved 16 December 2014.

[43] Robertson, Lloyd (1972-11-09). "Anik A1 launching: bridging the gap". CBC English TV. Retrieved 2007-01-25.

[44] Ezell, Linda N. (22 January 2010). "NASA - ATS". *Nasa.gov*. NASA. Retrieved 1 July 2014.

[45] Long Distance Television Reception (TV-DX) For the Enthusiast, Roger W. Bunney, ISBN 0900162716

[46] "Ekran". *Astronautix.com*. Astronautix. 2007. Retrieved 1 July 2014.

[47] "Ekran".

[48] Feder, Barnaby J. (15 November 2002). "Taylor Howard, 70, Pioneer In Satellite TV for the Home". *New York Times*. Retrieved 19 July 2014.

[49] Public Service Broadcasting in the Age of Globalization, Editors: Indrajit Banerjee, Kalinga Seneviratne. ISBN 9789814136013

[50] Wade, Mark. "Gorizont". Encyclopedia Astronautica. Retrieved 2008-06-29.

[51] The "Glory Days" of Satellite

[52] Browne, Ray (2001). *The Guide to United States Popular Culture*. Madison, Wisconsin: Popular Press. p. 706. ISBN 9780879728212. Retrieved 1 July 2014.

[53] Giarrusso, Michael (28 July 1996). "Tiny Satellite Dishes Sprout in Rural Areas". *Los Angeles Times* (Los Angeles: Los Angeles Times). Retrieved 1 July 2014.

[54] Keating, Stephen (1999). "Stealing Free TV, Part 2". *The Denver Post* (Denver, CO: The Denver Post). Retrieved 3 July 2014.

[55] Stein, Joe (1989-01-24). "Whatta dish : Home satellite reception a TV turn-on". *Evening Tribune*. p. C-8.

[56] "Earth Station Is Very Popular Dish". *Reading Eagle* (Kansas City, Missouri). 21 December 1980. Retrieved 21 July 2014.

[57] Brooks, Andree (10 October 1993). "Old satellite dish restrictions under fire New laws urged for smaller models". *The Baltimore Sun* (Baltimore, MD: The Baltimore Sun). Retrieved 1 July 2014.

[58] Nye, Doug (14 January 1990). "SATELLITE DISHES SURVIVE GREAT SCRAMBLE OF 1980S". *Deseret News* (Salt Lake City: Deseret News). Retrieved 30 June 2014.

[59] *Ku-Band Satellite TV: Theory, Installation and Repair*. Frank Baylin et al. ISBN 9780917893148.

[60] Stecklow, Steve (1984-07-07). "America's Favorite Dish". *The Miami Herald*. Knight-Ridder News Service. p. 1C.

[61] Reibstein, Larry (1981-09-27). "Watching TV Via Satellite Is Their Dish". *The Philadelphia Inquirer*. p. E01.

[62] Dawidziak, Mark (1984-12-30). "Satellite TV Dishes Getting Good Reception". *Akron Beacon-Journal*. p. F-1.

[63] "Broadband Cable 10th Anniversary". TinyPic. Retrieved 5 May 2013.

[64] "Industry History". *sbca.com*. Satellite Broadcasting and Communications Association. 2014. Retrieved 5 June 2014.

[65] Stecklow, Steve (1984-10-25). "Research Needed in Buying Dish: High Cost Is Important Consideration for Consumer". *Wichita Eagle*. Knight-Ridder News Service. p. 6C.

[66] Takiff, Jonathan (1987-05-22). "Satellite TV Skies Brighten As War With Programmers Ends". *Chicago Tribune*. Knight-Ridder Newspapers. Retrieved 2014-04-10.

[67] Wolf, Ron (1985-01-20). "Direct-Broadcast TV Is Still Not Turned On". *The Philadelphia Inquirer*. p. C01.

[68] Lyman, Rick; Borowski, Neill (April 29, 1986). "On The Trail Of 'Captain Midnight'". Philly. Retrieved May 20, 2014.

[69] Paradise, Paul R. (1 January 1999). *Trademark Counterfeiting, Product Piracy, and the Billion Dollar Threat to the U.S. Economy*. Westport, Connecticut: Greenwood Publishing Group. p. 147. ISBN 1567202500. Retrieved 3 July 2014.

[70] "Scrambled NBC Bad News for Satellite Pirates". *The San Francisco Chronicle*. United Press International. 1988-11-03. p. E3.

[71] Article STATUTE-106-Pg1460.pdf, *Cable Television Consumer Protection and Competition Act of 1992*, Act No. 1460 of 8 October 1992 (in English). Retrieved on 3 July 2014.

[72] "ASTRA 1A Satellite details 1988-109B NORAD 19688". N2YO. 9 July 2014. Retrieved 12 July 2014.

[73] Grant, August E. *Communication Technology Update* (10th ed.). Taylor & Francis. p. 87. ISBN 978-0-240-81475-9.

[74] Bell-Jones, Robin; Berbner, Jochen; Chai, Jianfeng; Farstad, Thomas; Pham, Minh (June 2001). "High Technology Strategy and Entrepreneurship" (PDF). *INSEAD journal* (Fontainebleau: INSEAD).

[75] Mirabito, M., and Morgenstern, B. (2004). *Satellites: Operations and Applications: The New Communication Technologies* (fifth edition). Burlington: Focal Press.

[76] Khaplil, Vidya R.; Bhalachandra, Anjali R. (April 2008). *Advances in Recent Trends in Communication and Networks*. New Delhi: Allied Publishers. p. 119. ISBN 1466651709. Retrieved 16 July 2014.

[77] "Rain fade: satellite TV signal and adverse weather". *Dish-cable.com*. Dish-cable.com. 2010. Retrieved 16 July 2014.

Bibliography

- Pattan, Bruno (31 March 1993). *Satellite Systems: Principles and Technologies*. Berlin: Springer Science & Business Media. ISBN 9780442013578. Retrieved 29 July 2014.

- Mott, William H.; Sheldon, Robert B. *Laser Satellite Communication: The Third Generation*. Westport, Connecticut: Greenwood Publishing Group. p. 235. ISBN 978-1567203295. Retrieved 30 July 2014.

- Tirró, S. (30 June 1993). *Satellite Communication Systems Design*. Berlin: Springer Science & Business Media. pp. 279–80. ISBN 978-0306441479. Retrieved 29 July 2014.

- Minoli, Daniel (3 February 2009). *Satellite Systems Engineering in an IPv6 Environment*. Boca Raton, Florida: CRC Press. ISBN 978-1420078688. Retrieved 29 July 2014.

- Dodd, Annabel Z. (2002). *The Essential Guide to Telecommunications* (5th ed.). Upper Saddle River, New Jersey: Prentice Hall. pp. 307–10. ISBN 0130649074. Retrieved 29 July 2014.

- Prentiss, Stan (1989). *TVRO Technology*. Prentiss Hall. ISBN 9780139333262. Retrieved 16 December 2014.

2.42.7　External links

History

- Steve Birkill's History of C-Band and Early Satellite TV

Channels and satellite fleets

- Lyngemark Satellite Charts

- Worldwide satellite locations

- Eutelsat satellite fleet

- Eutelsat TV channel guide

- SES fleet information and map

- SES guide to receiving Astra satellites

- SES guide to channels broadcasting on Astra satellites

- Linowsat PID-Lists and Videobitrate Charts

Tracking and utilities

- Online Satellite Calculations

- Online Satellite Finder Based on Google Maps

- Dish Alignment Calculator with Google Maps

General

- Satellite-TV/TVRO/ C-Band FAQ List

- General frequency allocation information, mainly for U.S.

2.43　Secret broadcast

A **secret broadcast** is, simply put, a broadcast that is not for the consumption of the general public. The invention of the wireless was initially greeted as a boon by armies and navies. Units could now be coordinated by nearly instant communications. An adversary could glean valuable and sometimes decisive intelligence from intercepted radio signals:

- messages that were not encrypted or poorly encrypted could be read

- order of battle and future intentions could be deduced by traffic analysis

- individual units could be located using direction finding

In the 1920s the United States was able to track Japanese fleet exercises even through fog banks by monitoring their radio transmissions.

A doctrine was developed of having units in the field, particularly ships at sea, maintain radio silence except for urgent situations, such as reporting contact with enemy forces. Ships in formation reverted to pre-wireless methods, including semaphore and signal flags, with signal lamps used at night. Communication from headquarters were sent by one-way radio broadcasts.

2.43.1　"Personal messages" on propaganda stations

During WWII, the BBC would include "personal messages" in its broadcasts of news and entertainment to occupied-Europe. Often they were coded messages intended for secret agents. Leo Marks attributes this idea to Georges Bégué, an agent for the Special Operations Executive who felt their use could eliminate a lot of the two-way radio traffic that often compromised agents. Such messages were also used to authenticate agents to sources of assistance in the field. The agent would arrange to have the BBC broadcast any short phrase the other person chose.

2.43.2　Numbers stations

Main article: Numbers station

In the mid-twentieth century, the High Frequency radio bands were used by numerous stations sending seemingly random Morse code, usually in five-letter groups. As more advanced communications methods, such as teleprinter and satellite, took over, the number of such stations diminished, but another type appeared that transmitted spoken and also seemingly random number and letter groups, the latter usually using words from a radio alphabet such as ICAO/NATO alphabet.

Though there has been no official confirmation (beyond a 1998 article in *The Daily Telegraph* which quoted a spokesperson for the Department of Trade and Industry as saying, "These [numbers stations] are what you suppose they are. People shouldn't be mystified by them. They are not for, shall we say, public consumption."[1]) there is little doubt that most of these numbers stations are primarily used to send messages to spies and other clandestine agents (additional possible uses include communication with embassies when a crisis might dictate destruction of cryptographic equipment and as a backup to normal command systems in wartime). Other intended recipients of secret broadcasts have faster and easier-to-use equipment at their disposal. But number stations are ideal for spies in that they require no special equipment, beyond a short-wave receiver. Morse code skills, once a staple of spy training, are no longer required.

2.43.3 Problems with secret broadcast

An issue in the past has been the limited bandwidth of the broadcast. Morse code was typically sent at 25 words per minute. Teleprinters could operate at or above 60 words per minute. The military uses a message precedence system to prioritize critical traffic, but all too often, senior commanders insisted on high precedence for lengthy messages lacking real urgency.

2.43.4 See also

- Letter beacon

- Numbers station

- Pirate radio - Piracy in amateur and two-way radio

- Traffic flow security

2.43.5 References

[1] "Salon People Feature | Counting spies". Salon.com. 1999-09-16. Retrieved 2010-08-26.

2.44 Shortwave bands

Shortwave bands are frequency allocations for use within the shortwave radio spectrum (the upper MF band and all

of the HF band). They are the primary medium for applications such as maritime communications, international broadcasting and worldwide amateur radio activity because they take advantage of ionospheric skip propagation to send data around the world. The bands are conventionally stated in wavelength, measured in metres. Propagation behavior on the shortwave bands depends on the time of day, the season and the level of solar activity.

2.44.1 International broadcast bands

The bands and frequencies below are derived from multiple sources, and different radios may have different frequency numbers. Most international broadcasters use amplitude modulation with 5 kHz steps between channels; a few use single sideband modulation.[1] The World Radiocommunication Conference (WRC), organized under the auspices of the International Telecommunication Union, allocates bands for various services in periodic conferences. The most recent WRC took place in 2012. At WRC-97 in 1997, the following bands were allocated for international broadcasting:

2.44.2 Amateur HF bands

Main article: Amateur radio bands

Amateur radio operators in many countries are allocated several shortwave bands for private, non-commercial use. Amateur radio is a communications service, educational tool and hobby. It is particularly useful in providing emergency communication where standard telecommunications infrastructure is compromised or nonexistent, such as a disaster area or remote region of the globe.

2.44.3 Marine, air, land mobile and fixed allocations

Designated bands in the shortwave spectrum are used for ships, aircraft, and land vehicles. Shortwave (HF radio) is used by transoceanic aircraft for communications with air-traffic control centers out of VHF radio range. Most countries with HF citizens'-band allocations use 40 or 80 channels between approximately 26.5 MHz and 27.9 MHz, in 10 kHz steps. Illegal "freeband" CB activity can be heard from 25 to 28 MHz,steps with operators generally using AM below 26.965 (US and European CB channel 1) and SSB above 27.405 (US and European CB channel 40). CB radio in the UK can be heard from 27.60125 to 27.99125 MHz in 10 kHz steps as well as the lower 26.965 to 27.405 MHz allocation. The UK and Ireland both operate Community Audio Distribution (CADS) in the UK or Wireless Public Address System (WPAS) in Ireland services in the 27.600 to 27.995 MHz portion, AM and FM mode, with two overlapping sets of 40 channels (27.60125 to 27.99125 MHz in 10 kHz steps, and 27.605 to 27.995 MHz in 10 kHz steps).[3] These transmissions are usually rebroadcasts of church services and can sometimes be heard hundreds or even thousands of km (miles). Part of the 11 m/27 MHz band was also allocated in many countries for early-model cordless phones. Due to antenna-length requirements and the band's long-distance propagation characteristics (undesirable in these cases), much land-mobile radio activity has moved to VHF or UHF and most cordless-phone use is at UHF or higher. Some segments of the HF spectrum are allocated for fixed services, providing point-to-point communication between sites with no access to wired communications.

2.44.4   Military HF use

In the US and Canada, as well as the Americas (ITU Region 2) as a whole, there are no pre-designated HF allocations for military use. Similar rules exist in Europe, where it has become necessary for European amateurs to police the bands due to overcrowding. Most military HF band incursions into the HF ham bands occur in Europe or Africa. Since the end of the Cold War specific military HF allocations have gradually disappeared from the HF bands, except for Africa and some parts of Asia. In Australia, the military shares the HF bands with civilian users; this is mainly due to low population density and relative under-use of the HF bands. The military in the Americas and Australia has tended to use the civilian fixed, maritime mobile and aeronautical mobile allocations on an *ad hoc* (non-interference) basis.

2.44.5   Industrial/Scientific/Medical (ISM) and other HF allocations

See also: ISM band

Above 10 MHz there are numerous frequencies set aside for radio astronomy, space research (FCC terminology) and standard- frequency-and-time services. RF diathermy equipment uses 27.12 MHz to heat bulk materials or adhesives for the purpose of drying or improving curing. The industrial use of the frequency suggested the use of the 11 m band for CB radio. About a dozen narrow ("sliver") allocations for ISM exist throughout the radio spectrum. These allocations are among the smallest in the HF band, with respect to national HF allocations.

2.44.6   See also

- Shortwave

- High frequency

- Medium frequency (for the unique case of the 120 m band)

- International broadcasting

- World Radiocommunication Conference

- World Administrative Radio Conference

- Citizen's Band Radio

2.44.7   References

[1] http://www.monitoringtimes.com/html/swb.html   Short wave broadcast bands, retrieved 2010 Nov 19

[2] Introduction on digital technology in the HFBC bands Accessed 2011-10-20. (Archived by WebCite® at http://www.webcitation.org/62aCbliW6

[3] http://stakeholders.ofcom.org.uk/binaries/consultations/cads_scheme/statement/statement.pdf

2.44.8   External links

- American Radio Relay League—the United States lobbying body for amateur radio and the body responsible for the ARRL Handbook

- Radio Amateurs of Canada—Canada's National Amateur Radio Society

- EiBi & DX - Complete list of International Broadcasting Stations worldwide, frequently updated.

- UnwantedEmissions.com - Radio spectrum allocations reference.

- short-wave.info - Easy to interrogate frequency schedules of short wave broadcasters.

- US Amateur Radio Bands chart

2.45   Shortwave broadcasting in the United States

Shortwave broadcasting in the United States is unique in that the United States allows private ownership of commercial and non-commercial shortwave stations that are not relays of existing AM/MW or FM radio stations, as are common in Africa, Europe, Asia, Oceania and Latin America. In addition to private broadcasters, the United States also has government broadcasters and relay stations for international public broadcasters. Most privately owned shortwave stations have been religious broadcasters, either wholly owned and programmed by Roman Catholic and evangelical Protestant charities or offering brokered programming consisting primarily of religious broadcasters. To better reach other continents of the world, several stations are located in far-flung US territories. Shortwave stations in the USA are not permitted to operate exclusively for a domestic audience; they are subject to antenna and power requirements to reach an international audience.

2.45.1 Non-religious private broadcasters

While many private shortwave broadcasters in the United States are operated by religious groups or carry mostly religious programming, there have also been attempts at starting non-religious shortwave stations.

Two such stations were WRNO in New Orleans and KUSW in Salt Lake City, both of them with a rock and roll music format. Both stations were well received by shortwave listeners, but could not make the format successful in the long run. KUSW was eventually sold to the Trinity Broadcasting Network and converted into religious broadcaster KTBN. WRNO kept its rock & roll format going for most of the 1980s but eventually switched formats to selling brokered airtime to political and religious broadcasts, suffered a damaged transmitter, and eventually ceased broadcasting following the death of its owner, Joe Costello. WRNO was acquired by Dr. Robert Mawire and Good News World Outreach in 2001. After installing a new transmitter, the station was within just days of returning to the air when Hurricane Katrina struck on August 29, 2005. The new transmitter was spared from flood waters, but the antenna was severely damaged by high winds. WRNO finally returned to broadcasting in 2009, operating 4 hours per day. On March 13, 2010, WRNO began transmitting a weekly religious broadcast in Arabic for a portion of its broadcast schedule.

A notable exception is WBCQ, a non-religious private station operated by Allan Weiner in Maine. WBCQ has been a success by brokering much of their airtime to religious programs like Brother Stair, while also carrying some music and entertainment programs. *Allan Weiner Worldwide*, which can be heard in most of North America, airs Fridays from 8:00 to 9:00 PM Eastern time and 7:00 to 8:00 PM central time on 7490 kHz.

2.45.2 Pirate radio

Numerous pirate radio stations have operated sporadically in or just outside the shortwave broadcast bands. Most are operated by hobbyists for the amusement of DX'ers with broadcasts typically only a few hours in length.

Few American pirates are political or controversial in their programming. Pirates have tended to cluster in unofficial "pirate bands" based on the current schedules of licensed shortwave stations and the retuning of amateur radio transmitters to operate outside the "ham" radio bands.

Most pirate activity takes place on weekends or holidays, Halloween and April Fool's Day being traditional favorites of pirates. Most broadcasts are only a few minutes to a few hours at a time. One notable exception was Radio Newyork International, a short-lived attempt to establish a permanent broadcasting station operating from international waters.

Some European nations have recently begun allowing privately owned shortwave stations on a far more limited scale.

2.45.3 Notable personalities

Preachers/Religious broadcasters

- Tony Alamo

- Kirby Anderson

- Mother Angelica

- Harold Camping (deceased)

- E. C. Fulcher

- Texe Marrs

- Robert Mawire

- Dr. Gene Scott (deceased)

- Melissa Scott (replaced Dr. Gene Scott)

- Brother Stair

- Peter J. Peters (deceased)

White Supremacists

- William Luther Pierce (deceased)

- Kevin Alfred Strom

- Hal Turner (FBI plant)

- Ernst Zündel

Commentators

- Jack Anderson (deceased) – was heard on AFRTS Radio in the 1980s

- Art Bell (a ham radio operator on 80 metres in his heyday) – via a Canadian affiliate's 49-metre shortwave relay service and WFLA's 11-metre relay

- Willis Conover (deceased)

- William Cooper (deceased)

- Mort Crim

- Chuck Harder

- Paul Harvey (deceased) – Paul Harvey News & Commentary/Rest Of The Story was carried on AFRTS Radio

- Glenn Hauser – World Of Radio

- Marie Lamb – DXing With Cumbre

- Rush Limbaugh – his show was carried on WRNO-Worldwide in the 1990s

- Dr Stan Monteith

- Jay Smilkstein (deceased) – WBCQ

- John Stadtmiller on WWCR – notorious for setting up Mark Koernke

- Allan Weiner – WBCQ

- John from Staten Island & Frank from Queens – hosting "The Right Perspective" on WWCR

2.45.4 Shortwave stations

Government broadcasters (USA)

- Voice of America

- WWV/WWVH

- Armed Forces Radio Network

- Radio Martí

Current privately owned US broadcasters

- KJES – "The Lord's Station" – Vado, New Mexico

- KNLS – World Christian Broadcasting – Anchor Point, Alaska

- KSDA – Adventist World Radio – Agat, Guam

- KTWR – Trans World Radio – Agana, Guam

- KVOH – "Voice of Hope" – Rancho Simi, California

- WBCQ – "The Planet" – Monticello, Maine

- WEWN – "Eternal Word Network" – Irondale, Alabama

- WHRI – "World Harvest Radio" – Furman, South Carolina

- WINB – "World International Broadcasting" – Red Lion, Pennsylvania

- WJHR – Milton, Florida

- WMLK – Assemblies of Yahweh – Bethel, Pennsylvania

- WRMI – "Radio Miami International" – Okeechobee, Florida

- WRNO – "WRNOradio.com" – New Orleans, Louisiana

- WTWW – "We Transmit World Wide" – Lebanon, Tennessee licensed June 30, 2009

- WWCR – "Worldwide Christian Radio" – Nashville, Tennessee

- WWRB – Manchester, Tennessee (successor to WGTG and WWFV)

Defunct broadcasters

- KAIJ – Dallas, Texas

- KFBS – Far East Broadcasting Company – Saipan, Northern Mariana Islands

- KGEI – San Francisco, California (studios and transmitter in Redwood City, California)

- KHBN – High Adventure Ministries Piti, Guam (now T8BZ in Palau)

- KHBI – Saipan, Northern Mariana Islands (Formerly KYOI)

- KIMF – Pinon, New Mexico (licensed, but never built)

- KRHO – Honolulu, Hawaii

- KSAI – Saipan, Northern Mariana Islands

- KTBN – Trinity Broadcasting Network – Salt Lake City, Utah

- KUSW – "Superpower" – Salt Lake City, Utah (station sold, and became KTBN)

- KWHR – "World Harvest Radio" – Naalehu, Hawaii

- KYOI – Saipan (1982-1989) "Super Rock" commercial station, later KHBI

- NDXE – (never built)

- WCSN – Maine -Operated by the Christian Science Monitor

- WGTG – McCaysville, Georgia

- WHRA – "World Harvest Radio" – Greenbush, Maine

- WJIE – Evangel World Prayer Center – Louisville, Kentucky

- WNRI: Bound Brook, New Jersey owned by NBC

- WNYW – "Radio New York Worldwide" – Scituate, Massachusetts

- WSHB – Furman, South Carolina

- WYFR – "Family Radio" – Okeechobee, Florida

- WWBS – Macon, Georgia

- Radio Newyork International – Pirate radio station operating from international waters

- WVOH and WTJC – Fundamental Broadcasting Network – Newport, North Carolina

New stations

- KTMI – Lebanon, Oregon – not yet on the air, construction permit

2.45.5   External links

- Official listing of active stations at the FCC.gov website. *CAUTION:* This FCC information is often out of date, and does not include domestic IBB relay stations.

- shortwavesites – The Shortwave Transmitter Site Archive – The shortwave transmitter site archive of current and historical shortwave transmitter site information, includes data on North American shortwave broadcasters' transmitter sites both past and present

- Glenn Hauser's World of Radio website

- SWDXER "The SWDXER" – with general SWL information and radio antenna tips

- Radio World special report on American shortwave stations

- International Broadcast Station KGEI: 1939–1994 History courtesy of FEBC International

2.46   Shortwave listening

A Sangean ATS-909 world band receiver.

Shortwave listening, or **SWLing**, is the hobby of listening to shortwave radio broadcasts located on frequencies between 1700 kHz and 30 MHz.[1] Listeners range from casual users seeking international news and entertainment programming, to hobbyists immersed in the technical aspects of radio reception and collecting official confirmations (QSL cards) that document their reception of distant broadcasts (DXing). In some developing countries, shortwave listening enables remote communities to obtain regional programming traditionally provided by local medium wave AM broadcasters. One 2002 estimate placed the number of shortwave listeners worldwide in the hundreds of millions.

The practice of long-distance radio listening began in the 1920s when shortwave broadcasters were first established in the US and Europe. Audiences discovered that international programming was available on the shortwave bands of many consumer radio receivers, and a number of magazines and listener clubs catering to the practice arose as a result. Shortwave listening was especially popular during times of international conflict such as World War II, the Korean War and the Persian Gulf War.

Listeners use inexpensive portable "world band" radio receivers to access the shortwave bands, and some advanced hobbyists employ specialized communications receivers featuring digital technology designed for optimum reception of shortwave signals, along with outdoor antennas to enhance performance.

With the advent of the internet, many international broadcasters have scaled back or terminated their shortwave transmissions in favor of web-based program distribution, while others are moving from traditional analog to digital broadcasting modes in order to allow more efficient delivery of shortwave programming. The number of organized shortwave listening clubs has diminished along with printed magazines devoted to the hobby; however, many enthusiasts continue to exchange information and news on the web.

2.46.1 History

The practice of listening to distant stations in the medium wave AM broadcast band was carried over to the shortwave bands. Frank Conrad, an early pioneer of medium wave broadcasting with KDKA in Pittsburgh, instituted some of the first shortwave broadcasts around 1921. Stations affiliated with General Electric and Crosley followed shortly after.

United States shortwave broadcasters began transmitting popular radio programs in an attempt to attract foreign audiences. During the 1930s, new shortwave receivers appeared on the market as well as popular shortwave magazines and clubs. Shortwave stations often offered unique QSL cards for DXers.

In Europe, shortwave broadcasts from Britain and the Netherlands such as Philips Radio's PCJJ began around 1927. Germany, Italy, the Soviet Union, Britain, and many other countries soon followed, and some classic shortwave broadcasters got their start. The BBC began on shortwave as the "BBC Empire Service" in 1932.[2] Its broadcasts were aimed principally at English speakers. Radio Moscow was broadcasting on shortwave in English, French, German, Italian and Arabic by 1939. The Voice of America (or VOA) began broadcasting in 1942 after its entry into World War II using the Yankee Doodle musical theme.

While technically minded shortwave listening hobbyists dwindled during the war years due in part to the demands of military service, casual listeners seeking war news from foreign broadcasters increased. Shortwave receiver manufacturers contributed to war production. Zenith launched the multi-band Trans-Oceanic series of radios in 1942. In some other countries, during the war, listening to foreign stations was a criminal offense. Established in 1939, 35-kilowatt Chinese shortwave station XGOY broadcast pro-

"The Voice of China" broadcast in 1942

gramming aimed at listening-restricted Japan. The station was often bombed by the Japanese.[3][4]

CBS began a shortwave listening program in September 1939, on an experimental basis, at the National Lawn Tennis Championships at West Side Tennis Club in Forest Hills, New York. Engineers installed equipment at the CBS booth when the location was found to have good reception, and monitors relayed European shortwave news to CBS headquarters in New York between tennis matches.[5] Throughout World War II, CBS captured Allied and enemy shortwave communications from more than 60 international stations via secretly located receivers. Translations of intercepted broadcasts were teletyped to all New York newspapers, Associated Press, United Press International and International News Service, and in turn disseminated to newspapers and radio stations throughout the United States. Major headline news frequently resulted, since big stories often broke first on radio.[6]

Shortwave listeners notified families of prisoners of war when studio announcers at stations in Axis powers countries, such as Germany and Japan, read prisoner-written messages. Allied monitors notified families, but many shortwave listeners were often first to report the broad-

CBS shortwave listening post (May 1941)

casts. Listeners in other countries also monitored POW messages.[7] Americans were actively discouraged from listening to these reports, however, since broadcasting the names of a few American prisoners was regarded as a propaganda trick to build up the listening audience for Axis radio programs. In May 1943 Jack Gerber, director of the CBS listening post, told journalist William L. Shirer that the International Red Cross was the only reliable source of information on prisoners, and expressed concern at receiving six or seven letters a week requesting transcripts of German broadcasts in which service members may have been mentioned:

The only reason the Nazis put on prisoner broadcasts is to get people justifiably anxious about relatives reported missing at the front to listen to their propaganda. Although many of the messages undoubtedly are true, they represent but a small fraction of our prisoners and we have no assurance that many of them are not faked from papers picked up on the battlefield. What concerns some of us is the consequences of listening to Nazi broadcasts unless you are a well-trained listener (and often, even if you are). Nazi arguments often sound plausible. A person may listen to them with all the skepticism in the world, knowing that every word is a lie. But if the content is sufficiently sensational (and it often is) the source may be forgotten in time, and out pops the Nazi lie, all unsuspecting.[8]

New Zealand shortwave listeners reported POW voices broadcast over Radio Peking during the Korean War.[7]

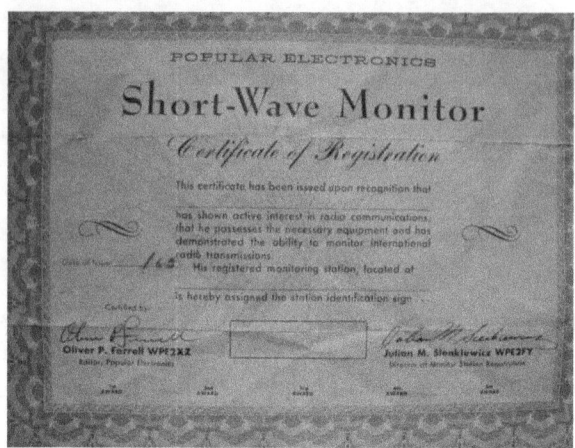

WPE shortwave monitor registration certificate circa 1963

In the 1950s and 60s, shortwave DX columns in US magazines such as *Popular Electronics'* "Tuning the Short Wave Bands" and *Electronics Illustrated'*s "The Listener" became news sources for serious radio listeners. *Popular Electronics'* "WPE Monitor Registration" program, begun in 1959, even offered callsign-like identifiers to hobbyists. A number of specialty radio clubs such as the *Newark News Radio Club* also arose during these decades and provided hobbyists with an exchange of DX news and information. When Popular Electronics and similar magazines expanded coverage of new electronics topics in the 1970s, this led to the cancellation of several long-time shortwave listening columns.[9]

Beginning with *Sweden Calling DXers* on Radio Sweden in 1948[10] (there was a slightly earlier short-lived program from Radio Australia), many shortwave radio stations began programs providing news. Some of the other prominent DX programs were Radio Netherlands' *DX Jukebox* (which became *Media Network*), the *SWL Digest* on Radio Canada International, and the *Swiss Shortwave Merry-go-round* on Swiss Radio International.

An example of notable shortwave programming was the *Happy Station Show*, popularly called the "world's longest-running shortwave radio program". The show originated on Philips Radio's PCJJ shortwave station in 1928, continuing until 1940. After World War II Radio Netherlands broadcast the show from 1946 until it terminated in 1995. Producer and presenter Keith Perron "resurrected" Happy Station on March 12, 2009. Although no longer associated with Radio Netherlands, the new effort proclaims itself as "transmitted globally via shortwave, podcasting and Internet streaming radio".[11]

During the Persian Gulf War in the 1990s, many Americans tuned into foreign news broadcasts on shortwave. Some electronics retailers even reported a "run" on portable short-wave receivers due to the increased interest at the time.[12]

2.46.2 Practices

Listening to shortwave broadcast stations for news and information programming is common, but for many shortwave listeners (abbreviated as "SWLs"), the goal is to receive as many stations from as many countries as possible, also known as DXing. "DXers" routinely test the limits of their antenna systems, radios and radio propagation knowledge. Specialized interests of shortwave listeners may include listening for shortwave utility, or "ute", transmissions such as shipping, sailing, naval, aviation, or military signals, listening for intelligence signals (numbers stations), or tuning in amateur radio stations.[1]

A Radio Moscow QSL card from 1969.

Listeners often obtain QSL cards (which confirm contact) from ham operators, broadcasters or utility stations as trophies of the hobby. Traditionally, listeners would send letters to the station with reception reports and requests for schedules. Many stations now accept E-mails or provide reception report forms on their Web sites. Reception reports give valuable information about propagation and interference to a station's engineers.[1]

There have been several publications dedicated to providing information to shortwave listeners, including the magazines *Popular Communications* (now a "digital supplement" to CQ Amateur Radio magazine), *Monitoring Times* (now defunct), and The Spectrum Monitor, a digital-only publication, in the United States, and the annual publications *Passport to World Band Radio* (now defunct) and the *World Radio TV Handbook* (WRTH). In addition, stations can provide broadcast schedules through the mail or E-mail. There are also shortwave radio programs dedicated to shortwave listening and DXing, such as the U.S.-based *World of Radio* and *DXing With Cumbre*, but recently these programs have been curtailed or dropped by many international broadcasters. As of 2007, Radio Habana Cuba still hosts a program called DXers Unlimited.

There are estimated to be millions of shortwave listeners. In 2002, according to the National Association Of Shortwave Broadcasters, for estimated numbers of households with at least one shortwave set in working order, Asia led with a large majority, followed by Europe, Sub Saharan Africa, and the former Soviet Union, respectively. The total estimated number of households worldwide with at least one shortwave set in working order was said to be 600,000,000.[13] SWLs are varied, with no common age or occupation. David Letterman is an admitted fan of the British Broadcasting Corporation (BBC).[1]

Some developing countries use shortwave as a means of receiving local and regional programming. China and Russia retransmits some domestic channels on shortwave that target listeners in far off provinces. Shortwave listening is also used as an educational tool in classrooms.[14] Poor sound reproduction, unreliable signal quality, and inflexibility of access are seen as disadvantages.[15]

Some humanitarian organizations like *Ears to Our World* distribute portable, self-powered shortwave radios to less developed parts of the globe, enabling people in remote, impoverished parts of the world to get educational programming, local and international news, emergency information and music. Recently, the group was involved in sending radios to Haiti so victims of the 2010 Haiti earthquake could stay abreast of local disaster recovery efforts.[11]

2.46.3 Equipment

Shortwave radio receivers

Radios for shortwave reception generally have higher performance than those intended for the local AM or FM broadcast band, since dependable reception of shortwave signals requires a radio with increased sensitivity, selectivity, and stability. Modern shortwave radio receivers are relatively inexpensive and easily accessible, and many hobbyists use portable "world band" receivers and built-in telescopic antennas.

Serious hobbyists may use expensive communications receivers and outdoor antenna located away from electrical noise sources, such as a dipole made from wire and insulators.

Features typical of modern solid state communications receivers:

- 500 kHz to 30 MHz frequency coverage

- Superheterodyne type - double, triple or quad conversion

- Multiple RF and IF stages

- A crystal controlled IF stage

- BFO product detector for SSB and CW reception

- Signal strength meter

- RF gain control; AVC/AGC adjustments

- Antenna tuner

- Bandwidth filters

- BFO tuning; audio limiters or attenuators.

- Frequency display dials - analog or digital.[16]

Three portable shortwave receivers

Older vacuum tube-based communications receivers are affectionately known as boatanchors for their large size and weight. Such receivers include the Collins R-390 and R-390A, the RCA AR-88, the Racal RA-17L and the Marconi Elettra. However, even modern solid-state receivers can be very large and heavy, such as the Plessey PR2250, the Redifon R551 or the Rohde & Schwarz EK070.[17]

A feature coming into wide use in modern shortwave receivers is DSP technology, short for digital signal processing. DSP is the use of digital means to process signals, and a primary benefit in shortwave receivers is the ability to tailor the bandwidth of the receiver to current reception conditions and to the type of signal being listened to. A typical analog-only receiver may have a limited number of fixed bandwidths, or only one, but a DSP receiver may have 40 or more individually selectable filters.[18]

Another important trend in modern shortwave listening is the use of "PC radios", or radios that are designed to be controlled by a standard personal computer. These radios as the name suggests are controlled by specialized PC software using a serial port connected to the radio. A PC radio may not have a front-panel at all, and may be designed exclusively for computer control, which reduces cost. In pure software defined radios, all filtering, modulation and signal manipulation is done in software, usually by a PC soundcard or by a dedicated piece of DSP hardware.[19]

2.46.4 Future of shortwave listening

The rise of the internet influenced many broadcasters to cease their shortwave transmissions in favor of broadcasting over the world wide web. When BBC World Service discontinued service to Europe, North America, Australasia, and the Caribbean, it generated many protests and activist groups such as the Coalition to Save the BBC World Service.[20] In the US, the shifting of resources from shortwave to Internet and television by the Broadcasting Board of Governors, which oversees U.S. international broadcasting, has also resulted in reduced broadcasting hours in the English language. Although most of the prominent broadcasters continue to scale back their analog shortwave transmissions or completely terminate them, shortwave is still very common and active in developing regions such as parts of Africa.

Some international broadcasters have turned to a digital mode of broadcasting called Digital Radio Mondiale for their shortwave outlets. One reason is that digital shortwave broadcasts using DRM can cover the same geographic region with much less transmitter power — roughly one-fifth the power — than traditional AM mode broadcasts, significantly reducing the electricity cost of operating a station. A traditional AM (analog) international shortwave station can have a power rating of 50 kilowatts to as much as one million watts per transmitter, with typical power levels in the 50–500 kilowatt range. Endorsed by the ITU, it has been approved as an international standard for digital broadcasts on the HF (shortwave) bands. A DRM broadcast rivals FM mono quality and can also send graphic images and web pages via a separate information channel.[21]

Shortwave listening also remains popular with some expatriates who tune in shortwave transmissions from their homeland. Additionally, a number of remotely controlled shortwave receivers located around the world are available to users on the web.[22] While radio hobbyists report that the number of shortwave listening clubs has diminished and printed magazines devoted to the hobby are few, enthusiasts such as Glenn Hauser and others continue to populate web sites, and originate podcasts dedicated to the pursuit.[23]

2.46.5 See also

- International broadcasting

- MW DX – Similar to SW DXing except on the MW (AM Radio) band

- List of American shortwave broadcasters

- *World War II Radio Heroes: Letters of Compassion*

2.46.6 References

[1] "Introduction To Shortwave Listening". *DXing.com*. Universal Radio Research. Retrieved 2007-11-21.

[2] Analysis: BBC's voice in Europe Jan Repa, BBC News Online: 25 October 2005

[3] *China Speaks Japanese*, Time Magazine, Dec. 28, 1942

[4] Jerome S. Berg (2007). *On the Short Waves, 1923–1945: Broadcast Listening in the Pioneer Days of Radio*. McFarland. pp. 234–. ISBN 978-0-7864-3029-1. Retrieved 30 September 2013.

[5] "Radio: Propaganda Pigeons." *Time*, September 7, 1942, pp. 65–66

[6] "24,000,000 'Stolen' Words Go to Library of Congress." *The Christian Science Monitor*, September 4, 1945

[7] Berg, Jerome S. (2008). *Listening on the Short Waves, 1945 to Today*. McFarland.

[8] Shirer, William L., "The Propaganda Front". *The Montana Standard*, May 4, 1943

[9] *On The Shortwaves, The "WPE" Monitor Registration Program*

[10] "About Sweden Calling DXers - MediaScan/Sweden Calling DXers". Sverigesradio.se. 2001-07-17. Retrieved 2012-08-31.

[11] Osterman, Fred. "Newsroom." *DXing Newsroom*. 2004. Universal Radio Research. 6 April 2010.

[12] *Tuning Into The World Via Shortwave, New York Times*, June 3, 1992.

[13] NATIONAL ASSOCIATION OF SHORTWAVE BROADCASTERS, Inc. October 2002

[14] Hawkins, Ralph G. (March 1989). "A learning experience via short wave radio". Tech Trends, Volume 34, Number 2. Retrieved 15 April 2010.

[15] Wipf, Joseph A. "Shortwave Radio and the Second Language Class." *The Modern Language Journal*. 68.1 (Spring 1984): 7–8. JSTOR. 3 March 2010.

[16] *Passport To WorldBand Radio*

[17] Osterman, Fred (1998). *Shortwave Receivers Past & Present: Communications Receivers 1942–1997*. Universal Radio Research, Reynoldsburg (USA).

[18] H. Ward Silver (1 June 2006). *The ARRL Ham Radio License Manual: All You Need to Become an Amateur Radio Operator. Technician]. Level 1*. American Radio Relay League. pp. 3–. ISBN 978-0-87259-963-5. Retrieved 7 May 2013.

[19] Ziff Davis, Inc. (19 January 1999). *PC Mag*. Ziff Davis, Inc. pp. 56–. ISSN 0888-8507. Retrieved 7 May 2013.

[20] Save the BBC World Service.

[21] *Digital Radio Mondiale*

[22] Petruzzellis, Thomas (2007). *22 Radio and Receiver Projects for the Evil Genius*. McGraw-Hill Professional. ISBN 978-0-07-148929-4.

[23] van de Groenendaal, Hans (January 7, 2009). "Is there a future for shortwave listening as a hobby?". EE Publishers. Retrieved 20 April 2010.

2.46.7 Further reading

- World Radio TV Handbook WRTH, ISBN 3-87463-356-X.

- Passport to World Band Radio, www.passband.com, ISBN 0-914941-61-5 (2007 ed.) ISBN 978-0-914941-66-8 (2008 ed.) ISBN 978-0-914941-80-4 (2009 ed.)

- Jerome S. Berg (2008). *Listening on the short waves, 1945 to today*. McFarland. ISBN 0-7864-3996-3.

- *Popular Communications* monthly magazine published by CQ Communications.

- *Monitoring Times* -- Monthly publication has ceased, but Grove Enterprises in Brasstown NC maintains some features at its website.

2.46.8 External links

- Shortwave Radio at DMOZ

2.47 Shortwave relay station

ALLISS antenna as viewed underneath

Shortwave relay stations are transmitter sites used by international broadcasters to extend their coverage to areas that cannot be reached easily from their home state, for example the BBC operates an extensive net of relay stations.[1]

These days the programs are fed to the relay sites by satellite, cable/optical fiber or the Internet. Frequencies, transmitter power and antennas depend on the desired coverage. Some regional relays even operate in the medium wave or FM bands.[2]

Relay stations are also important to reach listeners in countries that practice Radio jamming. Depending on the effect of the Shortwave dead zone the target countries can jam the programs only locally, e.g. for bigger cities. For this purpose Radio Free Europe/Radio Liberty with studios in Munich/Germany operated a relay station in Portugal, in the extreme west of Europe, to reach the then communist Eastern Europe.[3]

2.47.1 Variations in design

One and only one broadcasting technology couples all of the components of a traditional shortwave relay station into one unit: the ALLISS module. For persons totally unfamiliar with the concepts of how shortwave relay stations operate this design may be the most understandable.

The ALLISS module is a fully rotatable antenna system for high power (typically 500 kW only) shortwave radio broadcasting—it essentially is a **self contained shortwave relay station**.

Most of the world's shortwave relay stations do not use this technology, due to its cost (15m EUR per ALLISS module: Transmitter + Antenna + Automation equipment).

2.47.2 Planning and design

A traditional shortwave relay station—depending on how many transmitters and antennas that it will have—may take up to two years to plan. After planning is completed, it may take up to five years to construct the relay station.

The historically long design and planning cycle for shortwave relay stations ended in the 1990s. Many advanced software planning tools (not related to the relay station design proper) became available. Choosing a series of possible sites for a relay station is about 100 times faster using Google Earth, for example. With the modern graphical version of Ioncap, simplified propagation studies can completed in less than a week for any chosen site.

In some cases, existing relay stations can have their designs more or less duplicated, thus speeding up development time. However, there is one general exception to this: the ALLISS Module. From initial planning to deployment of ALLISS Modules may take a mere 1.5 years to 9 months depending on the number of modules deployed at one time in a particular sector of a country.

Graphic examples

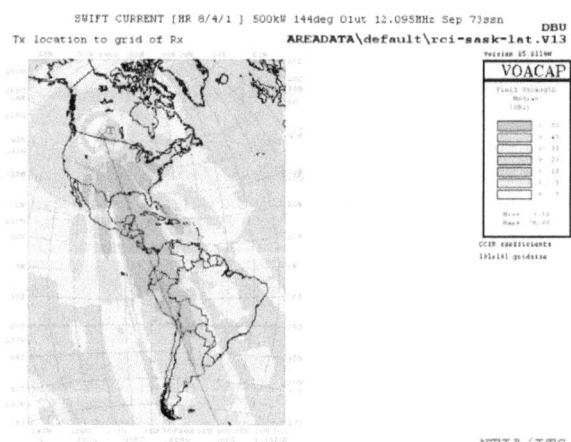

Example of a theoretical shortwave relay station in central Canada, single antenna with single transmitter

How relay stations operate

These are considered general operating parameters:

- 20 hours per day, but geopolitical reasons may dictate some stations run 24 hours per day (a 168-hour week)

- Generally 360 days per year, depending on the number of redundant transmitters and antennas

- Relay Stations generally consume from 250 kilowatts (kW) to 10 megawatts (MW)

- A single 100 kW SW transmitter consumes 225 kW RMS as a general rule

- A single 300 kW SW transmitter consumes 625 kW RMS as a general rule

- Modulator efficiency: Class-B modulators have about a 65% efficiency level, but digital (PDM or PSM or hybrid variants) modulators have about an 85% efficiency level as a general rule (for Amplitude Modulation)

- Broadcast times and frequencies are under ITU regulation

How relay stations are designed

General requirements of shortwave relay stations:

- Road access (fairly universal)

- HVAC mains access building or transformer in the transmitter building itself

- Staff quarters (if the relay station is not fully automated)

- Incoming audio processing centre, but since the mid-1980s this has evolved into one to five rack units

- Transmitter hall (50 kW, 100 kW, 250 kW, 300 kW, 500 kW shortwave transmitter)

- Switch matrix (but these are not typically used by ALLISS modules)

- Baluns (but their use is not always required nor universal)

- antenna tuners (sometimes called ATUs or roller coasters because of their appearance)

- Feeder lines (coax cable and open feeder lines are the most common feeders in use)

- HRS-type antennas, or occasionally log-periodic (horizontal)

- In parts of the developing world log-periodic (horizontal) antennas are used to provide less directional gain to a target area.

Where the broadcast programs go

- generally to target areas that are more than 300 km from the transmitter site

- most shortwave relay station target areas are 1500 km to 3500 km from the transmitter site

Mobile relay stations

The IEEE Book series "The History of International Broadcasting" (Volume I) describes mobile shortwave relay stations used by the German propaganda ministry during WWII, to avoid them being located by radio direction finding and bombed by the Allies. They consisted of a generator truck, transmitter truck and an antenna truck, and are thought to have had a radiated power of about 50 kW. Radio Industry Zagreb (RIZ Transmitters) currently produces mobile shortwave transmitters.

2.47.3   Notable sites : Issoudun

- RFI at Issoudun

- Volga ALLISS Module

- Ganges ALLISS Module

- Former RFI Issoudun Relay station feeders and curtain arrays

- Former RFI Issoudun Relay curtain arrays

The International broadcasting center of TDF (Télédiffusion de France) is at Issoudun/Ste Aoustrille. As of 2011, Issoudun is utilized by TDF for shortwave transmissions. The site uses 12 rotary ALLISS antennas fed by 12 transmitters of 500 kW each to transmit shortwave broadcasts by Radio France International (RFI), along with other broadcast services.

2.47.4   See also

- Broadcast relay station

- Imperial Wireless Chain

2.47.5   References

[1] BBC Cyprus relay. Retrieved 2011-04-01.

[2] How to listen to the BBC. Retrieved 2011-04-01.

[3] Radio Free Europe. Retrieved 2011-04-01.

2.48 SINPO code

For other uses, see Sinpo (disambiguation).

SINPO, acronym for **signal, interference, noise, propagation, and overall**, is a Signal Reporting Codes used to describe the quality of radio transmissions, especially in reception reports written by shortwave listeners. Each letter of the code stands for a specific factor of the signal, and each item is graded on a 1 to 5 scale (where 1 stands for nearly undetectable/severe/unusable and 5 for excellent/nil/extremely strong).

The code originated with the CCIR (a predecessor to the ITU-R) in 1951, and was widely used by BBC shortwave listeners to submit signal reports, with many going so far as to mail audio recordings to the BBC's offices.[1] It has been expanded in some places to a SINPFEMO code which includes rating the station's modulation and other audio qualities, but the expanded code is rarely used in practice.

Both SINPO and SINPFEMO are the official signal reporting codes for international civil aviation.[2]

The use of the SINPO code can be subjective and may vary from person to person. Not all shortwave listeners are conversant with the SINPO code and prefer using plain language instead.

2.48.1 Code explained

S (Signal strength) The relative strength of the transmission.

I (Interference) Interference from other stations on the same or adjacent frequencies (man-made noise).

N (Noise) The amount of atmospheric noise.

P (Propagation) Whether the signal is steady or fades from time to time.

O (Overall merit) An overall score for the listening experience under these conditions.

Each category is rated from 1 to 5 with 1 being 'unusable' or 'severe' and 5 being 'perfect' or 'nil'. MANY raters misunderstand the code and will rate everything either 55555 or 11111 when in reality both extremes are unusual in the extreme. '55555' essentially means 'perfect reception akin to a local station' while that is occasionally possible, when talking about long-distance short-wave reception, it is almost never the case.

Another common mistake in rating is presenting an 'O' higher than any previously rated element. By definition, a station cannot present 'perfect' reception if there is any Noise or Interference or Fading present. In other words, it is NOT 'perfect local quality' reception if any of those things are present.

2.48.2 Examples of SINPO code applied

In responding to a shortwave reception, the SINPO indicates to the transmitting station the overall quality of the reception.

The SINPO code in normal use consists of the 5 rating numbers listed without the letters, as in the examples below:

54554 - This indicates a relatively clear reception, with only slight interference; however, nothing that would significantly degrade the listening experience.

33433 - This indicates a signal which is moderately strong, but has more interference, and therefore deterioration of the received signal.

Generally, a SINPO with a code number starting with a 2 or lower would not be worth reporting, unless there is no noise, interference or loss of propagation, since it would be likely the signal would be unintelligible.

Although the original SINPO code established technical specifications for each number (i.e., a number 3 in the P column meant a fixed number of fades per minute), these are rarely adhered to by reporters. The 'S' meter displays the relative strength of the received RF signal in decibels; however, this should not be used as the sole indication of signal strength, as no two S meters are calibrated exactly alike, and many lower-priced receivers omit the S meter altogether. References to a "SINFO" code may also be found in some literature. In this case, the 'F' stands for Fading, instead of 'P' for Propagation, but the two codes are interchangeable. It was presumed that the average listener would be more familiar with the meaning of "fading" than "propagation". A simple way to insure the rating applied is useful is to rate the "O" column first based on the intelligibility of the station. If you can understand everything easily, the station will rate a 4 or higher. If you have to work hard, but can understand everything '3' is the appropriate rating. If you cannot understand everything although you put great effort into it, a '2' is appropriate, and if you cannot understand the programming at all '1' is the appropriate rating.

Some listeners may not know how to distinguish between the 'I' which indicates interference from adjacent stations, and the 'N' which describes natural atmospheric or man-made noise; also for some listeners, the rating for 'Propagation' may not be completely understood. As a result of this confusion, many stations suggest the SIO code -- a simpler code which makes the limitations noted above not relevant. Despite this, some books and periodicals maintain the SINPO code is the best for DX reporters.[3]

2.48.3 See also

- RST code

- QSL card

2.48.4 References

[1] "BBC Engineering Division Monnograph, Number 43: September 1962" (PDF).

[2] "Procedures for Air Navigation Services: ICAO Abbreviations and Codes" (PDF). Retrieved 27 April 2015.

[3] "Reception Reports: Reporting Code - Writing Useful Reception Reports". Retrieved 27 April 2015.

2.48.5 External links

- itu.int: SM.1135 - Sinpo and sinpfemo codes - ITU

2.49 Skip zone

"Zone of silence" redirects here. For the purported anomaly area in Mexico, see Mapimí Silent Zone.

A **skip zone**, also called a **silent zone** or **zone of silence**,

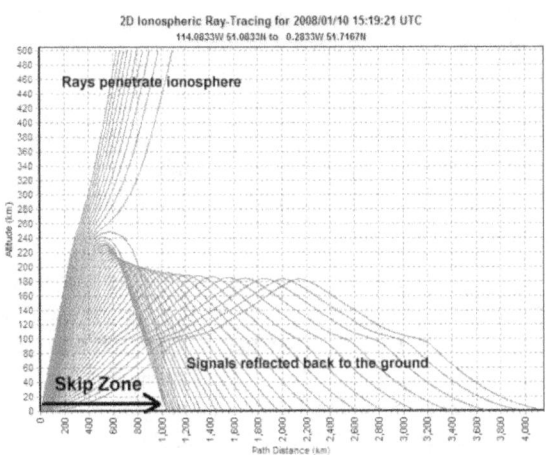

Formation of a skip-zone using Proplab-Pro 3.

is a region where a radio transmission can not be received. The zone is located between regions both closer and farther from the transmitter where reception is possible.

When using medium to high frequency radio telecommunication, there are radio waves which travel

both parallel to the ground, and towards the ionosphere, referred to as a ground wave and sky wave, respectively. A skip zone is an annular region between the farthest points at which the ground wave can be received and the nearest point at which the refracted sky waves can be received. Within this region, no signal can be received because, due to the conditions of the local ionosphere, the relevant sky waves are not reflected but penetrate the ionosphere.

The skip zone is a natural phenomenon that cannot be influenced by technical means. Its width depends on the height and shape of the ionosphere and, particularly, on the local ionospheric maximum electron density characterized by critical frequency f_oF_2. It varies mainly with this parameter, being larger for low f_oF_2. With a fixed working frequency it is large by night and may even disappear by day. Transmitting at night is most effective for long distance communication but the skip zone becomes significantly larger. Very high frequency waves and higher normally travel through the ionosphere wherefore communication via skywave is exceptional. A highly ionized Es-Layer that occasionally may appear in Summer may produce such an example.

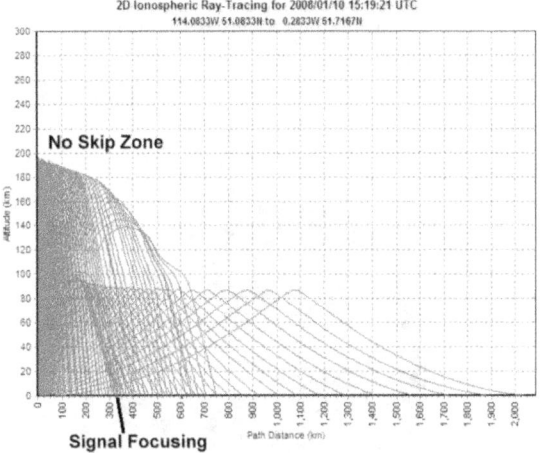

If the radio wave frequency is decreased, a point is reached where all waves (even vertically incident waves) are reflected back to the Earth.

Another method of decreasing the skip zone is by decreasing the frequency of the radio waves. Decreasing the frequency is akin to increasing the ionospheric width. A point is eventually reached when decreasing the frequency results in a zero distance skip zone. In other words, a frequency exists for which vertically incident radio waves will always be refracted back to the Earth. This frequency is equivalent to the ionospheric plasma frequency and is also known as the ionospheric critical frequency, or f_oF_2.

Skip zone is the subject of a film 'SKIPZONE' made in 1992 by UK artist, Peter Lee-Jones. It refers to areas in Scottish Highlands where it is difficult to obtain radio and

TV reception.

2.49.1   See also

- Skip (radio)

- Skywave

- Shortwave

- Ionosphere

2.49.2   Sources

- This article incorporates public domain material from the General Services Administration document "Federal Standard 1037C" (in support of MIL-STD-188).

2.50   Skywave

For the satellite terminal company in Ottawa, see SkyWave Mobile Communications.

In radio communication, **skywave** or **skip** refers to the

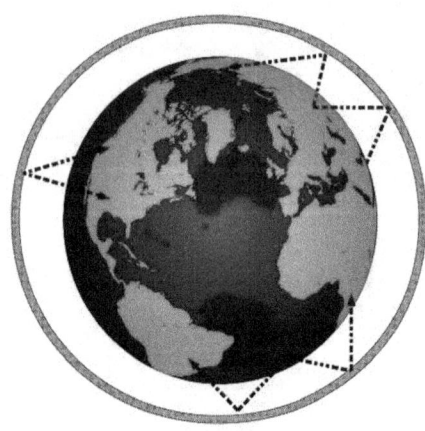

Radio waves (black) reflecting off the ionosphere (red) during skywave propagation.

propagation of radio waves reflected or refracted back toward Earth from the ionosphere, an electrically charged layer of the upper atmosphere. Since it is not limited by the curvature of the Earth, skywave propagation can be used to communicate beyond the horizon, at intercontinental distances. It is mostly used in the longwave frequency bands.

As a result of skywave propagation, a signal from a distant AM broadcasting station, a shortwave station, or—during sporadic E propagation conditions (principally during the summer months in both hemispheres)—a low frequency television station can sometimes be received as clearly as local stations. Most long-distance shortwave (high frequency) radio communication—between 3 and 30 MHz—is a result of skywave propagation. Since the early 1920s amateur radio operators (or "hams"), limited to lower transmitter power than broadcast stations, have taken advantage of skywave for long distance (or "DX") communication.

Skywave propagation is distinct from:

- groundwave propagation, where radio waves travel near Earth's surface without being reflected or refracted by the atmosphere—the dominant propagation mode at lower frequencies,

- line-of-sight propagation, in which radio waves travel in a straight line, the dominant mode at higher frequencies.

2.50.1   Explanation

The ionosphere is a region of the upper atmosphere, from about 80 km to 1000 km in altitude, where neutral air is ionized by solar photons and cosmic rays. When high frequency signals enter the ionosphere obliquely, they are back-scattered from the ionized layer as scatter waves.[1] If the midlayer ionization is strong enough compared to the signal frequency, a scatter wave can exit the bottom of the layer earthwards as if reflected from a mirror. Earth's surface (ground or water) then diffusely reflects the incoming wave back towards the ionosphere. Consequently, like a rock "skipping" across water, the signal may effectively "bounce" or "skip" between the earth and ionosphere two or more times (multihop propagation). Since at shallow incidence losses remain quite small, signals of only a few watts can sometimes be received many thousands of miles away as a result. This is what enables shortwave broadcasts to travel all over the world.

If the ionization is not great enough, the scatter wave is initially deflected downwards, and subsequently upwards (above the layer peak) such that it exits the top of the layer slightly displaced. Sky wave propagation occurs in the waveguide formed by the ground and ionosphere, each serving as reflectors. With a single "hop," path distances up to 3500 km may be reached. Transatlantic connections are mostly obtained with two or three hops.[2]

The layer of ionospheric plasma with equal ionization (the reflective surface) is not fixed, but undulates like the surface of the ocean. Varying reflection efficiency from this changing surface can cause the reflected signal strength to change, causing "*fading*" in shortwave broadcasts.

Depending on the transmitting antenna, signals below approximately 10 MHz during the day and 5 MHz at night, entering the ionosphere at a steep angle (near-vertical incidence) may be back-scattered down to Earth within a short range. Alternatively, signals beamed close to the horizon enter the ionosphere at a shallow angle and return to Earth over medium to long distances.

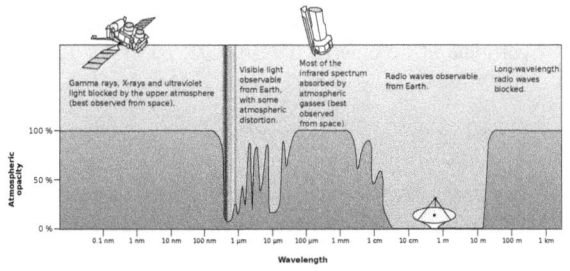

Rough plot of Earth's atmospheric transmittance (or opacity) to various wavelengths of electromagnetic radiation, including radio waves.

2.50.2   Other considerations

VHF signals with frequencies above about 30 MHz usually penetrate the ionosphere and are not returned to the Earth's surface. E-skip is a notable exception, where VHF signals including FM broadcast and VHF TV signals are frequently reflected to the Earth during late Spring and early Summer. E-skip rarely affects UHF frequencies, except for very rare occurrences below 500 MHz.

Frequencies below approximately 10 MHz (wavelengths longer than 30 meters), including broadcasts in the mediumwave and shortwave bands (and to some extent longwave), propagate most efficiently by skywave at night. Frequencies above 10 MHz (wavelengths shorter than 30 meters) typically propagate most efficiently during the day. Frequencies lower than 3 kHz have a wavelength longer than the distance between the Earth and the ionosphere. The maximum usable frequency for skywave propagation is strongly influenced by sunspot number.

Skywave propagation is usually degraded—sometimes seriously—during geomagnetic storms. Skywave propagation on the sunlit side of the Earth can be entirely disrupted during sudden ionospheric disturbances.

Because the lower-altitude layers (the E-layer in particular) of the ionosphere largely disappear at night, the refractive layer of the ionosphere is much higher above the surface of the Earth at night. This leads to an increase in the "skip" or "hop" distance of the skywave at night.

2.50.3   History

Discovery of skywave propagation

Amateur radio operators are credited with the discovery of skywave propagation on the shortwave bands. Early long-distance services used surface wave propagation at very low frequencies,[3] which are attenuated along the path. Longer distances and higher frequencies using this method meant more signal attenuation. This, and the difficulties of generating and detecting higher frequencies, made discovery of shortwave propagation difficult for commercial services.

Radio amateurs conducted the first successful transatlantic tests[4] in December 1921, operating in the 200 meter mediumwave band (1500 kHz)—the shortest wavelength then available to amateurs. In 1922 hundreds of North American amateurs were heard in Europe at 200 meters and at least 30 North American amateurs heard amateur signals from Europe. The first two-way communications between North American and Hawaiian amateurs began in 1922 at 200 meters. Although operation on wavelengths shorter than 200 meters was technically illegal (but tolerated as the authorities mistakenly believed at first that such frequencies were useless for commercial or military use), amateurs began to experiment with those wavelengths using newly available vacuum tubes shortly after World War I.

Extreme interference at the upper edge of the 150-200 meter band—the official wavelengths allocated to amateurs by the Second National Radio Conference[5] in 1923—forced amateurs to shift to shorter and shorter wavelengths; however, amateurs were limited by regulation to wavelengths longer than 150 meters (2 MHz). A few fortunate amateurs who obtained special permission for experimental communications below 150 meters completed hundreds of long distance two way contacts on 100 meters (3 MHz) in 1923 including the first transatlantic two way contacts.[6] in November 1923, on 110 meters (2.72 MHz)

By 1924 many additional specially licensed amateurs were routinely making transoceanic contacts at distances of 6000 miles (~9600 km) and more. On 21 September several amateurs in California completed two way contacts with an amateur in New Zealand. On 19 October amateurs in New Zealand and England completed a 90-minute two-way contact nearly halfway around the world. On October 10, the Third National Radio Conference made three shortwave bands available to U.S. amateurs[7] at 80 meters (3.75 MHz), 40 meters (7 MHz) and 20 meters (14 MHz). These were allocated worldwide, while the 10-meter band (28 MHz) was created by the Washington International Radiotelegraph Conference[8] on 25 November 1927. The 15-meter band (21 MHz) was opened to amateurs in the United States on 1 May 1952.

Marconi

In June and July 1923, Guglielmo Marconi's transmissions were completed during nights on 97 meters from Poldhu Wireless Station, Cornwall, to his yacht Ellette in the Cape Verde Islands. In September 1924, Marconi transmitted during daytime and nighttime on 32 meters from Poldhu to his yacht in Beirut. Marconi, in July 1924, entered into contracts with the British General Post Office (GPO) to install high speed shortwave telegraphy circuits from London to Australia, India, South Africa and Canada as the main element of the Imperial Wireless Chain. The UK-to-Canada shortwave "Beam Wireless Service" went into commercial operation on 25 October 1926. Beam Wireless Services from the UK to Australia, South Africa and India went into service in 1927.

Far more spectrum is available for long distance communication in the shortwave bands than in the long wave bands; and shortwave transmitters, receivers and antennas were orders of magnitude less expensive than the multi-hundred kilowatt transmitters and monstrous antennas needed for long wave.

Shortwave communications began to grow rapidly in the 1920s,[9] similar to the internet in the late 20th century. By 1928, more than half of long distance communications had moved from transoceanic cables and long wave wireless services to shortwave "skip" transmission and the overall volume of transoceanic shortwave communications had vastly increased. Shortwave also ended the need for multimillion-dollar investments in new transoceanic telegraph cables and massive long wave wireless stations, although some existing transoceanic telegraph cables and commercial long wave communications stations remained in use until the 1960s.

The cable companies began to lose large sums of money in 1927, and a serious financial crisis threatened the viability of cable companies that were vital to strategic British interests. The British government convened the Imperial Wireless and Cable Conference[10] in 1928 "to examine the situation that had arisen as a result of the competition of Beam Wireless with the Cable Services". It recommended and received Government approval for all overseas cable and wireless resources of the Empire to be merged into one system controlled by a newly formed company in 1929, Imperial and International Communications Ltd. The name of the company was changed to Cable and Wireless Ltd. in 1934.

2.50.4 See also

2.50.5 References

[1] Sony Corporation. (1998). *Wave Handbook*. p.14. OCLC 734041509.

[2] K.Rawer: *Wave Propagation in the Ionosphere*. Kluwer Acad.Publ., Dordrecht 1993. ISBN 0-7923-0775-5.

[3] Stormfax. Marconi Wireless on Cape Cod

[4] "1921 - Club Station 1BCG and the Transatlantic Tests". Radio Club of America. Retrieved 2009-09-05.

[5] "Radio Service Bulletin No. 72". Bureau of Navigation, Department of Commerce. 1923-04-02. pp. 9–13. Retrieved 2009-09-05.

[6] Archived November 30, 2009 at the Wayback Machine

[7] "Recommendations for Regulation of Radio: October 6-10, 1924". Earlyradiohistory.us. Retrieved 2012-08-31.

[8] http://www.twiar.org/aaarchives/WB008.txt

[9] "Full text of "Beyond the ionosphere : fifty years of satellite communication"". Archive.org. Retrieved 2012-08-31.

[10] Cable and Wireless Pl c History

2.50.6 Further reading

- Davies, Kenneth (1990). *Ionospheric Radio*. IEE Electromagnetic Waves Series #31. London, UK: Peter Peregrinus Ltd/The Institution of Electrical Engineers. ISBN 0-86341-186-X.

2.50.7 External links

- Navy - Propagation of Waves

- Radio wave propagation basics

- HFRadio Propagation forums

- Rare gamma-ray flare disturbed ionosphere

- Articles on sporadic E and 50 MHz Radio Propagation

- Radio propagation overview Details of many forms of radio propagation

2.51 SR International – Radio Sweden

Radio Sweden (Swedish: *Sveriges Radio International*) is Sweden's official international broadcasting station. Swedish Radio is a non-commercial and politically independent public service broadcasting company.

2.51.1   History

SR International is part of Sveriges Radio (SR), Sweden's non-commercial public-radio broadcasting organization. The service was founded in 1938, at the approach of World War II, as a way of keeping Swedes living abroad informed of happenings in Sweden and of Swedish opinion. Programming was at first in Swedish only, but in 1939 English- and German-language broadcasts were added.

After the war, further language services were added: in French, Portuguese, Russian, and Spanish. At the close of the Cold War, the services in French, Portuguese, and Spanish were gradually phased out and replaced by new services in Estonian and Latvian. The latter services were withdrawn once Estonia and Latvia had developed their own independent media and joined the European Union.

Until recently Radio Sweden also operated a service in Belarusian but this was closed in 2010.

In the 1990s Radio Sweden was merged with SR's Immigrant Languages Department to form the SR International channel. For a while, immigrant language services, such as those in Arabic and Kurdish, were additionally carried on shortwave.

2.51.2   Current programming in English

The Radio Sweden English Service seeks to provide a window on the diverse perspectives and issues in Sweden today.

Its programs and website offer a "smörgåsbord" of news and current affairs, science and technology, lifestyle, and culture.

2.51.3   Former programs

Sweden Calling DXers

One of the most popular programs on Radio Sweden was *Sweden Calling DXers*, founded in 1948 by Arne Skoog. He reasoned that shortwave listening or DXing was a very young hobby, and that by providing information in a weekly program for shortwave listeners about their hobby, Radio Sweden was teaching its own audience about how to listen better. While the first program was based solely on Arne's own listening, listeners were encouraged to write in with their own news, and soon virtually all of the program was based on listener's letters (an early example of interactvity).

The program was carried on Tuesdays in all of Radio Sweden's services except Swedish.

When Arne Skoog retired in connection with the program's 30th anniversary in 1978, the program was taken over by

George Wood, a member of the Radio Sweden English Service. After a number of years, as media changed, it began to cover less about shortwave, and more about satellite radio and television. Later coverage was extended to the Internet, the focus was shifted to concentrating on Swedish media, and the name of the English version of the program was changed to "MediaScan". "MediaScan" was the first English-language radio program in Europe (and the second overall in Europe) to post audio on the Internet (on the ftp sites ftp.funet.fi and ftp.sunet.se, as well as via Internet Multicasting).

In 2001 the program was discontinued, but remains in a somewhat sporadic form on the Radio Sweden website.

The Saturday Show

Another popular Radio Sweden program was the *The Saturday Show*, with Roger Wallis and Kim Loughran, which ran from 1967 to 1981. The program was launched to showcase Swedish rock and pop music in a world dominated by American and British rock. Using Radio Sweden's relatively high-powered medium wave transmitter on 1179 kHz, the entire program was 90 minutes in length, and featured many satirical sketches, often political and sometimes controversial. A 30-minute segment of the entire broadcast was the Radio Sweden shortwave program on Saturdays.

2.51.4   End of shortwave broadcasting

On October 20, 2010, Radio Sweden ceased broadcasts on shortwave and medium wave, as Swedish Radio's management decided that the Internet had matured enough to support international broadcasts. At the same time the English Service was extended to national broadcasts on FM, but programming on weekends was discontinued. Services in English also continue on satellite, both using Swedish Television's satellite and the World Radio Network.

2.51.5   See also

- External program hours - Comparison with some other external radio broadcasters

- Sveriges Radio, the Swedish publicly funded radio broadcaster

2.51.6 References

2.51.7 External links

- SR International - Radio Sweden website (Arabic) (English) (German) (Kurdish) (Persian) (Romani) (Russian) (Somali)

Coordinates: 59°20′5″N 18°6′5″E / 59.33472°N 18.10139°E

2.52 Stand Up to Cancer

SU2C logo

Stand Up To Cancer (S↑2C) is a charitable program of the Entertainment Industry Foundation (EIF). SU2C aims to raise significant funds for translational cancer research through online and televised efforts. Central to the program is a telethon that was televised by three major broadcast networks (ABC, NBC, CBS) in over 170 countries on September 5, 2008. SU2C raised over $100 million after that evening's broadcast.[1]

2.52.1 History

Stand Up To Cancer started in autumn 2007, by a group of women who had been affected by cancer. The group believes by merging the recourse of the media and entertainment industries into a single operation they would be able to fight against this disease, in a more profound way.[2]

Stand Up To Cancer was formally launched on May 27, 2008. Current members of the SU2C Council of Founders and Advisors (CFA) include Katie Couric, Sherry Lansing, Kathleen Lobb, Lisa Paulsen, Rusty Robertson, Sue Schwartz, Pamela Oas Williams, and Ellen Ziffren.[3] All current members of the CFA were co-producers of the 2012 televised special. The late co-founder Laura Ziskin executive produced both the September 5, 2008 and September

10, 2010 broadcasts. SU2C was formally launched on May 27, 2008. Sung Poblete, Ph.D., R.N., has served as SU2C's president and CEO since 2011. In the United Kingdom, Channel 4, along with Cancer Research UK, launched its own version of *Stand Up To Cancer* in October 2012 and again in 2014.[4]

2.52.2 Initiative

The Stand Up To Cancer initiative aims to raise awareness and bring about an understanding that everyone is connected by cancer. The stat used most often by SU2C is from the American Cancer Society: one out of every two men and one out of every three women will be diagnosed with cancer in their lifetime,[5] meaning everyone is affected in some way, or will be.

The campaign has featured televised moments during World Series baseball games where fans literally "stand up to cancer" by rising and holding signs inscribed with the names of friends and loved ones who have struggled with the disease.[6] As of October 2015, these signs have not treated or cured a single case of cancer.

- Fans "stand up to cancer" at a Phillies game on August 22, 2014

- Major League Baseball umpires participate in a tribute to Stand Up to Cancer at a Philadelphia Phillies' game on August 22, 2014

2.52.3 USA

Four telethons have been broadcast in USA and are made available to more than 190 countries. To date, more than $261 million has been pledged to support SU2C's innovative cancer research programs. The first Stand Up to Cancer event was held on September 5, 2008, raising over $100 million.[1]

The second was held on 10 September 2010.[7] The third was held on 7 September 2012, and the fourth was held on 5 September 2014,[8] presented by Gwyneth Paltrow and Katie Couric from Los Angeles. The averaged combined ratings for the 2014 telethon were 1.8 from 2.1 in 2012.[9] The 2014 telethon was broadcast, as before, on ABC, CBS, and NBC. This time, however, Fox, Ion Television, ABC Family, E!, VH1, Bravo, Audience Network, TNT, Oxygen, LMN, The Cooking Channel, ESPNews, MLB Network, FX Movie Channel, Pivot, National Geographic Channel, HBO, HBO Latino, Showtime, Starz, Encore, Epix, and Logo TV joined in the broadcast.

In Canada, the 2014 telethon aired September 5, 2014 in simultaneous substitution (in many areas) with the American telethon, simultaneously on English language networks CBC Television, City, CTV and Global, as well as CHCH-DT in Hamilton, Ontario, CHEK-DT Victoria, British Columbia, and specialty channels TLN, Fight Network, Gusto TV, AMI-TV and the Hollywood Suite channels. The Canadian broadcast of the telethon benefits EIF Canada, the Canadian arm of the Entertainment Industry Foundation, with all donations benefiting cancer research in Canada.[10]

2.52.4   UK

Main article: Stand Up to Cancer UK

On 19 October 2012, the Channel 4 network in the United Kingdom aired their first telethon for Stand Up To Cancer hosted by Davina McCall and Alan Carr along with Dr. Christian Jessen. A second telethon was broadcast on 17 October 2014, again hosted by McCall and Carr along with Dr. Jessen.

Stand Up To Cancer's fundraising activities in the UK include the London 3 Peaks Challenge, which involves participants running up the stairs of Heron Tower, 30 St Mary Axe and 200 Aldersgate before abseiling down the outside of 200 Aldersgate.[11]

2.52.5   References

[1] Keveney, Bill (2008-09-08). "Stars 'Stand Up to Cancer,' raise $100M - USATODAY.com". Usatoday30.usatoday.com. Retrieved 2015-07-26.

[2] "Stand Up To Cancer — FAQ". Standup2cancer.org. 2008-05-27. Retrieved 2015-07-26.

[3] "Stand Up To Cancer — SU2C Leadership". Standup2cancer.org. Retrieved 2015-07-26.

[4] "Stand Up To Cancer". *Channel 4*. Retrieved 26 September 2012.

[5] ACS Cancer Facts and Figures 2008. Retrieved on 27 May 2008.

[6] "MLB stands up to cancer during Game 4 of Series". Major League Baseball. Retrieved 17 October 2014.

[7] "Friday, Sept. 10 TV highlights: 'Stand Up for Cancer' unites networks". Washingtonpost.com. 2010-09-10. Retrieved 2015-07-26.

[8] "Stand Up To Cancer — Dine Out with MasterCard and Support a Priceless Cause: Fighting Cancer". Standup2cancer.org. 2015-07-13. Retrieved 2015-07-26.

[9] Harp, Justin (2014-09-06). "Friday ratings: Stand Up to Cancer down from 2012 - US TV News". Digital Spy. Retrieved 2015-07-26.

[10] "SU2C Canada - Press Releases". Standup2cancer.ca. 2014-07-09. Retrieved 2015-07-26.

[11] "Have you got what it takes to tackle London's Three Peaks Challenge (aka running up three skyscrapers)? – Now. Here. This. – Time Out London". Now-here-this.timeout.com. Retrieved 2015-07-26.

2.52.6   External links

- Stand Up To Cancer website

- Stand Up To Cancer Canada website

2.53   Stephen Williams (Radio Luxembourg)

Not to be confused with Stephen Williams (1900-1957), a music and drama critic who was also a frequent broadcaster..

For other people of the same name, see Stephen Williams (disambiguation).

Stephen Williams (31 March 1908 - 23 November 1994) was a British radio announcer, presenter and producer, and a pioneer of commercial radio for the UK.

Born in London and educated at Trinity College, Cambridge, as a young boy he was already, in his words, a "wireless fanatic": he said, "I was able to listen proudly to the debut of the BBC on 14 November 1922, and from the moment I heard the announcer say "This is 2LO calling, 2LO, the London station of the British Broadcasting Company", I was seized with an ambition to have a job like his."

During his university vacation in 1928 he got a job as announcer on a "broadcasting yacht" sponsored by the Daily Mail newspaper group. This vessel went round the coast of Britain, transmitting music on records and advertisements for the Daily Mail, from just outside territorial waters, an early precursor of the 1960s "pirate radio" ships..

The broadcasting yacht had been the idea of the paper's Circulation and Publicity Director, Valentine Smith, who soon transferred to the Sunday Referee, where he gave Williams a job with the idea of involving the paper in commercial broadcasting. They were soon in contact with Captain Leonard F. Plugge, who was starting the International Broadcasting Company (IBC).[1]

At the beginning of 1932, Williams was sent to France to join Max Staniforth at the IBC's new Radio Normandy service, broadcasting to the south of England.[2]

Recordings of two commercials read by Stephen Williams on Radio Normandy in 1932 are said to survive.[3]

Stephen Williams moved on to a rival company to IBC, Radio Publicity, which started broadcasting to the UK from Radio Paris, then a much more powerful station than Radio Normandy, which rapidly gained a large British audience. But French listeners complained that their most popular station was now too dominated by English programmes, so Radio Publicity turned to the newly created Radio Luxembourg, where it was able to gain the sole concession for English programmes.[4]

In December 1933 Williams thus became the first English radio presenter in Luxembourg as well as serving as manager of the station.

Shortly before his death in 1994, Williams gave an interview to Roger Bickerton about his work at Radio Luxembourg which was published in *The Historic Record and AV Collector*, Issues 39,40 and 41, April, June & October 1996.[5]

Radio Luxembourg closed down on the outbreak of World War II in September 1939. Stephen Williams resumed his duties there when the station restarted in 1946. He left Luxembourg in 1948, and worked until 1975 as a freelance broadcaster.

When, on 1 January 1992, Radio Luxembourg's English service closed down as a terrestrial radio station, the last words heard, "Good luck, good listening ... and goodbye" were spoken by 83-year-old Stephen Williams, who had been the first person ever to say on the air "This is Radio Luxembourg" over 58 years earlier.[6]

Williams was awarded the Order of Merit by the Grand Duke of Luxembourg in 1992.

He collected a large archive of material related to Radio Luxembourg, much of which his widow has donated to the Centre for Luxembourg Studies at Sheffield University.[7]

2.53.1 References

[1] Stephen Williams, "Pioneering commercial radio the D-I-Y way", *European Journal of Marketing*, October 1987. ISSN: 0309-0566 (journal title and date wrongly cited on web page).

[2] Roy Plomley, *Days Seemed Longer: Early Years of a Broadcaster*, London, 1980, p.123. ISBN 0-413-39730-0

[3] Sean Street, "Radio For Sale: Sponsored Programming in British Radio during the 1930s", in *Sound Journal*, Bournemouth University, 1999.

[4] Williams, *op. cit.*

[5] Sean Street, "Recording Technologies and Strategies for British Radio Transmission Before the 2nd World War", in *Sound Journal*, Bournemouth University, 2002.

[6] "Last hours on 208m", aircheck available at playlist No 40 on *The Station of the Stars: Audio Archive*.

[7] Department of Germanic Studies, University of Sheffield

2.54 The Filipino Channel

The Filipino Channel (or **TFC**) is a global brand of premium cable/satellite/iptv/mobile/digital/video on demand television network based in Redwood City, California. It is owned by the Filipino media conglomerate ABS-CBN Corporation and is targeted towards Filipino expatriates and their families.

It broadcasts a 24-hour line-up of shows imported from ABS-CBN, a national television network in the Philippines as well as some originally produced programming. The Filipino Channel is also backed by several other international direct-to-home services like TFC Direct of DirecTV, TFC IPTV of Sprint, Filipino On Demand of Comcast, and online on TFC.tv. Today, The Filipino Channel has over two million subscribers worldwide accounting to over 8 million viewership.[1] As of 2015, TFC has over 3 million subscribers worldwide most of which are in United States, Middle East, Europe, Australia, Canada, and Japan.[2]

2.54.1 History

The old logo of The Filipino Channel (2000s-2011).

On September 24, 1994, then ABS-CBN Broadcasting Corporation (now ABS-CBN Corporation) through its newly established subsidiary ABS-CBN International signed a historic deal with the PanAmSat to bring the first trans-Pacific Asian programming service to some two million Filipino immigrants in the United States using the then-newly launched PAS 2 satellite.[3]

The first headquarters of TFC was built in a garage in Daly City, California with only eight employees doing all the tasks from managing the phones, the computers, and the likes.[4] By 1995, TFC has grown to 25,000 subscribers in the United States.

In 1998, TFC Direct! was launched, an independently operated direct-to-home television service that incorporates the TV channels Sarimanok News Network (now ABS-CBN News Channel), Pinoy Blockbuster Channel (now Cinema One), Pinoy Central (Renamed as Kapamilya Channel, then replaced by Bro, and now replaced by ABS-CBN Sports + Action), and the radio channels DZMM Radyo Patrol 630 and WRR 101.9 For Life! (now MOR 101.9).[5]

By 2004, TFC had grown to 250,000 subscribers in the United States. This growth led to the expansion of TFC to other territories in the world.[6]

In 2005, ABS-CBN International signed an affiliation agreement with DirecTV, one of the leading DTH providers in the United States. Under the deal, DirecTV has the exclusive right to distribute the TFC package on its DTH platform. In return, DirecTV will pay license fees to ABS-CBN and to ABS-CBN International.[7] Later that year, the now defunct internet television service TFC Now! was launched. TFC Now! was later replaced by TFC.tv video streaming website. In this year, ABS-CBN International acquired San Francisco International Gateway from Loral Space & Communications. SFIG is a telecommunications port company based in Richmond, California. SFIG provides satellite communications services through its 2.5 acre (1 hectare) facility consisting of 19 satellite dish antennas and 9 modular equipment buildings. Also in this year, ABS-CBN International opened its state-of-the-art studio and office in Redwood City, California.[8] ABS-CBN International received Federal Communications Commission licensing approval in April 2006. In 2006, SFIG successfully handled the pay per view distribution to In Demand and DirecTV for the Manny Pacquiao vs. Oscar Larios super featherweight championship title fight. SFIG's customers include Discovery Communications, CBS, ESPN, Playboy among others. SFIG is a member of the World Teleport Association.[9]

In 2007, ABS-CBN International launched myx (now Myx TV), the first and only television channel in the United States that is targeted to the Asian-American youth audience.[10] As of 2011, TFC had over 2.47 million subscribers worldwide.[1] As of 2015, The Filipino Channel has over three million subscribers worldwide most of which are in United States, Middle East, Australia, Japan, Europe, and Canada.

2.54.2 Programming

Main article: List of programs broadcast by The Filipino Channel

The programming of The Filipino Channel consists mostly of imports from a national television network ABS-CBN. The line-up include delayed telecast of shows from the Philippines as well as previously aired shows, films, and live sports events from the Philippines. Original programming is also produced by ABS-CBN International – mostly news and talk shows.

2.54.3 Studio TFC

- Jing Reyes, substituting for Gel Santos-Relos in *Balitang America*.

- *Adobo Nation*, an original program taping from Studio TFC.

- A show for Myx TV in Studio TFC.

2.54.4 References

[1] Valisno, Jeffrey (14 June 2012). "Pinoy TV goes international". *BusinessWorld*. Retrieved 18 June 2012.

[2] P. Valdueza, Rolando (April 24, 2015). 2014 Annual Report (17-A) (Report). *Philippine Stock Exchange*. Retrieved April 27, 2015.

[3] Jessel, Harry; Taishoff, Lawrence (2010). "Television and Radio". *Encyclopædia Britannica Ultimate Reference Suite*. Encyclopædia Britannica, Inc.

[4] "A Journey of Triumph of the Filipino Spirit" (PDF). *ABS-CBN International*. 2009. Retrieved 18 June 2012.

[5] "The Filipino Channel: Bringing overseas Pinoys closer to home". *The Philippine Star*. 24 March 2003. Retrieved 18 June 2012.

[6] Katigbak, Antonio (11 April 2004). "ABS-CBN's The Filipino Channel marks a decade of steady growth". *The Philippine Star*. Retrieved 18 June 2012.

[7] Villanueva, Paul Michael (31 December 2011), *ABS-CBN 17-A 2011*, *Scribd.com*, p. 8, retrieved 9 August 2012

[8] "Customer Showcase: ABS-CBN". Advanced Systems Group.

[9] "ABS-CBN International". *World Teleport Association*. 14 December 2011. Retrieved 18 June 2012.

[10] "StudioTFC". *ABS-CBN International*. Retrieved 21 June 2012.

2.54.5 External links

- www.abs-cbnglobal.com

2.55 The Flattery Show

The Flattery Show was the first English-language radio chat show aired in the south west of France and a flagship Sunday evening programme broadcast on Radio Coteaux in the Gascony region. The show was presented by Irish radio presenter John Slattery with American co-host Patricia McKinnes. It ran for twenty-one weeks and proved to be very popular with expatriates living in France.

2.55.1 The show

Slattery and McKinnes regularly began the show at 5pm and finished at 7pm with a handover to Patrick Martinez. *The Flattery Show* was a music-based programme that included light-hearted banter and whimsical jokes by the hosts. Listeners' views and comments were discussed and there were rotating segments such as "Song of The Week", "Ask a Frenchie" and "Life in the South-West". To keep listeners on their toes, random topics were picked from a big "Bag O' Topics".

2.55.2 Invited guests

The show ran interviews with regular and/or rotating guests, dealing with subjects of both local interest and of general interest to the expatriate community. The first people to be interviewed on the show were the English sisters Marcha King and Sian Cash, former proprietors of the landmark café "Bar Memory" in the Hautes-Pyrénées town of Castelnau-Magnoac. They recounted humorous stories of their years behind the bar at the French café. The interview proved enormously popular with English-speaking listeners in France and it served to create a pattern of jesting and buffoonery that would become an underlying trait of later interviews.

2.55.3 Regular contributors

The show relied heavily on e-mailed material sent in by listeners that was then read out on air. Most of the regular contributors were local expatriates living in the Gascony region.

2.55.4 The end

The show ran for twenty-one weeks in the early months of 2011. When it ceased, Radio Coteaux's Sunday evening slot was filled by a new show called *The Gascony Show*, also hosted by John Slattery.

2.55.5 External links

- *The Flattery Show* home page

- *The Gascony Show* home page

- Radio Coteaux

- «So british» sur radio Coteaux *La Dêpeche*, 19 March 2011

2.56 Tropospheric propagation

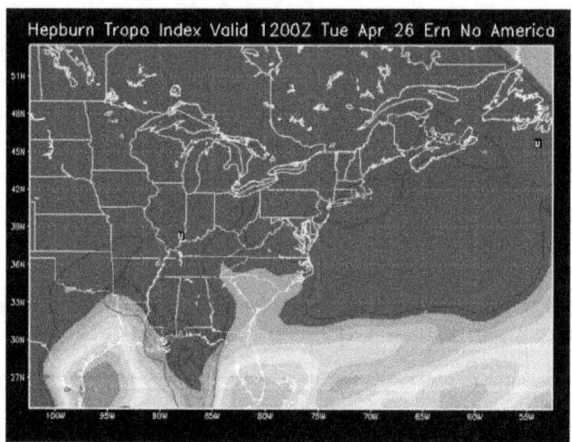

Meteorologist William Hepburn's forecast maps often provide early indications of potential tropospheric DX (long distance) openings.

Tropospheric propagation describes electromagnetic propagation in relation to the troposphere. The service area from a VHF or UHF radio transmitter extends to just beyond the optical horizon, at which point signals start to rapidly reduce in strength. Viewers living in such a "deep fringe" reception area will notice that during certain conditions, weak signals normally masked by noise increase in signal strength to allow quality reception. Such conditions are related to the current state of the troposphere.

Tropospheric propagated signals travel in the part of the atmosphere adjacent to the surface and extending to some 25,000 feet (7,620 m). Such signals are thus directly affected by weather conditions extending over some hundreds of miles. During very settled, warm anticyclonic weather (i.e., high pressure), usually weak signals from distant transmitters improve in strength. Another symptom during such conditions may be interference to the local transmitter resulting in co-channel interference, usually horizontal lines or an extra floating picture with analog broadcasts and break-up with digital broadcasts. A settled high-pressure system gives the characteristic conditions for enhanced tropospheric propagation, in particular favouring signals which travel along the prevailing isobar pattern (rather than across it). Such weather conditions can occur at any time, but generally the summer and autumn months are the best periods. In certain favourable locations, enhanced tropospheric propagation may enable reception of ultra high frequency (UHF) TV signals up to 1,000 miles (1,600 km) or more.

The observable characteristics of such high-pressure systems are usually clear, cloudless days with little or no wind. At sunset the upper air cools, as does the surface temperature, but at different rates. This produces a boundary or temperature gradient, which allows an inversion level to form – a similar effect occurs at sunrise. The inversion is capable of allowing very high frequency (VHF) and UHF signal propagation well beyond the normal radio horizon distance.

The inversion effectively reduces sky wave radiation from a transmitter – normally VHF and UHF signals travel on into space when they reach the horizon, the refractive index of the ionosphere preventing signal return. With temperature inversion, however, the signal is to a large extent refracted over the horizon rather than continuing along a direct path into outer space.

Fog also produces good tropospheric results, again due to inversion effects. Fog occurs during high-pressure weather, and if such conditions result in a large belt of fog with clear sky above, there will be heating of the upper fog level and thus an inversion. This situation often arises towards night fall, continues overnight and clears with the sunrise over a period of around 4 – 5 hours.

2.56.1   Tropospheric ducting

Tropospheric ducting is a type of radio propagation that tends to happen during periods of stable, anticyclonic weather. In this propagation method, when the signal encounters a rise in temperature in the atmosphere instead of the normal decrease (known as a temperature inversion), the higher refractive index of the atmosphere there will cause the signal to be bent. Tropospheric ducting affects all frequencies, and signals enhanced this way tend to travel up to 800 miles (1,300 km) (though some people have received "tropo" beyond 1,000 miles / 1,600 km), while with tropospheric-bending, stable signals with good signal strength from 500+ miles (800+ km) away are not common

This example of 1,340-mile (2,160 km) tropospheric ducting reception shows Auckland, New Zealand 175.25 MHz ch4 TV received by Robert Copeman, Sydney, Australia.

when the refractive index of the atmosphere is fairly high.

Tropospheric ducting of radio and television signals is relatively common during the summer and autumn months, and is the result of change in the refractive index of the atmosphere at the boundary between air masses of different temperatures and humidities. Using an analogy, it can be said that the denser air at ground level slows the wave front a little more than does the rare upper air, imparting a downward curve to the wave travel.

Ducting can occur on a very large scale when a large mass of cold air is overrun by warm air. This is termed a temperature inversion, and the boundary between the two air masses may extend for 1,000 miles (1,600 km) or more along a stationary weather front.

Temperature inversions occur most frequently along coastal areas bordering large bodies of water. This is the result of natural onshore movement of cool, humid air shortly after sunset when the ground air cools more quickly than the upper air layers. The same action may take place in the morning when the rising sun warms the upper layers.

Even though tropospheric ducting has been occasionally observed down to 40 MHz, the signal levels are usually very weak. Higher frequencies above 90 MHz are generally more favourably propagated.

High mountainous areas and undulating terrain between the transmitter and receiver can form an effective barrier to tropospheric signals. Ideally, a relatively flat land path between the transmitter and receiver is ideal for tropospheric ducting. Sea paths also tend to produce superior results.

In certain parts of the world, notably the Mediterranean Sea and the Persian Gulf, tropospheric ducting conditions can become established for many months of the year to the extent that viewers regularly receive quality reception of signals over distances of 1,000 miles (1,600 km). Such conditions are normally optimum during very hot settled summer weather.

Tropospheric ducting over water, particularly between California and Hawaii, Brazil and Africa, Australia and New Zealand, Australia and Indonesia, Strait of Florida, and Bahrain and Pakistan, has produced VHF/UHF reception ranging from 1000 to 3,000 miles (1,600 – 4,800 km). A US listening post was built in Ethiopia to exploit a common ducting of signals from southern Russia.

Tropospheric signals exhibit a slow cycle of fading and will occasionally produce signals sufficiently strong for noise-free stereo, reception of Radio Data System(RDS) data, and solid locks of HD Radio streams on FM or noise-free, color TV pictures.

Virtually all long-distance reception of digital television occurs by tropospheric ducting (due to most, but not all, DTV stations broadcasting in the UHF band).

2.56.2   Notable tropospheric DX receptions

- On October 18, 1975, Rijn Muntjewerff, the Netherlands, received UHF channel E34 Pajala, Sweden, at a distance of 1,150 miles (1,851 km).[1]

- On June 13, 1989, Shel Remington, Keaau, Hawaii, received several 88-108 MHz FM signals from Tijuana, Mexico, at a distance of 2,536 miles (4,081 km).[2]

- Throughout the 1990s, Fernando Garcia, located at what could be considered an ideal tropospheric DX location near Monterrey, Mexico, received numerous 1,000+ mile (1,600+ km) stations via tropospheric propagation, both over the Gulf of Mexico and past land. Among his receptions are WGNT-27 from Portsmouth, Virginia, at a distance of 1,608 miles (2,588 km) and low-power (LPTV) station W38BB from Raleigh, North Carolina, at a distance of 1,460 miles (2,350 km)[3]

- On June 24, 2001, a Romanian engineer Ioan Albesteanu received Russian ORT television on channel 31 from the Babadag hills in the Russian city Назрань, Nazran. The reception was made at a distance of 1,290 kilometres (802 mi).[4]

- On May 11, 2003, Jeff Kruszka, living in south Louisiana, received a few UHF DTV signals from 800+ miles. The longest of these was WNCN-DT, channel 55, Goldsboro, North Carolina, at a distance

of 835 miles (1,344 km) (at the time, the record for UHF DTV).[5]

- On the late evening of June 19, 2007 and into the early morning hours of June 20, 2007, three DXers in eastern Massachusetts, Jeff Lehmann, Keith McGinnis, and Roy Barstow, received FM signals from southern Florida via tropo. All three logged WEAT 104.3 West Palm Beach, Florida, and WRMF 97.9 Palm Beach, Florida, at distances of around 1,200 miles (1,931 km), and Barstow logged WHDR 93.1 Miami, Florida, at a distance of 1,210 miles (1,947 km).[6]

- On December 3, 2007 Bulgarian dxer "FMDXBG" received Radio Militsaysk, 105.5 MHz via tropo near Gurgulica chalet in eastern Rila, at a distance of 1,312 kilometres (815 mi).[7]

- On December 17, 2007 Polish dxer Maciej Lugowski received 93,7 BBC Radio Scotland from Keelylang Hill transmitter in Gora Kalwaria, Poland. The distance from his site to Orkney Islands is 1,745 km (1,084 mi). BBC Scotland reception lasted for next two days, as extreme tropo ducting was built over Baltic and Northern Sea.[8]

- On November 3, 2008 Swedish Radio Amateur Kjell Jarl SM7GVF contacted Russian Radio Amateur RA6HHT at a distance of 2,315 km (1,438 mi) on 144Mhz.[9]

- On April 23, 2009, a San Antonio-area DXer received WFTS-TV 28's digital signal from Tampa, Florida, at a distance of 995 miles (1,601 km).[10]

- On the late evening of August 24 into the afternoon of August 25, 2009, a DX'er in Burnt River, Ontario, Canada, received several FM radio stations via tropo from Arkansas, Illinois, Iowa, Kansas, Michigan, Missouri, Ohio, Oklahoma, Pennsylvania, and Wisconsin.[11]

- On September 11, 2010, Daniel Albu (Bucharest, Romania) received Radio TRT-FM from Amasya, Turkey at a distance of 922 km.

- On August 9, 2012, Greek dxer Peter "p15able" (Pyrgos, Greece) received Alger Chaîne 2 on 97.5 MHz from Doukhane, Algeria at a distance of 1,228 km.

- On October 7, 2012, Aleksandr (Poltava, Ukraine) received 90.8 MHz - BNR Hristo Botev from Bulgaria (Tsarevo, Burgas Province) at a distance of 980 km during the tropospheric propagation.[12]

2.56.3 See also

- Atmospheric duct

- MW DX

- Skywave

- Radio propagation

- Tropospheric scatter

- Velocity of propagation

- Thermal fade

- Clear-channel station

- Federal Standard 1037C

- Looming and similar refraction phenomena

2.56.4 References

[1] "Rijn Muntjewerff's 1961-2005 TV DX". *Todd Emslie's TV DX Page*. Retrieved August 29, 2005.

[2] http://home.twcny.rr.com/nordquistsyr/FMRECDIS.htm *(dead link)*

[3] http://home.twcny.rr.com/nordquistsyr/tvrecdis.htm *(dead link)*

[4] http://img24.imageshack.us/img24/944/blankrusia1din27august2.png *(dead link)*

[5] Jeff Kruszka's Record-Breaking DTV Tropo

[6] TROPO MA to FL!!!! 97.9 WRMF, *WTFDA Forums*, June 19, 2007

[7] FMDXBG Youtube Page

[8] http://www.fmdx.pl/thebest.htm *(dead link)*

[9]

[10] WA5IYX DTV Screen Captures

[11] Aug 24-25 2009 Es (er, Tropo), *WTFDA Forums*, August 28, 2009

[12] "[Tropo] 90.8 MHz - BNR Hristo Botev - Tsarevo, Burgas Province, Bulgaria - (980 km)".

- "DXing FAQ". *Worldwide TV-FM DX Association*. Retrieved April 25, 2005.

- "William Hepburn's VHF / UHF Tropospheric Ducting Forecast". *William Hepburn's TV & Radio DX Information Centre*. Retrieved June 12, 2006.

- Tropospheric Ducting YouTube Channel of FMDXUA -

- "Bellevue, NE DX Photos". *Matthew C. Sittel's DX Page*. Retrieved April 26, 2005. *(dead link)*

- "Jeff Kadet, K1MOD's TV DX Photos". *oldtvguides.com*

2.57 Tsunami Aid

Tsunami Aid: A Concert of Hope was a worldwide benefit held for the tsunami victims of the 2004 Indian Ocean earthquake. It was broadcast on NBC and its affiliated networks of USA Network, Bravo, PAX, MSNBC, CNBC, Sci-Fi, Trio, Telemundo and other NBC Universal stations and was heard on any Clear Channel radio station. The benefit was led by the actor George Clooney on January 15, 2005, and was similar to *America: A Tribute to Heroes* (set up after the September 11th, 2001 attacks). Digital Media innovator Jay Samit enabled viewers to purchase digital downloads of the performances as a new way to raise money for the cause; including live recordings by Elton John, Madonna, Sheryl Crow, Eric Clapton and Roger Waters. Taking a cue from Bob Geldof (the man who had organized the Live Aid concerts for African famine relief), it consisted of famous Hollywood entertainers and former American presidents George H.W. Bush and Bill Clinton. It was two hours long with stories and entertainment from a huge array of Hollywood popstars notables that include Brad Pitt, Donald Trump, and much more. It was estimated to raise at least five million dollars by the end of the broadcast.

2.57.1 External links

- *Tsunami Aid* at the Internet Movie Database

2.58 Utility station

The term **utility station** is used to describe fixed radio broadcasters disseminating signals that are not intended for reception by the general public (but such members are not actively prohibited from receiving). Utility stations, as the name suggests, do broadcast signals that have an immediate practical use, by means of analog or usually digital modes; most often utility transmissions are of a "point-to-point" nature, intended for a specific receiving station. Utility stations are most prevalent on shortwave frequencies, though they are not restricted to the shortwave frequencies.

2.58.1 Examples of utility station and modes

One common use of utility stations is disseminating weather information. Weather information is often broadcast using RTTY and sending synoptic codes, or weather charts are sent using radiofax, which are used by mariners and others. Airports make voice weather broadcasts on HF, known as VOLMET. Some examples include New York Radio, which broadcasts weather information for locations in the eastern United States, or Shanwick Radio, which does the same for Europe.

HF frequencies are still often used for trans-oceanic air traffic control. News agencies previously used RTTY for news stories, and, less commonly, radiofax for the images, although this is no longer done. Satellite communications and the Internet have replaced HF for this application.

Many maritime radio services are often known as utility stations, including as ship-to-shore and vice versa telephony and error-controlled radioteletype such as SITOR.

Military use of shortwave is also common, but nearly all transmissions are encrypted, with voice encrypted using modes such as ANDVT. Data transmission may make use of encrypted RTTY, use Link-11 for radar tracking data, or use of Automatic link establishment modes to set up communication links automatically.

Some utility stations are on other frequency bands, including NOAA Weather Radio, traveler information stations, and the like; other utility-type signals are piggybacked on FM broadcast subcarriers.

2.58.2 See also

- Coast radio station

- Shortwave listening

2.58.3 External links

- Utility World with various sound samples of utility modes

- KB9UKD Digital Modes with mostly non-HF utility modes samples

- Shortwave Utility Radio with information and links to utility sites, information, and products

- Police Scanner Information frequencies, audio feeds and links related to police scanners

- Global Frequency Database

2.59 Voice of America Bethany Relay Station

The National Voice of America Museum of Broadcasting is the main building and campus of the original VOA, Bethany Relay Station.

The Voice of America's Bethany Relay Station was located in Butler County, Ohio's Union Township about 25 miles (40 km) north of Cincinnati, adjacent to the transmitter site of WLW. Starting in 1944 during World War II it transmitted American radio programming abroad on shortwave frequencies, using 200,000-watt transmitters built by Crosley engineers under the direction of R.J. Langley. The site was developed to provide 'fallback' transmission facilities inland and away from the East Coast, where transmitters were located in Massachusetts, on Long Island in New York, and in New Jersey, all close to the ocean, subject to attack from German submarines or other invading forces.

Programming originated from studios in New York until 1954, when VOA headquarters moved to Washington.

The station operated until 1994. The facility took its name from the Liberty Township community of Bethany, which was about two miles north of the facility.

In 1943, the United States government bought nearly all of Section 12 of Township 3, Range 2 of the Symmes Purchase, the northeasternmost section of Union Township. From Hazel Beckley, 170 acres (688,000 m²) were purchased; from Philip Condon, 143 acres (579,000 m²); from Lola Gray Coy, 100 acres (405,000 m²); from John Miller, 69 acres (279,000 m²); and from Suzie Steinman, 142 acres (575,000 m²). The site was chosen for its elevation and its shallow bedrock and is today bounded by Tylersville Road on the south, Cox Road to the west, Hamilton-Mason Road to the north, and Butler-Warren County Line Road.

The transmitters were built by Powel Crosley Jr.'s Crosley Broadcasting Corporation about one mile west of the company's tower for WLW-AM in Mason. The Office of War Information began broadcasting in July 1944 and Adolf Hitler is said to have denounced the "Cincinnati liars". Following the war with the OWI abolished, the facility was taken over by the State Department in 1945. It became part of the newly created United States Information Agency in 1953. The Crosley Broadcasting Corporation operated the facility for the government until November 1963, when the Voice of America assumed direct control.

At its peak the facility had three transmitters broadcasting with 250 kW, three broadcasting with 175 kW, and two transmitting with 50 kW.

The facility was closed on November 14, 1994; because of changing technologies, the transmissions shifted to satellites. The towers were brought down from December 1997 to February 1998. Most of the land was turned over to the county and township for use as a park. Part in the southwest corner was sold to developers who have erected a shopping center called the Voice of America Centre. The Miami University Voice of America Learning Center opened on the site in January 2009.

Today that building is being transformed into an historical center that will not only explain the importance of what happened there in the past, but how technology, honesty and the creative spirit, the guiding principles of The Voice of America, are still relevant today in spreading truth and providing encouragement globally to those seeking information without political bias . The VOA mantra, "Tell the truth and let the world decide" transcends the shortwave broadcasts of the 1940s and 50s and continues to ring true in the digital era of iPads, iPhones and all things digital. The National Voice of America Museum of Broadcasting will transform the former Bethany Station, a 30,000 sq. ft. building originally designed as one of the world's most powerful radio transmission facilities into an educational resource for the region and the nation. Using a blueprint developed by the acclaimed Jack Rouse and Associates, the facility will use state of the art displays and interactive experiences to relate the story of the Voice of America. Incorporating other re-lated collections from Media Heritage and the Gray History of Wireless Museum, the new facility will have content of interest to a wide demographic segment and age groups.

The development will be phased over several years as support is generated and construction and program design work is completed. Phase One calls for the construction of a convocation center at the museum designed as a place to host meetings, lectures, and traveling displays. The 2400 sq. ft. multi-purpose room will have a new ADA compliant entrance and modifications will be made to the building allowing access to the exhibit and meeting areas barrier-free. Two modern rest rooms will also be added.

Phase Two will modify the main transmitter room which once housed the six high power short wave transmitters. This area will be the main exhibition concourse. The space will be restored to the ambience of the 1940s, complete with the original second level observation platform located off the main lobby. The main exhibit space will feature and 8,000 + open area and another 5,000 sq. ft. of smaller display rooms. Archival and support offices will occupy the second floor of the main building. There will still be some 3000 sq. ft. of undeveloped space. Funding is currently being sought for Phase One of the plan.

2.59.1 References

[1] Staff (2008-04-15). "National Register Information System". *National Register of Historic Places*. National Park Service.

2.59.2 Further reading

- Jim Blount. *The 1900's: 100 Years In the History of Butler County, Ohio*. Hamilton, Ohio: Past Present Press, 2000.

- Virginia I. Shewalter. *A History of Union Township, Butler County, Ohio*. [West Chester, Ohio?]: The Author, 1979.

- Stern, David & Banks, Michael, *CROSLEY: Two Brothers and a Business Empire that Transformed the Nation* Cincinnati, Ohio: Clerisy Press, 2006.

2.59.3 External links

- AmateurLogic Special Presentation on the VOA Museum

- Bob Heil from Ham Nation takes a tour of the VOA Museum

2.60 Willis Conover

Willis Clark Conover, Jr. (December 18, 1920 – May 17, 1996) was a jazz producer and broadcaster on the Voice of America for over forty years. He produced jazz concerts at the White House, the Newport Jazz Festival, and for movies and television. By arranging concerts where people of all races were welcome, he is credited with desegregating Washington D.C. nightclubs.[2] Conover is credited with keeping interest in jazz alive in the countries of Eastern Europe through his nightly broadcasts during the Cold War.[3]

2.60.1 Youth

As a young man Conover was interested in science fiction, and published a science fiction fanzine, *Science Fantasy Correspondent*. This brought him into contact with horror writer H. P. Lovecraft. The correspondence between Lovecraft, who was at the end of his life, and the young Conover, has been published as *Lovecraft at Last* (Carrolton-Clark, 1975; reprint 2004).

Conover's father had intended for him to attend The Citadel and follow his family's tradition of military service. Instead, he attended the Maryland State Teacher's College at Salisbury, Maryland, and became a radio announcer for WTBO in Cumberland, Maryland.

He later moved to Washington, D.C., and focused on jazz in his programming, especially the Duke Ellington hour on Saturday nights. His guests on this program and Saturday morning shows included many important artists, such as Boyd Raeburn.

2.60.2 Voice of America

Conover's first arrival in Poland (1959)

Conover came to work at the Voice of America, and eventually became a legend among jazz lovers, primarily due to the hour-long program on the Voice of America called *Voice of America Jazz Hour*. Known for his sonorous baritone voice, many would argue that he was the most important presenter on Voice of America. His slow delivery and the use of scripts written in "special English" made his programmes more widely accessible and he is said to have become the first teacher of English to a whole generation of East European jazz lovers.[4] Conover was not well known in the United States, even among jazz aficionados, as the Voice of America did not broadcast domestically except on shortwave, but his visits to Eastern Europe and Soviet Union brought huge crowds and star treatment for him. On a trip to Moscow a taxi driver recognized him by his distinctive deep-toned voice. He was a celebrity figure in the Soviet Union, where jazz was very popular and the Voice of America was a prime source of information as well as music.

In 1956, Conover conducted a series of interviews with jazz luminaries like Duke Ellington, Billie Holiday, Stan Getz, Peggy Lee, Stan Kenton, Benny Goodman, and Art Tatum. His interview with Tatum is noted as "the only known in-depth recorded interview with the pianist". These interviews were selected by the Library of Congress as a 2010 addition to the National Recording Registry, which selects recordings annually that are "culturally, historically, or aesthetically significant".[5]

2.60.3 Death

He died of lung cancer. He had been a smoker for 57 years.[6]

2.60.4 Legacy

In 2015, the University of North Texas announced its Willis Conover Collection would make digitized copies of Conover's programs available online.[7]

2.60.5 References

[1] Thomas, Jr., Robert McG. (May 19, 1996). "Willis Conover, 75, Voice of America Disc Jockey". *New York Times*. p. 35. Retrieved December 12, 2010.

[2] Robert McG. Thomas Jr., "Willis Conover Is Dead at 75; Aimed Jazz at the Soviet Bloc", *New York Times*, May 19, 1996. Retrieved February 4, 2010.

[3] *Willis Conover: Broadcasting Jazz To The World*, by Terence M. Ripmaster (born 1933), iUniverse (2007); OCLC 180237422

[4] Alexei Yurchak, *Everything Was Forever, Until It Was No More: The Last Soviet Generation*, Princeton University Press, Princeton, 2006, pp. 180–181.

[5] "The National Recording Registry 2010". Library of Congress. Retrieved April 10, 2011.

[6] James Lester, "Willis of Oz", *Central Europe Review*, Vol. 1, No. 5, July 26, 1999. Retrieved February 4, 2010.

[7] Ramsey, Doug. "The Willis Conover Archive Is Online". *ArtsJournal*. Rifftides. Retrieved September 11, 2015.

2.60.6 External links

- Profile for Willis Conover at Find-A-Grave

- Doug Ramsey, "Willis Conover", Rifftides, October 29, 2005.

- Article about Willis Conover from the Voice of America web site

- Willis Conover introduces a band of jazz musicians during a live performance

- Willis Conover tells his story

- Voices of Freedom: A Celebration of VOA Jazz and Willis Conover

- Farber, Gary (June 29, 2008). "If It Ain't Got That Swing". *Amygdala*. Article about Conover's interest in science fiction and fantasy, including his correspondence with H. P. Lovecraft.

- The Willis Conover Collection at the University of North Texas

- David Brent Johnson, "Conover's Coming Over: Willis Conover and Jazz at the VOA", Indiana Public Media (NPR program), November 12, 2007.

2.61 World Administrative Radio Conference

The **World Administrative Radio Conference** (WARC) was a technical conference of the International Telecommunication Union (ITU) where delegates from member nations of the ITU met to revise or amend the entire international Radio Regulations pertaining to all telecommunication services throughout the world. The conference was held in Geneva, Switzerland, with preparatory conferences held in Panama City, Panama.

In 1992 at an *Additional Plenipotentiary Conference* in Geneva the ITU was restructured and as a result from 1993 the conference became known as the World Radiocommunication Conference or WRC.[1]

2.61.1 Conferences list

This list is incomplete; you can help by expanding it.

- ITU Preparatory to World Administrative Radio Conference Panama 1979

- ITU World Administrative Radio Conference Geneva 1979 (WARC-79)

- ITU World Administrative Radio Conference Geneva 1984 (WARC-84)

- ITU World Administrative Radio Conference Geneva 1992 (WARC-92)

2.61.2 See also

- Amateur radio

- Radio

- Telecommunications

- World Radiocommunication Conference

2.61.3 References

[1] ITU.int

2.62 World Radio TV Handbook

"WRTH" redirects here. For the radio station, see WRTH (FM).

The **World Radio TV Handbook**, also known as **WRTH**, is a directory of virtually every radio and TV station on Earth, published yearly. It was started in 1947 by Oluf Lund Johansen (1891–1975) as the *World Radio Handbook* (WRH).[1] The word "TV" was added to the title in 1965, when Jens M. Frost (1919–1999) took over as editor.[2] It had then already included data for television broadcasting for some years. After the 40th edition in 1986, Frost handed over editorship to Andrew G. (Andy) Sennitt.[3]

The first edition that bears an edition number is the 4th edition, published in 1949. The three previous editions appear to have been:

- the 1st edition, marked "Winter Ed. 1947" on the cover and completed in November 1947

- the 2nd edition, marked "1948 (May-November)" on the cover and completed in May 1948

- the 3rd edition, marked "1948-49" on the cover and completed in November 1948.

Summer Supplements appear to have been issued from 1959 through 1971. From 1959 through 1966 they were called the Summer Supplement. From 1967 through 1971 they were called the Summer Edition.

Through the 1969 edition, the WRTH indicated the date on which the manuscript was completed.

Issues with covers in Danish are known to have been available for the years 1948 May-November (2d ed.), 1950-51 (5th ed.; cover and 1st page in Danish, rest in English, most ads in Danish), 1952 (6th ed.; cover and 1st page in Danish, rest in English, most ads in Danish), and probably others. The 1952 English ed., which is completely in English, has an extra page with world times and agents, and ads in English which are sometimes different from the ads in the Danish edition. Also, the 1953 ed. mentions the availability of a German edition.

Oluf Lund Johansen published, in conjunction with Libreria Hispanoamericana of Barcelona, Spain, a softbound Spanish-language version of the 1960 WRTH. The book was printed in Spain and called *Guia Mundial de Radio y Television*, and carried the WRTH logo at the time as well as all the editorial references contained in the English-language version.

Hardbound editions are known to have been available for the years 1963 through 1966, 1968, 1969, and 1975-1978, and probably others.

2.62.1 Publications

- Gilbert, Sean; Nelson, John; Jacobs, George, *World Radio TV Handbook 2007*, Watson-Guptill, 2006. ISBN 0-9535864-9-9.

2.62.2 References

[1] O. Lund-Johansen, presented by OZ6GH.

[2] DSWCI Member of Honour: Jens M. Frost

[3] Andy Sennitt, own presentation.

2.62.3 External links

- http://www.wrth.com/

Chapter 3

Reference LIsts

3.1 List of international television channels

This is a **list of channels that are distributed in more than one country**.

3.1.1 Worldwide distribution

This is a list of television channels with worldwide distribution via satellite or cable. Due to Internet television, some localized channels might have worldwide distribution. These channels should only be listed only if they also broadcast via cable or satellite.

Other international programs aired from other programs [ex. Temptation of Wife - a Korean drama never aired on GMA Pinoy TV] due to copyright issues. CNA Channel News Asia, Singapore news distributed in major cities around Asia.

This list is incomplete; you can help by expanding it.

- ABS-CBN-TFC (The Filipino Channel) - international television network, based in the Philippines;. A service of ABS-CBN Global, a fully owned subsidiary of ABS-CBN.

- Abu Dhabi TV - worldwide Arabic-language channel

- ABP News is a Indian international channel

- Aaj Tak this Indian channel is last 35 year is number 1

- AFN American Forces Network, broadcasts worldwide via satellite, only available to US Forces and their families, but is available in South Korea.

- Al Jazeera - based in Qatar, 2 separate channels available in Arabic and English

- Arirang - Korea international Broadcasting Foundation, based in the Republic of Korea, English language

- ANI Is a Indian channel this channel is coming world 195 country

- Awesome TV - based in USA, distributed worldwide

- BBC - British Broadcasting Corporation, based in the United Kingdom. Operating BBC World News news and information service, BBC Entertainment, BBC Lifestyle, BBC Knowledge, CBeebies, and BBC HD. BBC programming available in North America through local channels- BBC America, BBC Canada and BBC Kids.

- BBC World Service - The Worlds oldest International Radio Station, programmes are broadcast in over 50 languages.

- BFBS - British Forces Broadcasting Service, broadcast to British Armed Forces bases/ships around the world via satellite.

- Cartoon Network - based in USA, distributed worldwide

- CCTV - the major state television broadcaster in mainland China

- CNN - several channels devoted to news broadcasting

- DD News is a Indian international channel is a Highly pay channel in Asian country

- Deutsche Welle, DW, (pronounced deh-veh) - based in Germany, also in English and Spanish

- Discovery Channel - several channels devoted to science, technology, health, history, etc.

- Disney Channel, based in the United States

- FOX, available everywhere in the world, FOX owns channels like National Geographic Channel, FX and BabyTV which is produced worldwide.

- EWTN - Catholic channel broadcasting across the world from Alabama.

- Fox News - 24-hour news channel

- KTV5 International and AksyonTV International - the television network in the Philippines which is a subsidiary of PLDT and covers Africa, Middle East and Europe

- France 24, new 24-hour news network in English and French based in France

- TV Globo Internacional - Brazilian broadcaster, worldwide distribution

- i24news - Based in Israel, available in English, French and Arabic.

- India TV his a Indian channel this channel Ability in all world his number one news channel in Asia his headquarters New Delhi

- India Today Television This is Indian channel his cover in Europe

- MTV - based in USA but with regionalized versions around the world (MTV2).

- NDTV 24x7 his a interested English based news channel in India his hindquarters in Noida India

- NHK World - International version of Japanese national television NHK, broadcasts in English but focused on news

- NHK World Premium - Broadcasts a mixture of news, sports and entertainment in Japanese worldwide via satellite as a subscription service

- Press TV, 24-hour news network in English and based in Iran

- PTC News his a panjabi lunges interested international news channel his hindquarters is Chandigarh India

- Rai Italia - Italian state broadcaster, worldwide distribution

- RTP Internacional - Portuguese state broadcaster, worldwide distribution except Lusophone African countries

- RT - Russian news channel broadcasting in English 24/7 in over 100 countries spread over five continents, available on cable, satellite and online

- KBS World - South Korea based news and entertainment channel broadcast internationally.

- Sky News - British based news channel broadcast internationally.

- Sun Channel Tourism Television - 24-hour Tourism network in Spanish

- Sony Entertainment Television Asia - Indian entertainment channel broadcasting in Hindi.

- TBN - Christian television network based in the USA, worldwide distribution

- Times Now is a Indian Top English language news channel This channel is going on All Asian country without Pakistan

- Televisa - Mexican broadcaster, worldwide distribution

- TGN - Owned and operated by TV5 (Thailand)

- TV5MONDE - French language, worldwide distribution

- Telefe - Argentinian broadcaster, worldwide distribution

- TVE Internacional - Spanish state broadcaster, worldwide distribution

- Voice of America - is a multimedia international broadcasting service funded by the U.S. Government through the Broadcasting Board of

- Zee News Asia most popular and powerful news channel his available on all Indian language as (Hindi, English, tamil, telghu, Bangali, Chinese, Thai, Uruguay, Malaysia, Bhutani) his hindquarters in }Mumbai

3.1.2 Distribution on several continents

This is a list of television channels with distribution on several continents via satellite or cable. Keep in mind that almost all European satellite channels can be seen in parts of Asia and Africa and vice versa. Such channels should not be listed.

This list is incomplete; you can help by expanding it.

- ABC (American Broadcasting Company) - based in USA, but also broadcasts into Canada and Mexico, parts of the Caribbean, and Guam.

- ABC Asia Pacific (Australian Broadcasting Corporation) - based in Australia, broadcasting to the neighbouring region.

- Africa Independent Television (AIT) - based in Lagos, Nigeria and broadcasts across the United Kingdom and United States.

- Al Jazeera and Al Jazeera English - Middle Eastern satellite TV station broadcasting to North America via Galaxy 25.

- ANT1 - Greek language, broadcasts into Macedonia, Albania, parts of Italy and Turkey; international ANT1 channels broadcast ANT1 programming to North America, Europe & Australia.

- ART News - English language, broadcasts into Sri lanka, parts of Italy and Turkey; international ANT1 channels broadcast ANT1 programming to North America, Europe & Australia.

- ARY Digital - Pakistani channel broadcasting in Urdu available in USA, Europe, Middle East and South Asian markets.

- Asian Food Channel

- TV Azteca - Also known as Azteca America in the United States, based in Mexico, broadcasts throughout Latin America, United States, Canada.

- AzTV - Azeri national broadcaster. Distribution to Europe, the Middle East, and North America.

- BFBS - British Forces Broadcasting Service, operating BFBS Television in Germany, Cyprus, former Yugoslavia, Iraq, Falkland Islands, Belize - only available to HM Forces and their families, except in Falkland Islands, and on cable in Ralston, Alberta (home of CFB Suffield, the British Forces' Canadian home)

- CBC - based in Canada, but also broadcast into part of USA, Bermuda, the Caribbean and probably Greenland.

- CBS - based in USA, but also broadcasts into Canada, Mexico, parts of the Caribbean, and Guam.

- Channel NewsAsia - based in Singapore with broadcasts in over 20 Asian territories.

- CaribVision - based in Barbados, the international broadcaster caters mostly to the English-Speaking Caribbean-community throughout the world.

- Cubavision International - International broadcaster from Cuba

- The CW - Launched in the United States in September 2006 as a merger of two networks, UPN and The WB. It is also seen in Puerto Rico and the U.S. Virgin Islands.

- DD India - Government-run Indian international station.

- Dua channel - broadcasting to the Middle East in English, Malay, French and Arabic

- ERT World, Greek language/Greek state broadcaster, available in North America, Europe & Australia

- ESPN, ESPN2 - Sports channel, based in USA

- Euronews - news television, mainly broadcast in Europe

- Eurosport- Sports channel, the sports equivalent of Euronews (although unrelated)

- Fox, based in USA but also broadcasts into Canada, Mexico, parts of the Caribbean American Samoa, and Guam. It also has a worldwide distribution.

- Fox Sports Net - several regional channels devoted to sports broadcasting including national Fox Sports 1 and Fox Sports 2 in the United States.

- France 2 and France 3 - Major television network in France, also available on satellite in Europe, Africa and the Americas

- Geo TV - Pakistani channel broadcasting in Urdu available in USA, Europe, Middle East and South Asian markets.

- GMA Pinoy TV - Popular channel from the Philippines owned by Filipino broadcaster GMA Network; now available throughout the world

- Hadi TV - Pakistani Islamic channel available worldwide

    - Hadi TV 1 (Urdu, English) - Available in South Asia, USA, Canada, Australia, New Zealand and Europe

    - Hadi TV 2 (Malay, English, Thai, Arabic) - Available in South East Asia and Middle East

    - Hadi TV 3 (Azeri, Turkish, Kurdish, English) - Available Azerbaijan, Turkey, Kurdistan and Iran

    - Hadi TV 4 (English, Pashto, French, Farsi, Malay) - Available in Pakistan, Afghanistan, North Africa and Malaysia

    - Hadi TV 5 (English, French, Hausa, Swahili) - Available in Africa

    - Dua Channel (English, Malay, French, Arabic) - Available in Malaysia, North Africa and Middle East

- Hauraki TV - Available in New Zealand, Australia & South America

- History Channel

- Indus Music - Pakistani music channel broadcasting in Urdu available in USA, Europe, Middle East and South Asian markets.

- Indus Vision - Pakistani general entertainment channel broadcasting in Urdu available in USA, Europe, Middle East and South Asian markets.

- ION Television - Formerly PAX then renamed *i*: Independent Television. Based in USA but also broadcasts into Canada, Mexico, and parts of the Caribbean. It also has a worldwide distribution.

- The Israeli Network, diffusion of Ch. 3 IL, Ch. 7 IL, and Ch. 10 IL, Mainly Hebrew language

- Living Asia Channel - Channel broadcast from the Philippines to Asia, Middle East, Oceania, Europe, and North America, primarily in English

- MCOT World - Owner and Operated by MCOT (Former Names is *MCOT2* and *Asean TV*), This is one of two International TV Stations by Government of Thailand

- Mega Cosmos, Greek Language available in North America & Australia

- MRTV3 - International channel from Myanmar.

- NBC - based in USA but also broadcasts into Canada, Mexico, and parts of the Caribbean, American Samoa, the Northern Mariana Islands, Sisimiut, Greenland, and Guam. It also offers a business channel called CNBC

- NBT World - Owner and Operated by National News Bureau of Thailand. and Sister TV Station with NBT

- NDTV 24x7 - Indian news channel broadcasting in English.

- NDTV Imagine - Indian entertainment channel broadcasting in Hindi.

- Nickelodeon, owned by Viacom, based in the United States, see Nickelodeon around the world.

- Nigerian Television Authority (NTA) based in Nigeria and also broadcasts in the United Kingdom via BEN Television, in North America FTA and throughout Africa.

- Polsat 2 International, Polish private network available in North America and Australia

- Rusiya Al-Yaum - Russian TV news channel broadcasting in Arabic via satellites for the audience in the Arab world and European states as well as other regions.

- SIC Internacional - Portuguese commercial broadcaster available in North America

- STAR Gold - Hindi language movie channel from India, owned by News Corp.

- STAR News - Hindi language news channel from India, owned by News Corp.

- STAR One - Hindi language entertainment channel from India, owned by News Corp.

- STAR Plus - Hindi language entertainment channel from India, owned by News Corp.

- TF1 - French language television network based in France with international distribution

- TVB - Based in Hong Kong but can be viewed in Canada, Malaysia, Singapore, UK, US, and Australia via satellite.

- TV Chile - Chilean channel with international distribution.

- TVK - Cambodian station with international distribution.

- TVRi - Romanian state broadcaster, international distribution

- UK.TV - 'Best of British' entertainment channel in Australia and New Zealand, carrying BBC, Thames Television and ITV programming

- Voice of America - is a multimedia international broadcasting service funded by the U.S. Government through the Broadcasting Board of Governors. Programs are produced in 45 languages.

- VMTV - Volksmusik TV is a German folc music station with European and American proragmming versions

- Zee Cinema - Indian movie channel broadcasting in Hindi.

-

3.1.3 Former/defunct international television

- Jetix (originally Fox Kids), replaced by Disney XD

3.1.4   See also

- List of shortwave radio broadcasters

3.2   List of shortwave radio broad-casters

3.2.1   References

[1] http://yle.fi/uutiset/yle_ends_short_wave_broadcasts/
     5756519

[2] http://www.hrt.hr/273790/satellite-broadcasting/
     the-voice-of-croatia-2

Chapter 4

Text and image sources, contributors, and licenses

4.1 Text

- **International broadcasting** *Source:* https://en.wikipedia.org/wiki/International_broadcasting?oldid=701667200 *Contributors:* Hephaestos, Infrogmation, GABaker, Liftarn, Glenn, Mxn, Dysprosia, Radiojon, Topbanana, JorgeGG, TMC1221, Scriptwriter, Anoop, PedroPVZ, Clngre, Rhombus, DHN, David Edgar, Cyrius, Mboverload, 159753, Formeruser-81, CaribDigita, Comandante, Andy Christ, Cab88, Maestrosync, Rich Farmbrough, Vague Rant, Pmsyyz, Xezbeth, Gerry Lynch, Loren36, Kiand, Cwolfsheep, Mithent, Swarve, Mavros, Ianblair23, Gene Nygaard, Mosesofmason, Jakes18, Weisbrod, Scriberius, Mu301, Eyreland, Graham87, Haikupoet, MarcosR~enwiki, Rjwilmsi, Koavf, IRT.BMT.IND, Duagloth, Tlitic, RattusMaximus, Alcot, Pigman, Gaius Cornelius, Oenomel, Bachrach44, Veledan, Thiseye, Bestofmed, Formeruser-82, Jfdunphy, Rearden9, SIGURD42, Curpsbot-unicodify, Thomas Blomberg, SmackBot, WikiWookie, Jeffreykopp, Iamajpeg, Mauls, Chris the speller, Endroit, Pinots, Colonies Chris, A. B., Jaellee, Colin41, Mitchumch, Asiaradio, FunkyFly, Breno, Beetstra, Aquarelle, AxG, KirrVlad, Hu12, RekishiEJ, Yosy, Eastlaw, CmdrObot, Wws, Adam JW, Siberian Husky, Soniamcewan, Maximilian Schönherr, Seaphoto, SkagitRiverQueen, Thevaran~enwiki, Sanskritkanji, Harryzilber, Enigmaticland, Desertsky85451, Geniac, Radiobroadcasting, Drkirby, Avicennasis, Cgingold, Dipper2, Welshleprechaun, Zir, Simon.spanswick, Funandtrvl, Noble Caraqueño, TreasuryTag, Ai4ijoel, Adrian two, Marknagel, TXiKiBoT, BaronVonchesto, Kaiketsu, Resurgent insurgent, Biscuittin, Pinoysurfer, Itisa, ImageRemovalBot, PipepBot, Kathleen.wright5, Kotalampi, Trivialist, DisambiguationGuy, Life of Riley, Twitherspoon, Good Olfactory, Addbot, Quantock, Yobot, Mahmudmasri, Xqbot, Capricorn42, FrescoBot, ImageTagBot, Full-date unlinking bot, Dilbilimci61, Zollerriia, Newsjessore, Erianna, Юлия Леонтьева, Noodleki, AleeeexisSCL, Cagoul, Theopolisme, MerlIwBot, Helpful Pixie Bot, Jyg1093, 4throck, Khazar2, Itbeso, Mogism, YiFeiBot, Mo2010, EditorMakingEdits, ElCommandanteVzl, Peacock.hero123, Punhal PK, AnonAnnu, KasparBot and Anonymous: 109

- **Shortwave radio** *Source:* https://en.wikipedia.org/wiki/Shortwave_radio?oldid=702309591 *Contributors:* Timo Honkasalo, The Anome, Peterlin~enwiki, TomCerul, Heron, Camembert, Edward, Michael Hardy, GABaker, Kwertii, Tregoweth, Curtisweyant~enwiki, Julesd, Glenn, Pakkio, Palmpilot900, Charles Matthews, Reddi, Dysprosia, Radiojon, Bloodshedder, Chris Rodgers, Jerzy, M1fcj, Denelson83, Twang, Cdang, Blainster, Pmcray, Magic Window, Timvasquez, Graeme Bartlett, Karn, Wiki Wikardo, Peter Ellis, K7jeb, Beland, Gaul, Kramer, Trilobite, Canterbury Tail, Alistair1978, Bender235, Evice, Ce garcon, Irrawaddy, Sparkgap, Sukiari, 119, Andrewpmk, Zippanova, Pjacklam, Wtmitchell, Mavros, Wtshymanski, Joeva3eo, Ringbang, Axeman89, Feline1, Unixxx, Scriberius, Bratsche, Rjairam, Eyreland, Mandarax, BD2412, Haikupoet, BorgHunter, Pmj, Rjwilmsi, Koavf, IRT.BMT.IND, Vegaswikian, FlaBot, Soredewa, Ossington2, Awotter, Sherubtse, Fragglet, Mskadu, Subversive, Chobot, Richard-L-James, Burnte, Bartleby, YurikBot, JarrahTree, Bhny, Kibbitzer, Dforest, Johantheghost, Matticus78, Voidxor, Mysid, Bota47, Dddstone, Caerwine, Maddog Battie, Ninly, Arthur Rubin, Junglecat, Groyolo, DocendoDiscimus, SmackBot, Fireworks, Jeffreykopp, Reedy, TBH, Sea diver, Gilliam, Carl.bunderson, Cabe6403, Chris the speller, Bidgee, Colonies Chris, A. B., Beatgr, OrphanBot, Kevinpurcell, Adamantios, Hateless, EdGl, Harryboyles, Chazchaz101, Sambot, Andrewjuren, GCW50, Nagle, Beetstra, EEPROM Eagle, Kvng, Hu12, דניאל צבי, Toddsschneider, Stereorock, Chetvorno, CmdrObot, Psycadelc, Wws, Outriggr (2006-2009), AndrewHowse, Altaphon, Djg2006, Josef Serf, Ward3001, WxGopher, Thijs!bot, Kablammo, Allquestions, Dfrg.msc, Dawnseeker2000, AntiVandalBot, Chill doubt, JAnDbot, Harryzilber, CosineKitty, Boguslinks, Ggugvunt, Jahoe, No more bongos, Dsergeant, Jerome Kohl, Steve Hosgood, CodeCat, Ashishbhatnagar72, 0612, Burgh House, Maurice Carbonaro, 8hhaggis, Mrceleb2007, Plasticup, Americandxer, Fluteboy, RJASE1, VolkovBot, Athletes Foot, Ai4ijoel, Adrian two, Bobbetts, The Original Wildbear, Trashbag, Sankalpdravid, Argument~enwiki, Cabezon-raven, NW7US, Cosprings, WereSpielChequers, Coati123, Miniapolis, Gregory.hand, Iceman63976, Altzinn, Dlrohrer2003, Serialdownloader, ClueBot, Trojancowboy, Binksternet, Badger Drink, Jelf64, Kathleen.wright5, Kotalampi, Sv1xv, Nobidicus, Kitsunegami, Rcooley~enwiki, Eeekster, Dxinginfo, Ysrc, Arjayay, Cexycy, XLinkBot, Twitherspoon, Rio de oro, Addbot, Quantock, Riadismet, Avobert, Vchorozopoulos, Damiens.rf, Download, Kicior99, Angrense, Lightbot, Dellium, Luckas-bot, Yobot, Rccoms, Donfbreed, Alfonso Márquez, ShrtWave, AnomieBOT, Andrewrp, Erud, JFY, Light,Love and Law, Kernel.package, Demigodgodd, Eugene-elgato, GliderMaven, FrescoBot, Surv1v4l1st, Jc3s5h, Citation bot 1, PigFlu Oink, Sibian, AstaBOTh15, DrilBot, Broadcasttransmitter, GrapedApe, AAT17, Hsnmoom, Mjs1991, PhillyDelphia, Mickeylove73, NameIsRon, Jackehammond, Steve03Mills, Graeme 2, John of Reading, Duncan952, GoingBatty, Ovidcaput, G7cnf, Dreamer26, Alpha Quadrant (alt), Noodleki, Bryanmaupin, ChuispastonBot, Mongoosander, A120068020, Ivolocy, ClueBot NG, Reify-tech, Danim, Billgrove, Helpful Pixie Bot, Jyg1093, BG19bot, Neøn, Jeremy112233, JoBaWik, Cyberbot II, 313 TUxedo, Dexbot, Wetrace, Redd Foxx 1991, JakeWi, YiFeiBot, Dough34, Procrastinatingpersona, Cgs17, AnonAnnu, Dxreport and Anonymous: 236

177

- **Agência Brasil** *Source:* https://en.wikipedia.org/wiki/Ag%C3%AAncia_Brasil?oldid=700315736 *Contributors:* Tedernst, RickK, Whisper-ToMe, Radiojon, Hajor, Robbot, Centrx, Carlosar~enwiki, Abu badali, PFHLai, Karl-Henner, Mavros, Ceyockey, Wikivandal1993, Leonar-doRob0t, KnightRider~enwiki, SmackBot, Mairibot, Bluebot, Rrtaddei, Victor Lopes, ThurnerRupert, CapitalR, Namiba, Ingolfson, Cgingold, Haddiscoe, LeadSongDog, Gits (Neo), Fadesga, TarzanASG, Trivialist, Addbot, דוד55, ComputerHotline, Yobot, AnomieBOT, DrilBot, Chat-fecter, Pierpao, EmausBot, MrFawwaz, Gabriel Yuji, Egeymi, Giso6150, Vitorabdala, VitorHeisenberg, Ms Tepre Saanws, Manuelle1133 and Anonymous: 7

- **Agência Estado** *Source:* https://en.wikipedia.org/wiki/Ag%C3%AAncia_Estado?oldid=671769693 *Contributors:* Bearcat, Pharaoh of the Wizards, Drpickem, Yobot and Anonymous: 1

- **ALLISS** *Source:* https://en.wikipedia.org/wiki/ALLISS?oldid=693129260 *Contributors:* Skysmith, Eugene van der Pijll, Mike Rosoft, Loganberry, ThomasK, CanisRufus, Ivansanchez, Pearle, GL, Gene Nygaard, Henrik, Eyreland, Royan, BD2412, Rjwilmsi, Missmarple, Dstudent, Ian Pitchford, Ground Zero, Mark83, Kolbasz, Conscious, Gaius Cornelius, Mikeblas, Banana04131, Bluebot, Fredvanner, Colonies Chris, TenPoundHammer, Gjp23, Rogerbrent, Iridescent, Rnb, CmdrObot, CMG, JodyB, JaGa, Jim.henderson, Cuddlyable3, Lightbot, Yobot, AnomieBOT, Jaydenillman, DrilBot, 6harts, ChrisGualtieri, Khazar2 and Anonymous: 21

- **Amateur radio** *Source:* https://en.wikipedia.org/wiki/Amateur_radio?oldid=702085749 *Contributors:* WojPob, Brion VIBBER, Vicki Rosen-zweig, Mav, Ap, Ik1tzo~enwiki, PierreAbbat, Waveguy, Artsygeek, Heron, Arj, Netcrusher88, DevilRaysFan, Michael Hardy, Altailji, Bdowd, CesarB, Nanshu, Glenn, Kimiko, Kwekubo, Cimon Avaro, Deisenbe, John K, Wfeidt, Dying, Charles Matthews, Andrevan, Reddi, Dysprosia, Lou Sander, Tpbradbury, Furrykef, SEWilco, Xyb, Mignon~enwiki, Bloodshedder, M1fcj, Frazzydee, Mrdice, Denelson83, Robbot, Cdang, KeithH, TMC1221, R3m0t, Scriptwriter, RedWolf, Kadin2048, Lowellian, Halibutt, TheLight, Hadal, Kd4ttc, Wikibot, Lupo, SpellBott, Jrash, Giftlite, DocWatson42, Yama, Laudaka, Sj, Oherrala, Nichalp, Dfrandin, Inter, Patrick-br, Ssd, Niteowlneils, Rdcole, Yekrats, BigHaz, Finn-Zoltan, Albany45, IrrelevantQuestionBoy, Peter Ellis, Wmahan, JeffyJeffyMan2004, Wleman, Piotrus, AlexanderWinston, AndrewTheLott, Icairns, Defenestrate, Sam Hocevar, Thparkth, N4zhg, Jakro64, Njh@bandsman.co.uk, ChrisRuvolo, Gachet, GoodStuff~enwiki, RossPatter-son, Discospinster, Guanabot, Somegeek, ArnoldReinhold, Flynns32547, User2004, Gerry Lynch, Bender235, Esn1d, PaulMEdwards, Peter-sam, Huntster, Nile, C1k3, Allyn, Adambro, Simon South, Nigelj, Ptemples, Smalljim, Cje~enwiki, Mink Butler Davenport, Cmdrjameson, Cmacd123, Brim, Sparkgap, Sukiari, Tractor~enwiki, Analogdemon, Flashweb, Jumbuck, Musiphil, Alansohn, Richard Harvey, Gblaz, Bee-Jay~enwiki, Stillnotelf, KB3JUV, Wtshymanski, Danhash, Wikicaz, CloudNine, Zoohouse, Ianblair23, SteinbDJ, Gene Nygaard, Kazvorpal, Firthy2002, Jakes18, Richard Weil, Johnwcowan, Feezo, Stemonitis, Flawiki, Bushytails, Linas, Pauley2483, Theloniouszen, Plaws, Rjairam, Df2dr, Shadyman, Andromeda321, Eyreland, Zzyzx11, Jon Harald Søby, MechBrowman, Kotoviski, Tslocum, Vu2ukr, Graham87, K3wq, BD2412, Jbarr, Sjakkalle, Rjwilmsi, Mdinan, Pyt, M1LCR, Linuxbeak, Vegaswikian, Ligulem, Gerard Hill, Yamamoto Ichiro, N0YKG, ZoeL, FlaBot, Mirror Vax, SchuminWeb, EnDumEn, Anonym1ty, RobyWayne, TeaDrinker, Alphachimp, Darranc, Flecom, Drakcap, Jit-tat~enwiki, Vchapman, Chobot, ShadowHntr, YurikBot, Wavelength, Borgx, RobotE, Huw Powell, Mukkakukaku, StuffOfInterest, Jeffthejiff, Kilowattradio, Epolk, Hmss007, Gaius Cornelius, CambridgeBayWeather, Randyholloway, Member, K.C. Tang, Deskana, 9cds, Brandon, Ed-mondo~enwiki, Gregburd, Mikeblas, Voidxor, Ma3nocum, Belayet, Brauhaus, Scottfisher, Mysid, Gadget850, Jeh, Jrbonica, Ke5crz, Dddstone, Gadget17, Yudiweb, Gat0r, Searchme, Erpingham, Light current, Ninly, Encephalon, Mike Selinker, David Jordan, MathGeek06, Rearden9, GutoAndreollo, EuroRaver, Practicalairsoft, Elliskev, Veinor, RupertMillard, SmackBot, BlakJak, Robfwb, Nsayer, Kipio, N3sgd, Kth, Gary Kirk, Goleson, Sailin, Grey Shadow, KVDP, TheDoctor10, Rjayres, Unforgettableid, Aij, Gilliam, Hmains, Skizzik, Kmarinas86, Jenhowse, Schmiteye, KD5TVI, Chris the speller, Pberrett, Kharker, Thumperward, Vees, RomaC, Dan Zimmerman, Zsinj, Dethme0w, Frap, Nixeagle, Neo3DGfx, MikeAus, PointyOintment, PetesGuide, Dreadstar, Kc2idf, Ras123, Pilotguy, WA3VJB, Drunken Pirate, Will Beback, Kb2jpd, Shane oneal, Harryboyles, KLLvr283, Jidanni, Rcasey, Dejudicibus, Andrewjuren, GCW50, 16@r, StanBrinkerhoff, SandyGeorgia, Mets501, Ryulong, Lenn0r, Ethertaxi, Iridescent, Paul Koning, CuteWombat, Mwhite66, Toptucker, Zero sharp, Stereorock, Ki4ihc, DJGB, HDCase, Wb9mcw, Sir Vicious, Tuvas, Makeemlighter, Smallpond, Requestion, Wingman358, KXL, Mike65535, ChardingLLNL, JFreeman, Palmiped, Julian Mendez, Starionwolf, Kozuch, Dmbaty, JodyB, Bolesjohnb, Nite owl, ZS5Z brad, Billth87, WillMak050389, Dfrg.msc, M0ffx, AgentPep-permint, Pcbene, Booshakla, Dawnseeker2000, Escarbot, Cvos, KidIncredible, JurgenG, CPWinter, Edit Centric, ErinHowarth, LuckyLouie, K0VIN, Jaredroberts, JAnDbot, Harryzilber, MER-C, CosineKitty, Afarhan, Captbryan, TAnthony, 7severn7, VoABot II, Skapare, Dsergeant, Think outside the box, TARBOT, Ling.Nut, BlakJakNZ, Robomojo, Ps2babyboy, WhatamIdoing, DodgerDean, March of the Ducks, LorenzoB, Edward321, Ekotkie, Misibacsi, TigerMo, PhantomS, Kf4yfd, Erazmus~enwiki, N734LQ, Microsloth, PrestonH, Theonlysilentbob, EdBever, AceNZ, J.delanoy, EscapingLife, Wa3frp, Wikip rhyre, Neon white, Little Professor, ZacharyWyman, Davandron, Cannibalicious!, Tarotcards, AntiSpamBot, Pianotech, Plasticup, NewEnglandYankee, Squidfryerchef, Finley Breese, BillyMassie, VolkovBot, John Darrow, Umalee, N2ueg, Nmewarlok, Noerrorsfound, Karl Shoemaker, Lradrama, AtaruMoroboshi, CodyGraves, Sultec, Jaqen, Andy Dingley, Haseo9999, Kmlengel1, BlueH2O, Synthebot, Expeditionradio, LittleBenW, ChrisZeddybear, Paloma Walker, Daveh4h, Billblyth25, VU3RDD, SieBot, Briefer, Eu-ryalus, Whimsley, Simmonds001, Gimili2, Pyroglyph, Oxymoron83, PamRivers8, Miniapolis, Lightmouse, Callidior, Navy.enthusiast, Diego Grez-Cañete, Svick, RedBlade7, Iknowyourider, Dodger67, Jacob.jose, MM3OXB, Omsk~enwiki, Nacarlson, Martarius, ClueBot, EoGuy, Ndenison, Kotalampi, Niceguyedc, Sv1xv, Excirial, Jusdafax, Crywalt, Papna, Rcooley~enwiki, EhJJ, Maniago, Another Believer, Doggydudu, Samson3000, InternetMeme, BarretB, XLinkBot, VK6DNA, 13 of Diamonds, Eliran Levi, Philsherrod, WikHead, Chuvaris, Cmr08, Leonarp, Thatguyflint, CalumH93, Addbot, Willking1979, Stuart lyster, DOI bot, Landon1980, Linespermillimter, Cst17, W4auv, Coasting, Glane23, Kc8nlr, Chzz, LinkFA-Bot, Jaydec, KALZOID-73-20METER, Krano, Jarble, Luckas-bot, Yobot, AnomieBOT, Ubuntujason, Jim1138, Ad-justShift, RandomAct, Citation bot, Ekconklin, EnamelWildcat, Almabot, ZimmerCircus, Qwertyzxcvbn, Surv1v4l1st, Fullautoglock, Ra-diomanPA, PigFlu Oink, Biker Biker, Sadaqa, Pinethicket, Jonesey95, A412, SpaceFlight89, PeopleString, FoxBot, Trappist the monk, Francis E Williams, Aiken drum, Jacers1, N4LXL, Mean as custard, RjwilmsiBot, NameIsRon, Mstern001, Dstone66, Kcuffel, DASHBot, EmausBot, John of Reading, Desertroad, Lunaibis, Dewritech, Slightsmile, Ὁ οἶστρος, Jstarx, Dgaddis, Sbmeirow, ClueBot NG, Codynosbig, FourLights, Taxilian, Karan Kamble, ChristophE, Paul Gaskell, HHaeckel, Arrknot, Widr, JordoCo, Wiki7373, Captain Klystron, Ham Radio Microphone, Dangerang, Xmike87, Sirhc808, Wp4oca, Rcunderw, Duxwing, NicholasJCole, TheUnnamedNewbie, Cqdx, Several Pending, Cyberbot II, Lucy34bell, Tow, Dexbot, Kk4kcu, Giancabr, Kf5kfj, Avianoutremont, Tentinator, Mw0rkb, M0SCU, Clientkill, Shrekogrelord, KK7PW, 2 Hertz, 32RB17, Wylieq, HHubi, MW6WOD, Monkbot, HMSLavender, Mtreit, Ruben The Handy Man, Paewiki, Pishcal, Super-gel, JopV, Comealongpond, KasparBot and Anonymous: 580

- **Arne Skoog** *Source:* https://en.wikipedia.org/wiki/Arne_Skoog?oldid=535204142 *Contributors:* Grafen, Colin41, Khazar, Shirt58, Waacstats, Tomas e, RjwilmsiBot and Radiowood

- **Association for International Broadcasting** *Source:* https://en.wikipedia.org/wiki/Association_for_International_Broadcasting?oldid= 647811759 *Contributors:* Xezbeth, FreplySpang, Alarob, Elonka, Adam sk, Cydebot, Simon.spanswick, Khairul hazim, Kathleen.wright5 and Anonymous: 2

- **British DX Club** *Source:* https://en.wikipedia.org/wiki/British_DX_Club?oldid=647807748 *Contributors:* Bearcat, Woohookitty, Hbdragon88, SmackBot, Trident13, Eastmain, MarshBot, Beachyboy, Hjal, Waacstats, Xnuala, Vanished user ikijeirw34iuaeolaseriffic, JL-Bot, Serialdownloader, LilHelpa, Birkonian, Antiqueight, ChrisGualtieri and Anonymous: 2

- **Digital Radio Mondiale** *Source:* https://en.wikipedia.org/wiki/Digital_Radio_Mondiale?oldid=700207164 *Contributors:* General Wesc, The Anome, Camembert, Liftarn, Ixfd64, Karada, Minesweeper, Glenn, Smack, Vanished user 5zariu3jisj0j4irj, Radiojon, Morwen, Thue, Hajor, JorgeGG, Denelson83, Bearcat, Robbot, Puckly, Apraetor, David Gerard, Corrados, Khalid hassani, Bobblewik, Peter Ellis, Shakata-GaNai, Maikel, DmitryKo, Grstain, GregDumbrell, Mindfsck, Rich Farmbrough, Cmdrjameson, CR7, Ashley Pomeroy, Mavros, Paul1337, Tedp, SteinbDJ, Gene Nygaard, LrdChaos, Armando, Maartenvdbent, Eyreland, Josh Parris, Bushido Hacks, FlaBot, Gurch, Windharp, Yurik-Bot, Ytrottier, Off!, Zwobot, Kewp, Ilmaisin, Closedmouth, RG2, Mardus, SmackBot, Chris the speller, Bidgee, Coinchon, ACupOfCoffee, Harumphy, Stuartfanning, Samuel.dellit, Hu12, DangerousPanda, CmdrObot, Wattyirl, WeggeBot, Altaphon, Thierry Bingen, Thijs!bot, Dark Serge, Gehtnich, JAnDbot, Harryzilber, CosineKitty, Cyclonius, Jahoe, Xb2u7Zjzc32, Steve Hosgood, Boffob, Japo, Vssun, Erpbridge, R'n'B, StewE17, Natobxl, VolkovBot, TreasuryTag, Rclocher3, Planish, The Original Wildbear, Agent452, McM.bot, Lillingen, Theaveng, Miniapolis, Lightmouse, Int21h, Chump Manbear, Mild Bill Hiccup, Niceguyedc, ElSaxo, Rcooley~enwiki, Dthomsen8, Addbot, RichardBond, Donfbreed, AnomieBOT, Mahmudmasri, Obersachsebot, Nikolaus2, RibotBOT, Sergedum, NoGo, Johnnie Rico, Young1sofparkesnswaust, I dream of horses, Broadcasttransmitter, Jmooredrm, DoobieFryer, John of Reading, Erianna, Zlatan8621, Matthiaspaul, Gnisten, Helpful Pixie Bot, BG19bot, Mark Arsten, Compfreak7, Bkjr3245bfv, ChrisGualtieri, YFdyh-bot, Spyglasses, Osbournehutch, Squizzler, Aetheradio, Tc.chair, Marcusmueller ettus and Anonymous: 124

- **Duga radar** *Source:* https://en.wikipedia.org/wiki/Duga_radar?oldid=702132372 *Contributors:* Malcolm Farmer, Maury Markowitz, Julesd, Smack, Shizhao, Secretlondon, M1fcj, Denelson83, Twang, Aliter, Cyrius, Alan Liefting, Timvasquez, D6, DanielCD, Brianhe, Rich Farmbrough, Rydel, Alistair1978, Kbh3rd, Loren36, Sergiek, W8TVI, Savvo, ליאור, Theodore Kloba, Gene Nygaard, Crosbiesmith, Bobrayner, Arru, Eyreland, Allen3, Graham87, Rjwilmsi, Ground Zero, Bottesford, Benlisquare, Brandon, Voidxor, Ninly, SIGURD42, DasBub, Jeff Silvers, SmackBot, Kuban kazak, PEHowland, Brossow, Ohnoitsjamie, Kharker, Hibernian, Rolypolyman, Daguero, Harumphy, Pettefar, EgorFiNE, LouScheffer, Easwarno1, Fuhghettaboutit, Rustypup49, A.R., Peaceduck, A5b, Charivari, JoshuaZ, Trounce, Meco, Siebrand, Iridescent, Chetvorno, Gholson, Cxw, Van helsing, Chrisahn, Cydebot, Tec15, Thijs!bot, SkonesMickLoud, N5iln, JustAGal, DagosNavy, Mwarren us, Bighousehx, LorenzoB, Cooper-42, R'n'B, Johnpacklambert, Necator~enwiki, DorganBot, 1812ahill, Nxsty, VolkovBot, TXiKiBoT, Kaiketsu, Michaeldsuarez, MaksKhomenko, 400Hz100V, AHMartin, VVVBot, Gerakibot, Cwkmail, Szater, Afernand74, ClueBot, Jackbootdash, Drmies, TheOldJacobite, Kathleen.wright5, Alexbot, SilvonenBot, Addbot, Nohomers48, TutterMouse, Bufar, Lightbot, Alkab, Luckasbot, Donfbreed, Chornobyl~enwiki, AnomieBOT, Obłuże, Xqbot, JimVC3, Pmlineditor, GrouchoBot, FrescoBot, Citation bot 1, RedBot, Belchman, Kelly2357, Jonkerz, EmausBot, Mgtm7m, Your Lord and Master, DefiningEternity, Exclusion-zone, Ingmar Runge, ClueBot NG, Helpful Pixie Bot, BG19bot, Original Token, Chrisrudge, MrPinks, Artyom Millachie, TortoiseWrath, Chernobyltour, K7DGI, Monkbot, Pripyatcanin, Jon Çobani, Олег Михайлович, Hammer5000, Vlan2 and Anonymous: 91

- **Duga-1 and Duga-2** *Source:* https://en.wikipedia.org/wiki/Duga-1_and_Duga-2?oldid=593564793 *Contributors:* Secretlondon, Alan Liefting, Crosbiesmith, Cxw, Tec15, Yobot, Mikhail The Russian and Gabiggio

- **DXing** *Source:* https://en.wikipedia.org/wiki/DXing?oldid=694842871 *Contributors:* Tim Ivorson, Mormegil, Discospinster, Shenme, Ceyockey, Woohookitty, Rjairam, BD2412, Rjwilmsi, Vegaswikian, Tbone, SystemBuilder, Anonym1ty, Mhking, YurikBot, StuffOfInterest, RussBot, Mysid, Jeh, Mike Selinker, SmackBot, Rutja76, Millifoo, Andy M. Wang, Bluebot, Neo-Jay, A. B., Jmlk17, Radagast83, Forster 06, ChrisCork, Thijs!bot, JurgenG, LuckyLouie, Alphachimpbot, Dman727, JAnDbot, Milonica, Harryzilber, MER-C, SiobhanHansa, VoABot II, STBot, Jim.henderson, Mange01, TomCat4680, Natobxl, AntiSpamBot, STBotD, Bonadea, Xnuala, VolkovBot, TXiKiBoT, N5na, Falcon8765, Djdubuque, 400Hz100V, SieBot, Igor.grigorov, EEMajor, TTQ07, Dlrohrer2003, ClueBot, TinyMark, Mild Bill Hiccup, Sv1xv, Cgord, Rcooley~enwiki, Dxinginfo, Mlaffs, Addbot, Quantock, SamatBot, Lightbot, Ivanov id, Zorrobot, Yobot, Cureden, RibotBOT, Asfarer, FrescoBot, Jcoltrane666, Neøn, Khazar2, Pyroray, Jjssfjfjd and Anonymous: 69

- **Eastern Bloc media and propaganda** *Source:* https://en.wikipedia.org/wiki/Eastern_Bloc_media_and_propaganda?oldid=690080966 *Contributors:* Bogdangiusca, Chowbok, Realsurreal, Megara, Piotrus, D6, Giraffedata, Saga City, Redvers, BD2412, Li-sung, Koavf, Ground Zero, Gaius Cornelius, Aeusoes1, Renata3, Gadget850, SmackBot, Incnis Mrsi, Mujanovic, Raymie, Rrburke, Derek R Bullamore, Shattered, CoolKoon, SimonD, J Milburn, AndrewHowse, Cydebot, Biruitorul, Zachwoo, Drazil91, Cloudz679, Maurice Carbonaro, Adamdaley, Ajfweb, Mycomp, Albanman, ImageRemovalBot, Iohannes Animosus, ChrisHodgesUK, Dthomsen8, WikHead, Addbot, Fyrael, Mosedschurte, Ivario, Luckas-bot, Yobot, Alexikoua, LilHelpa, I dream of horses, D.Albionov, Σ, Cheong Kok Chun, RjwilmsiBot, Gaubbi, Nkrita, H3llBot, Hazhk, Helpful Pixie Bot, Jzcool4455, Politbir, Jevoite, MartinZwirlein, Egeymi, Classafelonymonkey and Anonymous: 22

- **Electromagnetic interference** *Source:* https://en.wikipedia.org/wiki/Electromagnetic_interference?oldid=704445279 *Contributors:* Damian Yerrick, Heron, Olivier, Ahoerstemeier, Ronz, Poor Yorick, Omegatron, Bearcat, Hankwang, Altenmann, Stewartadcock, Khashmi316, DavidCary, Bkonrad, Glenn Koenig, Beland, Alistair1978, Bender235, Femto, Cmdrjameson, Kjkolb, Slambo, Hooperbloob, M00dimus, Atlant, Ddlamb, Wtshymanski, DV8 2XL, Nightstallion, Lkinkade, Linas, Pol098, CPES, ObsidianOrder, BD2412, Ground Zero, Neonil~enwiki, Lmatt, Srleffler, StuffOfInterest, Sasuke Sarutobi, Kirill Lokshin, Shaddack, Varnav, Mkill, Kkmurray, Samuel Blanning, Amux, Cadmium, Oli Filth, Nbarth, Frap, OrphanBot, Rrburke, Dougmc, Drphilharmonic, Checco, Flip619, NewTestLeper79, Beetstra, Zureks, ShelfSkewed, Requestion, Pushdgr8, Crum375, Thijs!bot, Sobreira, Roggg, Hcobb, Sbandrews, LuckyLouie, Tneedham, Steelpillow, .anacondabot, Avjoska, JamesBWatson, Budnick1, JaGa, Rwd999, Vanessaezekowitz, Jim.henderson, Glrx, STBotD, ICE77, GLPeterson, The Original Wildbear, Martin451, Lova Falk, Chronicpokerist, Tomaxer, Anoko moonlight, SieBot, TYLER, Danierrr, Bentogoa, Kuros0, NBS, Virusunknown, ClueBot, Wasami007, Thubing, Mild Bill Hiccup, Muhandes, EivindJ, WikiDao, ZooFari, Addbot, Metagraph, D0762, Semiwiki, Care, Luckas-bot, Alexkin, Hairhorn, Suludo, B137, Materialscientist, Futureseer, Benjachristensen, Chetankathalay, EMCarticles, Dougofborg, Pinethicket, Iceman797, PSUdesigner, WikiEMC123, Abm26, Dumitrescu Cristian, Callanecc, Rock3853277, Alfaisanomega, Erianna, Zukaz, ChuispastonBot, ClueBot NG, HAAlice, AeroPsico, Tylko, Snotbot, HHaeckel, JordoCo, BG19bot, Philsvision, Eleazarlazo, Xlicolts613, Loriendrew, BattyBot, Neshmick, Joerega, Mr. Wikipediania, Shaikh haque mobassir, Faizan, LiveRail, Hanthoec, Monkbot, Kyledonbaker, SageGreenRider, Sharma.ashoks, Samss86, KasparBot and Anonymous: 100

- **English-language radio** *Source:* https://en.wikipedia.org/wiki/English-language_radio?oldid=659520959 *Contributors:* Oneiros, SmackBot, Chris the speller, Iridescent, CmdrObot, Sgt Simpson, Prolog, Cgingold, Englishlanguageradio, Dick Kimball, Dravecky, Callmederek, Chris-Gualtieri, Polemicista and Anonymous: 17

- **Euronews** *Source:* https://en.wikipedia.org/wiki/Euronews?oldid=704244091 *Contributors:* Edward, Tim Starling, Haakon, Lee M, Radiojon, Jsanroman, Morven, MrWeeble, PedroPVZ, James Hatts, Admbws, DocWatson42, Pmaguire, Malcontent, Zoney, OldakQuill, 159753, Darksun, Trilobite, AndrewM1, CanisRufus, Kiand, Joaopais, Deathawk, Väsk, Shenme, Cmdrjameson, BAN NORM AND TEXTURE, Rjhatl, Paul-Hanson, Anthony Appleyard, Swarve, Bete, Ashlux, Woohookitty, Awostrack, Uncle G, RicJac, Matturn, Xizer, BD2412, Dueyfinster, Rjwilmsi, Nanami Kamimura, Tim!, Slac, SLi, FlaBot, LiangHH, Mhking, YurikBot, Wavelength, RussBot, THOMAS, UKWiki, DJ Bungi, DelftUser, Rbarreira, PGPirate, Chincheta~enwiki, Closedmouth, Arthur Rubin, Pejman, Luckystars, Airodyssey, AntL, Kungfuadam, Thomas Blomberg, Luk, SmackBot, Bormalagurski, Estoy Aquí, C.Fred, Eskimbot, Giandrea, Betacommand, Chris the speller, MK8, Hibernian, Alphathon, Chlew-bot, WorldWide Update, Azumanga1, Iam4Lost, Yulia Romero, LeonidasSpartan, Fatla00, WayKurat, JLogan, WikiWitch, NotMuchToSay, Cyzor, Kuru, Khazar, John, Euchiasmus, LUCPOL, Strainu, Davidvankemenade, -js-, AxG, Liski, Eve215454, Skinsmoke, Mike Doughney, Nehrams2020, Tomwood0, Octane, JLCA, Lee Stanley, DangerousPanda, CmdrObot, GMc, AndrewHowse, Pit-yacker, Cydebot, Peripitus, Tomta1, Thijs!bot, Wikid77, Bobblehead, Neilajh, Kaaveh Ahangar~enwiki, Nick Number, Natalie Erin, BokicaK, Sobaka, Superzohar, JAnD-bot, Tobibln, Maitre So, Rothorpe, Jay1279, .anacondabot, Pedro, Scanlan, Haku8645, Wrightaway, Gr1st, Adriaan, Sinigagl, Radical3, Sk wiki, Flrn, CommonsDelinker, Gammondog, Ssolbergj, DrFrench, J.delanoy, Herbythyme, Bot-Schafter, GenTV~enwiki, Peracute, Ajfweb, Jasmeet 181, Sphayros, GrahamHardy, Ciaranholahan, Glossologist, VolkovBot, Howth575, Gottago, Accountready, AlnoktaBOT, Gpeilon, TXiKiBoT, Griffindd, Yunaresuka, Predavatel, DieOfGoodLuck, Alan1989, Broadbot, Lokioak, Msw1002, AlleborgoBot, Refalm, Skytvfreak, Dan69en, SieBot, Pawebster, Yintan, Flyer22 Reborn, Oahiyeel, Nua eire, Momo san, SeoR, Aspects, Lightmouse, P.Marlow, LarRan, Laurentiu Popa, Radical-Dreamer, Andriy155, Snigbrook, Starkiller88, Nark64, Isebito, Abbaroodle, Mild Bill Hiccup, Babelia, Kathleen.wright5, Niceguyedc, ElSaxo, Bbb2007, PixelBot, Mfa fariz, Rui Gabriel Correia, Donegal92, DumZiBoT, A.h. king, Wertuose, Dthomsen8, Silvo-nenBot, MystBot, Addbot, Satbuff, Coryrob1979, MrOllie, Tassedethe, Faunas, Livestation, Luckas-bot, Yobot, Bearas, MacTire02, Exam-tester, AnomieBOT, Turkish Flame, Gold Wiz113, Jim1138, Homoatrox, Cmd2222, Mahmudmasri, ArthurBot, LilHelpa, Xqbot, J4lambert, Cthuang, Almabot, GrouchoBot, RibotBOT, Invest in knowledge, Lulu97417, Esfandiasil, LucienBOT, Mxipp, The GateKeeper07, Nima Farid, DrilBot, Mraandthebigbrother, TheToch, InMontreal, John123521, Cobija, WardMuylaert, Partialouest, Cheong Kok Chun, Heamsi, Artury-atsko, RjwilmsiBot, Saftorangen, Spacejam2, CTWPerfection, EmausBot, John of Reading, Spikethepunkrocker, G8crash3r, Ismukhammed, Polgraf, C0re1980, Eliolb97, Kkm010, ZéroBot, Dandublin93, Daky585, EWVGN3T0, H3llBot, Ruairimcaleer, Abtalion, TheA5, Michalis Melidonis, Rufflos, WPSamson, Sasukemadara, Widr, T-resh, BG19bot, Billy Liakopoulos, Christian2941, Modern Cromwell, Golstein, Ph-nomPencil, HIDECCHI001, Kath3169, Clacos9, Surtalnar, Wassim.Mansouri, Justincheng12345-bot, Pai Walisongo, Franckito73, Pratyya Ghosh, Tohandr009, Egeymi, Ekren, Mediaba, Cmdireland, Natig98, NightShadow23, Parix, Lemnaminor, RabeaMalah, Fabienamnet, RCP-TxDW, Mcollenrati, Meiræ, Monkbot, Ianfella, LBG80, Poveglia, Crystallizedcarbon, SLBedit, MediaWatch Anon, ScrapIronIV, Louiemer-cado2, XPanettaa, Prinsgezinde, Urgup-tur and Anonymous: 268

- **European Broadcasting Area** *Source:* https://en.wikipedia.org/wiki/European_Broadcasting_Area?oldid=689050871 *Contributors:* Rwalker, Eurosong, Mais oui!, Heitordp, Iridescent, Vanjagenije, Dispenser, PDFbot, CT Cooper, Leonard9ca, Addbot, Zorrobot, Rubinbot, Khajidha, MAXXX-309, Danish Expert, ChuispastonBot, HTML2011 and Anonymous: 3

- **FTA receiver** *Source:* https://en.wikipedia.org/wiki/FTA_receiver?oldid=641041013 *Contributors:* Ixfd64, Auric, Rhombus, Utcursch, Xez-beth, Thebrid, Sole Soul, Mazca, XP1, SchuminWeb, D.brodale, MMuzammils, RussBot, WikiY, Madlobster, PRehse, NetRolller 3D, Smack-Bot, C.Fred, Chris the speller, Lyran, Nakon, Gobonobo, Mihitha, Synergy, Mentifisto, Majorly, Harryzilber, Dipper2, R'n'B, Commons-Delinker, Jmccormac, J.delanoy, Arcanedude91, Funandtrvl, Signalhead, Finngall, SieBot, ImageRemovalBot, ClueBot, Snigbrook, Carnage99, SpikeToronto, Alikian, Robasap, Dekisugi, Mlaffs, Versus22, Bobthebugman, Jimmydean3012, XLinkBot, RoverRexSpot, HexaChord, Ad-dbot, Satbuff, WiiNie, Fluffernutter, Lightbot, Keeyon, Teknow, RottenDurian, Starbois, Killiondude, Bonusballs, I613, PSBsatellite, Grou-choBot, FrescoBot, Matchmanhattan, Kukdide, Quantanew, Oolloo2010, Diamondland, Sangyun357, YFdyh-bot, Ziggyhorse, K7L, Odoustar and Anonymous: 110

- **Gascony Show** *Source:* https://en.wikipedia.org/wiki/Gascony_Show?oldid=603322358 *Contributors:* Vonaurum, Deansfa, Dl2000, Commons-Delinker, SteveStrummer, Deveze, Kjell Knudde, Helpful Pixie Bot, Claviere and BattyBot

- **George Wood (Radio Sweden)** *Source:* https://en.wikipedia.org/wiki/George_Wood_(Radio_Sweden)?oldid=704230166 *Contributors:* BD2412, Ser Amantio di Nicolao, Waacstats, Griffindd, KathrynLybarger, Tomas e, XLinkBot, RjwilmsiBot, BG19bot and Radiowood

- **Glenn Hauser** *Source:* https://en.wikipedia.org/wiki/Glenn_Hauser?oldid=704497303 *Contributors:* Edward, Klemen Kocjancic, D6, Rich Farmbrough, Alai, Eyreland, Joe Decker, Jrtayloriv, RussBot, Jaxl, Caerwine, Aschmidt, SmackBot, Sluggoster, Chris the speller, Ohconfu-cius, JustAGal, Milonica, Waacstats, VolkovBot, Bolt, Mr.Z-bot, Phil Bridger, Miniapolis, Kumioko (renamed), Trivialist, Addbot, Lightbot, Kiddo27, Dana60Cummins, January, RjwilmsiBot, Hourtop, Helpful Pixie Bot, BattyBot, KellyLovato38, Finnusertop, KasparBot and Anony-mous: 10

- **Greenwich Time Signal** *Source:* https://en.wikipedia.org/wiki/Greenwich_Time_Signal?oldid=703666516 *Contributors:* Matthew Woodcraft, Zundark, The Anome, Tarquin, William Avery, Heron, Fonzy, Mintguy, Tregoweth, Angela, Cgs, [212], HappyDog, Mtcv, Denelson83, Pigson-thewing, Fredrik, Naddy, Chris Roy, TimR, Wereon, Alerante, Andrewphelps, Bkonrad, Markus Kuhn, Tom-, 159753, Geni, H1523702, PFH-Lai, Aerion, Jp347, Picapica, Geof, Pak21, MBisanz, Bobo192, Zachlipton, Philip Cross, ProhibitOnions, Paul1337, Gpvos, Pcpcpc, The JPS, Pol098, Eyreland, Harmonica~enwiki, Tim!, Destroyer of evil, Coolhawks88, Trainthh, GeeJo, DragonHawk, LibraryCJH, Multichill, Cleared as filed, Nick, Lcmortensen, Zr2d2, Tim Parenti, Lucas42, Saltmarsh, Zquack, SmackBot, JMiall, Colonies Chris, Aviageek, Ritchie333, PhilipB, PhotoJim, Charivari, Briantist, GothCabbage, DouglasCalvert, Neurillon, Fursday, W guice, MFlet1, Oliverdavison, AndrewHowse, Cydebot, Rrsmac, Hebrides, After Midnight, Andrew Clark, Davidhorman, Dfrg.msc, EdJohnston, TygerTyger, Majorly, AstroLynx, JAnDbot, Pedro, Tomcakes, Zahakiel, Zir, R'n'B, Slash, JonnyH, Swaddon1903, Sween64, Guns2006, Khairul hazim, Kevin Steinhardt, Synthebot, AHMartin, GoddersUK, SieBot, Lightmouse, SimonTrew, Mcbill88, Glennb28, Mr. Granger, Digital.Diablo, Islenska2007x, Piledhigheranddeeper, Triv-ialist, Bbb2007, DumZiBoT, XLinkBot, SP1R1TM4N, Al tally, Ka Faraq Gatri, CF1V8, Tassedethe, מבשׁע-זרם, Legobot, Luckas-bot, Yobot, Fraggle81, KBurchfiel, Radiopathy, Starbois, AnomieBOT, Ulric1313, Colinward1970, CyberneticianDave, Celuici, FrescoBot, Chemical2009,

Pjbpr1, Maarten2778, CarlosDude1337, Trappist the monk, Fxd25, Aw16, Perthshire2009, John of Reading, Exok, Ifore2010, Sebthedev, Basicporch, Wbm1058, Aisteco, ChrisGualtieri, KiwiNeko14, CoffeeWithMarkets, Osterforde, Froglich, Ginsuloft, SuperRafiuddin, Sharpglue, Dr. British12 and Anonymous: 127

- **Guglielmo Marconi** *Source:* https://en.wikipedia.org/wiki/Guglielmo_Marconi?oldid=704719626 *Contributors:* Tobias Hoevekamp, Derek Ross, Mav, Bryan Derksen, Andre Engels, Danny, DavidLevinson, Panairjdde~enwiki, Rsabbatini, Montrealais, Tedernst, Someone else, Renata, Shyamal, Mic, Ixfd64, GTBacchus, Delirium, Spliced, Looxix~enwiki, Ahoerstemeier, William M. Connolley, Bueller 007, Nikai, Merry, Jengod, Reddi, Magnus.de, Nataraja~enwiki, WhisperToMe, Zoicon5, Selket, Tpbradbury, Maximus Rex, Hyacinth, RayKiddy, Shizhao, Joy, Raul654, Dimadick, Riddley, Robbot, Pigsonthewing, Kizor, Securiger, Blainster, Timrollpickering, Mervyn, JackofOz, Mushroom, Tobias Bergemann, Alan Liefting, Hexii, Ancheta Wis, Giftlite, Wolfkeeper, Nunh-huh, Tom harrison, Everyking, No Guru, Curps, Duncharris, Gzornenplatz, Dingo~enwiki, Alexf, Antandrus, Mustafaa, PDH, SimonArlott, 1297, Beardless, Dcandeto, Picapica, Gcanyon, Lacrimosus, Corti, D6, TheBlueWizard, Ta bu shi da yu, Imroy, Noisy, Discospinster, Rich Farmbrough, Guanabot, Brutannica, Psd, Xezbeth, Mani1, Martpol, MarkS, SpookyMulder, Bender235, ESkog, Pissipo, Sfahey, Kwamikagami, Hayabusa future, Tom, Bobo192, Cymsdale, Kilclaren, Man vyi, Nk, JesseHogan, Knucmo2, Jumbuck, Danski14, Alansohn, Gary, TheParanoidOne, Anthony Appleyard, Arcenciel, Ben davison, Jeltz, Craigy144, Leonardo Alves, Riana, AzaToth, Ksnow, Wtmitchell, Hadlock, Plainclothedman, Wtshymanski, Mikeo, Gene Nygaard, Ghirlandajo, Richard Weil, Bastin, Smark33021, Richard Arthur Norton (1958-), Woohookitty, FeanorStar7, Camw, Barrylb, Jaavaaguru, Scjessey, Gimboid13, DESiegel, Stefanomione, Palica, Emerson7, Graham87, Kbdank71, Edison, Ketiltrout, Rjwilmsi, Koavf, Lockley, MZMcBride, Tawker, Vegaswikian, Kazrak, Ghepeu, Sferrier, Bhadani, Matt Deres, Algebra, FlaBot, Djrobgordon, GünniX, CaptainCanada, OrbitOne, Srleffler, OpenToppedBus, Smelendez, Gareth E. Kegg, Holmwood, Valentinian, Butros, King of Hearts, Chobot, Jared Preston, Gwernol, HJKeats, YurikBot, Crotalus horridus, Sceptre, StuffOfInterest, Anglius, Phantomsteve, RussBot, Sputnikcccp, Hack, Hede2000, Lexi Marie, SluggoOne, SpuriousQ, RadioFan2 (usurped), Chensiyuan, Million Little Gods, Gaius Cornelius, CambridgeBayWeather, Rsrikanth05, Kimchi.sg, Shanel, NawlinWiki, Bachrach44, Voyevoda, Tastemyhouse, Cholmes75, Dmoss, Jpbowen, Pyroclastic, PhilipC, Jbourj, Bucketsofg, Bota47, Tachs, Thomas H. White, Nick123, Wknight94, Ms2ger, Homagetocatalonia, Hnatiw, OtherDave, Closedmouth, Jwissick, Pietdesomere, Whobot, JLaTondre, Curpsbot-unicodify, Che829, Selkem, Meegs, Thomas Blomberg, NeilN, Huldra, Elliskev, Luk, Attilios, SmackBot, FocalPoint, Iacobus, KnowledgeOfSelf, KocjoBot~enwiki, Discordanian, Ozone77, Frymaster, Canthusus, Carbonix, Gilliam, Hmains, Tolivero, Bluebot, Dahn, BabuBhatt, Liamdaly620, MalafayaBot, StevenCole, WikiFlier, Zinneke, Cassivs, Sbharris, Trekphiler, Can't sleep, clown will eat me, Aleksandar Šušnjar, Ww2censor, TheKMan, Athene noctua~enwiki, Whpq, Stevenmitchell, E bruton, Nakon, Dreadstar, Kismetmagic, Where, DDima, Vina-iwbot~enwiki, SashatoBot, Aaronmack, Frade, HDarke, John, Carnby, Mike1901, J 1982, Jperrylsu, Shlomke, Henry Bradford, Dicklyon, Aditreeslime, Waggers, Doczilla, Neddyseagoon, Jayzel68, Kanon6996, Citicat, Hu12, Paul Koning, Spark, Shoeofdeath, Jamesonking, Delta x, Gil Gamesh, FairuseBot, Tawkerbot2, Chetvorno, CmdrObot, Dycedarg, Kburton@ptd.net, Drinibot, Karenjc, Chicheley, Estnyboer, CMG, Cydebot, Aodhdubh, Fl, TrevorWright, DMeyering, Flowerpotman, R-41, MWaller, Studerby, Trident13, Fifo, Tawkerbot4, Codetiger, DumbBOT, Optimist on the run, Omicronpersei8, CieloEstrellado, JamesAM, Thijs!bot, Epbr123, Kayag, Kablammo, HappyInGeneral, Martin Hogbin, N5iln, Jimbo L, Count-Dracula, Oliver202, Headbomb, Missvain, Studio503, Bunzil, Mikeeg555, 00666, Eleuther, AntiVandalBot, RobotG, Kramden4700, Akradecki, Seaphoto, Sobaka, QuiteUnusual, Dawz, SoumenRoy, Res2216firestar, JAnDbot, Harryzilber, Bernardmarcheterre, Robina Fox, Andonic, Greensburger, Trey314159, Mrmdog, Juggernaut0102, Magioladitis, Connormah, VoABot II, JamesBWatson, Froid, Avicennasis, Mwalimu59, Eiyuu Kou, LorenzoB, Canyouhearmenow, DerHexer, Shakirashakira92, Bayboy4, Faelomx, PhantomS, MartinBot, Mmoneypenny, Arjun01, Anarchia, R'n'B, CommonsDelinker, Fconaway, Lilac Soul, RockMFR, J.delanoy, Geezenstacks, Rgoodermote, AAA!, Numbo3, SuperGirl, Jiuguang Wang, Eliz81, WFinch, Hifliercanada, FruitMonkey, McSly, Pyrospirit, Mle-mot-dit, Fountains of Bryn Mawr, Pterre, DadaNeem, Ontarioboy, Polsi, Jobeto, Jrcla2, Juliancolton, Entropy, Cometstyles, STBotD, WJBscribe, Lucifero4, DorganBot, Dorftrottel, Bvasek1082, Fbarton, Idioma-bot, Kitchawan, Signalhead, Lights, Unia 1000, VolkovBot, CWii, Johan1298~enwiki, ABF, Pleasantville, Dappled Sage, Soliloquial, Chiarafumo, Station1, Philip Trueman, JuneGloom07, TXiKiBoT, Oshwah, Zamphuor, AceofdataBase, Jimmyeatskids, Abcvince10, Sankalpdravid, Qxz, Anna Lincoln, Clarince63, DennyColt, JOHNDOE5555, Don4of4, Broadbot, LeaveSleaves, Inventis, Jwy2k2, Cremepuff222, Spinninghead, Wiae, Alakazam138~enwiki, Saturn star, Tennisnutt92, Enigmaman, CoolKid1993, Falcon8765, Smileblame, Insanity Incarnate, Cbjohnny, Richc1977, NHRHS2010, AHMartin, M.V.E.i., SieBot, Jeanettephair, Harriyott, Tresiden, Hertz1888, Boy1jhn, SE7, GrooveDog, Keilana, Niall9, MF-Warburg, Radon210, Nix D, A. Carty, PolarBot, Mustluvkats94, Jimthing, Funnyhahaha, Antonio Lopez, Demack, ViennaUK, Techman224, Moletrouser, G.-M. Cupertino, Spitfire19, Vojvodaen, Torchwoodwho, JEZZAP321, TSUBI, Dabomb87, RS1900, Faithlessthewonderboy, Loren.wilton, Mbssbs, ClueBot, Binksternet, Pretzelfactory, The Thing That Should Not Be, All Hallow's Wraith, NotnventedHere, MikeVitale, Knepflerle, AusTerrapin, Niceguyedc, Blanchardb, OfficeBoy, Liempt, Shannon bohle, DragonBot, Excirial, Alexbot, Jusdafax, Solsticedhiver, NuclearWarfare, AndyFielding, Sensai Dan, WellsSt, Nobody of Consequence, Audaciter, Thingg, Cardinalem, Pspprogo, Subash.chandran007, Camboxer, Bletchley, Hotcrocodile, AgnosticPreachersKid, Bilsonius, 21stCenturyGreenstuff, Marconiwireless, Noctibus, Griffin3435, Good Olfactory, Bridgetfox, Kbdankbot, Sobelone, Addbot, Proofreader77, AVand, Some jerk on the Internet, Broletto, AkhtaBot, Ronhjones, Bender21435, CanadianLinuxUser, Cst17, Desyman44, Glane23, Favonian, FaithF, LemmeyBOT, AtheWeatherman, Zabmaru, Lemonade100, Dr.woxford, ACM2, Tyw7, Mycatchip93, Mr.Xp, VASANTH S.N., Tide rolls, Lightbot, Darkblood1899, Ben Ben, Legobot, Math Champion, Luckas-bot, Yobot, Worldbruce, Marcbela, THEN WHO WAS PHONE?, JEms123, Jean.julius, South Bay, NewJerseyRadioHistory, AnomieBOT, DemocraticLuntz, 1exec1, Galoubet, Profangelo, Pooopfreid101, Ulric1313, Materialscientist, Jcrct, Citation bot, Bob Burkhardt, Peachesman94, ArthurBot, LilHelpa, John Bessa, MauritsBot, Xqbot, WilliamturnerII, Ekconklin, Jayarathina, Capricorn42, Wperdue, Timmyshin, Davshul, Sante44, Tad Lincoln, Shojego, GrouchoBot, Patrick12211, Omnipaedista, Shirik, RibotBOT, Yoganate79, Brutaldeluxe, Trafford09, NJam101, Cheerchicka 2012, A.amitkumar, Captain-n00dle, Vietnamvat, FrescoBot, Magnagr, Tobby72, VS6507, JuniperisCommunis, KokkaShinto, HJ Mitchell, Ripo20, Bob dole28, Froster69er1, Sea King 27, WikiDude23, Pinethicket, I dream of horses, Brayalad, Anibar E, Plucas58, Hamtechperson, Geogene, Fat&Happy, Fumitol, Jauhienij, TobeBot, Silent Billy, Mr Mulliner, Etincelles, Number579d, Oracleofottawa, Vrenator, Viktor Laszlo, Diannaa, Suffusion of Yellow, DARTH SIDIOUS 2, Dancinfoool!, Necroking1, Tanneryvillage, Whisky drinker, Axxxion, RjwilmsiBot, Altes2009, DexDor, Meanstheatre, Grepman, Dfgsd45, Crazymarsman, Der Künstler, EmausBot, John of Reading, Immunize, Never give in, Coolychris911, Racerx11, IncognitoErgoSum, RA0808, Dhawalvankar007, Tommy2010, Wonderation, Wikipelli, Dcirovic, K6ka, C16sh, Hhhippo, Kkm010, PBS-AWB, Lemeza Kosugi, Fæ, Emily Jensen, Cooksmagooks, Ὁ οἶστρος, Wir9796, Frigotoni, Tolly4bolly, Akasseb, Paymanpayman, Ludovica1, KAWiggins, Donner60, Chewings72, Historydocumentation, ChuispastonBot, Forever Dusk, TYelliot, DASHBotAV, Xonqnopp, ClueBot NG, Rich Smith, Mrmaryjanelova, Jonnycomp, LogX, Engradio, Movses-bot, AveVeritas, Marconi1901, Snotbot, Frietjes, ScottSteiner, YellowFratello, Widr, Gizgalasi, Oddbodz, Helpful Pixie Bot, YborCityJohn, Mhreid, Juro2351, PearlSt82, MusikAnimal, Josvebot, Mark Arsten, Helevorn, Maupertius, Majorbolz, JSWHU, Rococo1700, JZCL, Wrath X, Jaqeli, Singhjeevan, Bagustris, Andrewgprout, Cloptonson, Cyberbot II, ChrisGualtieri, SD5bot, Drow69, JYBot, Aditya Mahar, Horation12, Mogism, Markdhewitt, Periglio, Jeremy507, Aries no Mur, Lugia2453, VIAFbot, Telfordbuck, Melanie0522, Da6dfri, Alyssagpenick, SilverHawk156, LaurentianShield, Cornishfern, Ghinozzi-nissim, NariceA, Owain Knight, Epic Failure, Sciencedude100, Ryanspleb, KnwonStealth, Bobby Martnen, Jonarnold1985, InfoDataMonger, Batmanlover9000, 115ash, Esplanade1, Anatelo, Ndstead, Poopscoop1232, Zazauly, KasparBot, JeremiahY, 5millionangry-hornets, Nosy thegazelle, DatGuy, Allthefoxes, I7624240 and Anonymous: 1002

- **International Broadcasting Act** *Source:* https://en.wikipedia.org/wiki/International_Broadcasting_Act?oldid=689969371 *Contributors:* Andrewman327, Bearcat, Rich Farmbrough, Ground Zero, Waggers, Hemlock Martinis, Cydebot, Pjoef, SoxBot, Yobot, AvicAWB, Sross (Public Policy), Stephmae, Jphill19 and Dsprc

- **International Broadcasting Convention** *Source:* https://en.wikipedia.org/wiki/International_Broadcasting_Convention?oldid=697288930 *Contributors:* Rpyle731, MakeRocketGoNow, Denniss, JeremyA, RichardWeiss, RussBot, Peter S., EngineerScotty, Joel7687, Eivind F Øyangen, Kvng, Stephen B Streater, CmdrObot, Chicheley, Cydebot, Trident13, Jvhertum, MartinBot, Verkle, Jpdavidson, Bonadea, HairyWombat, Kig8472, Kathleen.wright5, Addbot, Margin1522, Yobot, AnomieBOT, Aoidh, EmausBot, CLI, Singhmahendra20, Amr.rs, MUSEnvi, TheGeekette and Anonymous: 15

- **Interval signal** *Source:* https://en.wikipedia.org/wiki/Interval_signal?oldid=697575503 *Contributors:* GABaker, Timc, JB82, Hadal, Xyzzyva, Alexwcovington, Robert Weemeyer, Catdude, JulieADriver, Picapica, BarkingFish, Gpvos, Eyreland, Mandarax, Rjwilmsi, RobotE, IBook of the Revolution, Eddie.willers, Mlouns, Mysid, Bcshell, SIGURD42, Garion96, SmackBot, Endroit, Bluebot, WorldWide Update, SQB, BENNYSOFT, WinBot, Tony Myers, Albany NY, Klackalica, Mrceleb2007, Samslipknot, Wiendietry~enwiki, SieBot, Bayonett, Miniapolis, Harry the Dirty Dog, Kathleen.wright5, ElSaxo, SchreiberBike, XLinkBot, Ost316, Addbot, SpBot, Yobot, Citation bot, Olivier92, FrescoBot, Oldbeforehistime, Troublemaker1949, Mickeylove73, ZéroBot, Ericmetro, Helpful Pixie Bot, BG19bot, Hongtiezhu, ChrisGualtieri, Timdog13, Monkbot and Anonymous: 50

- **Joe Adamov** *Source:* https://en.wikipedia.org/wiki/Joe_Adamov?oldid=660304392 *Contributors:* Altenmann, Alan Liefting, Kingal86, Rich Farmbrough, Rugxulo, BD2412, Metaspheres, Rjwilmsi, RussBot, Conscious, SIGURD42, TA-ME, Hvn0413, Luna Santin, JhsBot, BOTijo, Reginald Perrin, Vershinin~enwiki, AllenHansen, Addbot, Lightbot, Timurite, AnomieBOT, Hovhannesk, Full-date unlinking bot, RjwilmsiBot, Kathi17, ChrisGualtieri, VIAFbot, Tiptoethrutheminefield, KasparBot and Anonymous: 4

- **John Robles** *Source:* https://en.wikipedia.org/wiki/John_Robles?oldid=624852531 *Contributors:* Mandarax, Wavelength, Petri Krohn, SmackBot, Tec15, Magioladitis, Waacstats, Parellic, ImageRemovalBot, Rowantoad, Interceptor369, Felyza, RjwilmsiBot and Anonymous: 4

- **Medium wave** *Source:* https://en.wikipedia.org/wiki/Medium_wave?oldid=697629819 *Contributors:* Timo Honkasalo, The Anome, RTC, GABaker, Spliced, Ellywa, Glenn, Palmpilot900, Wfeidt, Mulad, Radiojon, Jerzy, Twang, RedWolf, Baldhur, Phil1988, JTN, ArnoldReinhold, Ericamick, Gerry Lynch, ZeroOne, Evice, Nigelj, ProhibitOnions, Wtshymanski, Rugxulo, Stephan Leeds, DV8 2XL, Gene Nygaard, Jakes18, KelisFan2K5, Jpers36, Pol098, BD2412, Squideshi, RexNL, Chobot, 121a0012, YurikBot, SpuriousQ, Dddstone, Daniel C, IslandRogger973, SmackBot, Eskimbot, Kharker, Tghe-retford, A. B., AKMask, Suicidalhamster, Harumphy, Danikayser84, Zonk43, Mattdp, Bucksburg, Metre01, Stereorock, Dani 7C3, Chetvorno, Wa2ise, Flowerpotman, Dawnseeker2000, JAnDbot, Harryzilber, Rothorpe, JMyrleFuller, Ashishbhatnagar72, Glrx, Kostisl, R'n'B, CommonsDelinker, J.delanoy, Ohms law, Liveinthewire, RingtailedFox, TXiKiBoT, Sankalpdravid, Qxz, Cuddlyable3, SteveSTFA, SieBot, Hertz1888, Purbo T, Callidior, Rcooley~enwiki, Dxinginfo, Iohannes Animosus, Mlaffs, DumZiBoT, Thebestofall007, Addbot, M.nelson, Lightbot, Luckas-bot, Yobot, Dellant, GrouchoBot, Omnipaedista, RibotBOT, Maitchy, Sibian, DABenji, RedBot, Cnwilliams, FoxBot, Unbitwise, Zumbooruk2, Rssbro, A930913, Peterh5322, ChuispastonBot, Matthiaspaul, PoqVaUSA, Danim, MerllwBot, Helpful Pixie Bot, Cqdx, Justincheng12345-bot, Hasenburg, 313 TUxedo, FellowesCarl, Corn cheese, One Of Seven Billion, Monkbot, DXFinder and Anonymous: 90

- **Millennium Live** *Source:* https://en.wikipedia.org/wiki/Millennium_Live?oldid=681557212 *Contributors:* Tregoweth, Rich Farmbrough, Hailey C. Shannon, Tim!, SNIyer12, Tarmo Tanilsoo, RussBot, Hmains, TimBentley, AussieLegend, TenPoundHammer, AxG, Dl2000, Juan Cruz~enwiki, STBot, Wiendietry~enwiki, Snowbot, Bjoh249, Heracletus, Unbuttered Parsnip, Arjayay, AnomieBOT, Timmyshin, FrescoBot, Full-date unlinking bot, Erpert, PhnomPencil, Mogism, SummerPhDv2.0 and Anonymous: 11

- **Modulation** *Source:* https://en.wikipedia.org/wiki/Modulation?oldid=704321532 *Contributors:* AxelBoldt, The Anome, Bdesham, Michael Hardy, Ralmin, Stw, Glenn, Smack, Wikiborg, Joy, Denelson83, Robbot, Altenmann, Giftlite, Svenjissom, DavidCary, Qartis, Inkling, RobertYu, Ssd, Superborsuk, Cihan, Starx, DmitryKo, Danh, Mike Rosoft, EugeneZelenko, West London Dweller, One-dimensional Tangent, Simon South, Photonique, Towel401, Rabarberski, Argilo, DV8 2XL, Mahanga, Meodou, Dandv, Dtwitkowski, BD2412, Pleiotrop3, HappyCamper, FlaBot, Lmatt, Tedder, Chobot, Hatch68, Krishnavedala, Roboto de Ajvol, YurikBot, RussBot, Red Slash, Splash, Grafen, BOT-Superzerocool, Mysid, Gadget850, Bota47, Plamka, Yeryry, Light current, Hadipedia, MaratL, Yaco, Katieh5584, Finell, SmackBot, Bernard François, Gilliam, Chris the speller, Oli Filth, MalafayaBot, McNeight, Gutworth, Dreadstar, Daniel.Cardenas, Muadd, Dicklyon, EEPROM Eagle, Kvng, Lee Carre, Chetvorno, JohnTechnologist, CmdrObot, MC10, Thijs!bot, Siwiak, Nick Number, Escarbot, Three Laws of Robotics, AntiVandalBot, Seaphoto, Pranav v, Doktor Who, JAnDbot, Harryzilber, VoABot II, Hmo, Wksalar, Read-write-services, Technicolorcavalry, Vanwhistler, Mange01, Yonidebot, Philippe23, DD2K, JClark2906, Atropos235, DorganBot, Idioma-bot, Funandtrvl, Sam Blacketer, VolkovBot, ICE77, Alinja, Cbradiomagazine, TXiKiBoT, Rei-bot, MichaelStanford, Jpat34721, Canaima, Cuddlyable3, Doc James, Logan, Bluemouse2306, Howard-Morland, SieBot, ToePeu.bot, Gerakibot, Bentogoa, Yerpo, Berserkerus, Smshaner, Unknownx123, Alex.muller, Anchor Link Bot, KLuwak, Tuxa, ClueBot, Binksternet, The Thing That Should Not Be, Pyr0technician, Wispanow, Sepia tone, Aua, Karlhendrikse, Coralmizu, Abrech, Jotterbot, Muro Bot, Berean Hunter, Johnuniq, Tprentice, Analogkidr, Mahamahamaha, Stradivariusis, Rebmertaumer, Ixhotl, Avoided, Addbot, Jpmonroe, Tanhabot, Maziaar83, Leszek Jańczuk, Bfallik, Favonian, Lightbot, Wireless friend, Frehley, DrFO.Tn.Bot, Snaily, Yobot, Gsmcoupe, Playclever, AnomieBOT, Jim1138, Materialscientist, ArthurBot, MauritsBot, Xqbot, Nasnema, Almabot, GrouchoBot, Jhbdel, RibotBOT, GliderMaven, Zhouyuanxin, Oalp1003, Z3r0kw3l, RedBot, MastiBot, Bjarkef, TobeBot, Suffusion of Yellow, Alph Bot, Urfriendshailesh, EmausBot, Rw4ni, Solarra, Thecheesykid, Dweremeichik, Ashwink911, ChuispastonBot, ClueBot NG, Widr, JordoCo, Helpful Pixie Bot, BG19bot, Brian Tomasik, زكري, Frogging101, MatthewIreland, YFdyh-bot, BrightStarSky, Calumet96, Checkmeleon, Hssaha, Superkronos, Kahtar, JaconaFrere, Tracey1022, Trackteur, Richard Yin, PaulRamone2, Undolie, Asddfas and Anonymous: 248

- **NEXUS International Broadcasting Association** *Source:* https://en.wikipedia.org/wiki/NEXUS_International_Broadcasting_Association?oldid=649837325 *Contributors:* TUF-KAT, 159753, Pearle, Sdr, Alcot, Esprit15d, Garion96, Carbonix, Chris the speller, Sadads, Derek R Bullamore, Mack2, Maurice Carbonaro, TreasuryTag, Kathleen.wright5, Trivialist, AnomieBOT, DrilBot, Lesser Cartographies and Anonymous: 3

- **NORMOB** *Source:* https://en.wikipedia.org/wiki/NORMOB?oldid=601720166 *Contributors:* Pegship, SmackBot, Kvng, Alaibot, MarshBot, Gunngx, Huku-chan, Yobot, Erik9bot and Anonymous: 3

- **Olympic Broadcasting Services** *Source:* https://en.wikipedia.org/wiki/Olympic_Broadcasting_Services?oldid=613818699 *Contributors:* Stickguy, SmackBot, Mauls, Chris the speller, Davidhorman, CardinalDan, Sjones23, Dravecky, Flashart1, Trivialist, Eeekster, Addbot, Luckas-bot, Yobot, AnomieBOT, SassoBot, Alvin Seville, ZenithZealotry, Charriscoquitlam, ZéroBot, BornonJune8, BG19bot, Yamatochem, YFdyhbot, Einstein2, GabeIglesia and Anonymous: 13

- **Our World (TV special)** *Source:* https://en.wikipedia.org/wiki/Our_World_(TV_special)?oldid=698628756 *Contributors:* Kchishol1970, Zanimum, Naddy, RealGrouchy, IRelayer, Everyking, Cantus, Bumm13, TonyW, Oknazevad, Grm wnr, Philip Cross, Dtcdthingy, Oliphaunt, Kelisi, Hailey C. Shannon, GregorB, Zzyzx11, Graham87, BD2412, David Levy, Tim!, Gareth E. Kegg, Kerry Raymond, Househippie, Tony1, Davidpatrick, Bantosh, CapitalLetterBeginning, BorgQueen, Garion96, Sugar Bear, Kingboyk, SmackBot, Chris the speller, Jprg1966, Rick7425, Colonies Chris, Azumanga1, FairuseBot, Andreasegde, Otto4711, BetacommandBot, Mrmusichead, Cgingold, Abebenjoe, Geoboy, Martin451, Argcar5199, Decur, Oxymoron83, Jón, Fuddle, Kai-Hendrik, The Thing That Should Not Be, Babelia, Alexbot, Wkharrisjr, Qwfp, XLinkBot, Bilsonius, Addbot, Yobot, Bunnyhop11, Radiopathy, AnomieBOT, Lucas0707, 205ywmpq, Trust Is All You Need, DrilBot, RjwilmsiBot, Wearealmosthere, Lily0pop, Are You The Cow Of Pain?, Unreal7, Gimelgort, Omnitographer, Lowlova, BattyBot, Fromthevaults and Anonymous: 34

- **Pangea Day** *Source:* https://en.wikipedia.org/wiki/Pangea_Day?oldid=702377401 *Contributors:* Kalki, Theresa knott, Rpyle731, Johnfreez, D6, Gigano, Rjwilmsi, Tim!, Abu-Dun, SmackBot, Chris the speller, Bazonka, Brenden105, Slakr, Dl2000, BetacommandBot, Thijs!bot, JustA-Gal, Escarbot, Shawn in Montreal, Nattfodd, JL-Bot, Sassf, Timsdad, Swnewyork, DumZiBoT, Duffbeerforme, Sw2135, Addbot, Manymerry-menmakingmuchmoneyinthemonthofMay, Monstermorrin, Songsandpictures, Akersville, Tushaaa, SassoBot, Wikipelli, ClueBot NG, Helpful Pixie Bot, Dobie80, Earflaps and Anonymous: 24

- **Pirate radio** *Source:* https://en.wikipedia.org/wiki/Pirate_radio?oldid=704382060 *Contributors:* Mav, Ortolan88, Daniel C. Boyer, Waveguy, Mbecker, Ericd, Edward, Conrad.au, Michael Hardy, Isomorphic, Liftarn, (, CesarB, Ihcoyc, Ahoerstemeier, Monk (usurped), Rossami, Kwekubo, Lee M, Conti, Mulad, Vanished user 5zariu3jisj0j4irj, RickK, Reddi, Random832, Paul Stansifer, Wik, Bernd zh, Maximus Rex, David Shay, Nnh, Lunchboxhero, Bearcat, Robbot, Tim Ivorson, Puckly, Texture, Auric, Smb1001, Benc, HaeB, Rsduhamel, Alan Liefting, DocWatson42, MrSnow, Bobblewik, Rlquall, N2271, Terryhufc, Neutrality, Nick Boulevard, Hobart, D6, A-giau, Herzen, Rich Farmbrough, Guanabot, Xezbeth, Ntennis, EDGE, RoyBoy, Acanon, AJP, Ronnus, Chbarts, MPLX, Cunningham, Zetawoof, Pearle, Stabilo~enwiki, Justinc, Phyzome, Calton, Erik, Wtshymanski, RJFJR, Cfrjlr, Ceyockey, Richard Weil, TimMartin, Siafu, Bushytails, Woohookitty, Scriberius, Dandv, GregorB, Toussaint, Deansfa, Graham87, Haikupoet, The-mart, Rjwilmsi, Joe Decker, Jivecat, Quiddity, Vegaswikian, Wackelpudding, SchuminWeb, Ground Zero, Woozle, Clio64B, Mrschimpf, Ravenswing, YurikBot, Rapido, Hairy Dude, Bhny, DanMS, Reluctantpopstar, Caspian, TransUtopian, Daniel C, King Kool, Mais oui!, Junglecat, Kingboyk, SmackBot, Melchoir, DeMyztikX, TharkunColl, Rohnadams, Chris the speller, Bluebot, Kharker, Victorgrigas, Nbarth, Colonies Chris, Brinerustle, Zhinz, DJboutit, Frap, Manstaruk, Kerkyra~enwiki, Dantadd, The PIPE, Bpe3812, Chymicus, AThing, Tazmaniacs, SilkTork, Gobonobo, MonstaPro, IronGargoyle, Beetstra, TastyPoutine, Stergiousakis, Iridescent, PERSBUREAU AZP, Blakegripling ph, Shoeofdeath, Nfutvol, Unidyne, Deke42, CmdrObot, Mattbr, Nczempin, Ken Gallager, A876, Crossmr, ST47, Dave Rabbit, Pir8radio, Chris Henniker, Jmg38, Invitatious, RickinBaltimore, Jdquadcities, Cnota, Lucky-Louie, Ok!, Harryzilber, SDX, SiobhanHansa, VoABot II, Snaxorb, Nyttend, Bloovee, Amiller545, Bdotgates, Edward321, Hankhayes, R'n'B, Kfm1000, Fountains of Bryn Mawr, KD Tries Again, Lukifer, HighKing, CWii, RingtailedFox, Bolt, AMAPO, Ai4ijoel, DaRaeMan, Oelting, Fragilethreads, Darklife, Newell Post, Agent452, Melsaran, Sgbirch, Veggieburgerfish, Kay Bear, Djmckee1, SieBot, Yeza, Donalkrista, MarcinZmudzki, Amusedly, Umrguy42, Miniapolis, Manway, Stupidisgod, SpiderMum, ClueBot, Ukthingy, Bleedingshoes, L12ra, Radio canada, Mild Bill Hiccup, Thecharg, Rek4385, Alexbot, AnthonyUK, Goodbyebean, Rhododendrites, Imokurnotok, Mlaffs, Thingg, XLinkBot, Hotcrocodile, The Rationalist, Zarateles, Crispness, Addbot, Glane23, CarterBar, Radiocitybill, Lightbot, OlEnglish, Pietrow, Faunas, Peterpirateman, Jarble, Luckas-bot, Yobot, TaBOT-zerem, EmpireForever, Rsupon, Pirateradioman, Sadrice, AnomieBOT, Rubinbot, Rjanag, Wickedradio, Materialscientist, DJ Zath, Xqbot, Fredolph, Crookesmoor, Cold ethyl, Eugene-elgato, Colt .55, WolfJackman, FrescoBot, Shiki2, Sandcat01, Pinethicket, Tpyvvikky, Jafol, Yappy2bhere, Its all relatives, Andy Richardz, EmausBot, Orphan Wiki, Radioholic, Dewritech, Ryguy611, Mobilewifi, Wispad, Fm tx, Decodicil, Acro mega man, ClueBot NG, Primergrey, JordoCo, Helpful Pixie Bot, Fmi7323104, Khazar2, 93, MidnightRequestLine, Rcolbert80, Weighted Cube, Saectar, Firefighter 127452936294, Eric Auckland, Highway 231, Jbschev, AlVieri, FMIworldwide, Ultimaxx9 and Anonymous: 295

- **Radio frequency** *Source:* https://en.wikipedia.org/wiki/Radio_frequency?oldid=699649871 *Contributors:* WojPob, Zundark, The Anome, Fredbauder, Aldie, SimonP, Waveguy, Rcingham, Heron, Mintguy, Stevertigo, Kku, Prefect, Bogdangiusca, Palmpilot900, Wfeidt, Conti, Mulad, RadarCzar, Emperorbma, Reddi, Radiojon, Tero~enwiki, SEWilco, Jerzy, Denelson83, Robbot, Moriori, RedWolf, Arkuat, Stewartadcock, Blainster, Hadal, Danceswithzerglings, DocWatson42, Average Earthman, Fleminra, Cantus, Jfdwolff, Bobblewik, Utcursch, Beland, Ary29, Ukexpat, Klemen Kocjancic, Deglr6328, Jakro64, Discospinster, Guanabot, LindsayH, Harriv, Bender235, El C, PhilHibbs, Bobo192, Sparkgap, MARQUIS111, Haham hanuka, ClementSeveillac, Atlant, Wtmitchell, Wtshymanski, Cburnett, Suruena, DV8 2XL, Gene Nygaard, WojciechSwiderski~enwiki, Kenyon, Alex.g, Camw, Pol098, Plaws, Zilog Jones, Ch'marr, Isnow, MarkPos, Zpb52, LimoWreck, Kotukunui, Koavf, Misternuvistor, Mike Peel, Vegaswikian, Fred Bradstadt, Ground Zero, RexNL, Smileyrepublic, Alvin-cs, Srleffler, King of Hearts, Wavelength, Fabartus, Anonymous editor, Yyy, Giro720, Teb728, Wiki alf, Retired username, Mortein, Anetode, Hyandat, Voidxor, Shadowblade, Kermi3, User27091, Jules.LT, Aparna82, GraemeL, Alureiter, SmackBot, Rutja76, Aim Here, C.Fred, Tbonnie, Yamaguchi⬜⬜, Gilliam, Bluebot, Avin, EncMstr, Bazonka, Can't sleep, clown will eat me, Harumphy, Frap, JustUser, Adamantios, Khoikhoi, NoIdeaNick, A.R., Anlace, Stattouk, Jcoy, Dicklyon, Stijak, NetBMC, Dsongman, Chetvorno, JohnTechnologist, Daggerstab, Chrumps, Requestion, Cydebot, Kanags, The Ultimate Koopa, Tsenapathy, L7HOMAS, Jstuby, Epbr123, Headbomb, Rosarinagazo, John254, Dawnseeker2000, AntiVandalBot, Orionus, Nitrous231, Myanw, MagiMaster, JAnDbot, Harryzilber, Ph.eyes, Ccrrccrr, Erpel13, .anacondabot, Bubba hotep, Benmcgraw, DonVincenzo, Rettetast, Lcabanel, PrestonH, Cpiral, JA.Davidson, Osndok, Coppertwig, NewEnglandYankee, DAID, Ibrahimyu, Петър Петров, Radioactivebloke, Steel1943, Deor, ICE77, Metroccfd, Mill haru, Anonymous Dissident, Anna Lincoln, Onevim, BwDraco, BotKung, Spinningspark, Hertz1888, Caltas, Bentogoa, Hovev~enwiki, Callidior, OP8, Lucyjuice, Regushee, Vcaeken, ClueBot, Noabar, Dean Wormer, CounterVandalismBot, Duane-light, Auntof6, PhySusie, M.O.X, Cexycy, Elizium23, Thefirm96, Jonverve, HumphreyW, InternetMeme, Ean5533, BarretB, Hotcrocodile, TopherGZ, Mitch Ames, Tvargy, Alexius08, Addbot, Yoenit, Zhipengye, Fgnievinski, Thaejas, Ronhjones, GyroMagician, Redheylin, Xicer9, Tide rolls, מבעת-יזרם, Bssquirrel, Neilforcier, Dede2008, Cflm001, Galoubet, Bluerasberry, Materialscientist, Limideen,

45Factoid44, Madjar, DSisyphBot, Barkinfool, RadiX, Mathonius, Nedim Ardoğa, Fotaun, Darwinius, Prari, Jc3s5h, Vhann, Drew R. Smith, Hawkpride3000, -jem-, Esar100, Tbhotch, RHC3, RjwilmsiBot, Paulmasters, Orphan Wiki, Wikipelli, OnePt618, Milliemchi, Mohsen.1987, Coasterlover1994, Vietcuong1212, Doris Camire, Taylor10897, RockMagnetist, Weisspiloti, ClueBot NG, AeroPsico, Mclinch, AlagrecoN-JITWILL, Solanki.3108, JordoCo, MerllwBot, Helpful Pixie Bot, BG19bot, Dragon2531, Shaun, Jeanlyhautyetienne, MeanMotherJr, NPSao, Theo's Little Bot, EstonianMan, DetroitSeattle, Shivansh Chaudhary, Soham, Spyglasses, Hanthoec, Airwoz, Symphero, JaconaFrere, Monkbot, Thadlooms32, Ff9473, Hafsa1982, SrihariThalla, Mcstacheattack, KasparBot and Anonymous: 383

- **Radio propagation** *Source:* https://en.wikipedia.org/wiki/Radio_propagation?oldid=700544656 *Contributors:* The Anome, Waveguy, Patrick, Michael Hardy, Glenn, Marknew, Wfeidt, Andrevan, Reddi, Dysprosia, Radiojon, Denelson83, Twang, Robbot, Giftlite, Graeme Bartlett, Ssd, Albany45, Beland, MisfitToys, Ojw, Geof, Rich Farmbrough, Vsmith, Cacophony, Smalljim, Shenme, Cmdrjameson, Sparkgap, Munkymu, RoySmith, Atomicthumbs, Wtshymanski, Gene Nygaard, LOL, Bhamer, Plrk, Rjwilmsi, Vegaswikian, Nihiltres, Athantor, Compotatoj, Srleffler, Random user 39849958, Bgwhite, StuffOfInterest, RussBot, Splash, Bergsten, Brandon, Mikeblas, Ma3nocum, Dan Austin, Reyk, ArielGold, DasBub, Kingboyk, SmackBot, Timrb, Hmains, Kmarinas86, A. B., Harumphy, Wharron, Drkirkby, Frap, MitchellShnier, Andrewpayneaqa, Lambiam, Khazar, BDM, MonstaPro, SlayerK, Hetar, JoeBot, Civil Engineer III, G-W, Chetvorno, Nczempin, Requestion, Andkore, Cydebot, Nsaum75, After Midnight, Epbr123, Barticus88, Sean2074, Malvineous, Autocracy, Lperez2029, CosineKitty, SiobhanHansa, Otivaeey, WikiTraveller, NMarkRoberts, Logictheo, Highsand, Read-write-services, Kf4yfd, Sm8900, AntiSpamBot, Ale2006, Kn4lf, The Original Wildbear, HopsonRoad, AlleborgoBot, NW7US, Lohost, Hertz1888, Rjfry, Igor.grigorov, Aillema, Miniapolis, Fratrep, Susan118, Dabbdabb, Dp67, Binksternet, GorillaWarfare, Kathleen.wright5, Niceguyedc, Sv1xv, PixelBot, Rwestafer, Solterdisp, Cowpip, JediSaint, Addbot, Fgnievinski, TutterMouse, Yobot, Ptbotgourou, GateKeeper, AnomieBOT, Jim1138, Nedim Ardoğa, Stiepan Pietrov, Veganacity, FrescoBot, Raise-the-Sail, BenzolBot, Kmarawer, 2A4Fh56OSA, Bejinhan, MaxDel, NZ4O, Dimitrisouza, Marie Poise, N0nbh, GoingBatty, Donner60, Orange Suede Sofa, ClueBot NG, Cwmhiraeth, Jimbo1qaz, Kkddkkdd, ChristophE, Frietjes, Helpful Pixie Bot, Sirhc808, ChrisGualtieri, Bnland, Sanya7901, Spyglasses, Jacob Gotts, VE3BMV, KasparBot and Anonymous: 111

- **Radio spectrum** *Source:* https://en.wikipedia.org/wiki/Radio_spectrum?oldid=704313035 *Contributors:* Mulad, Treutwein, DavidCary, Nomad~enwiki, Gerrit, Maaf, Sparkgap, Alansohn, Wtshymanski, Thryduulf, Tabletop, SDC, MarkPos, Vegaswikian, TheAnarcat, Wavelength, Ninly, Sardanaphalus, SmackBot, Kmarinas86, Chris the speller, Harumphy, Bjankuloski06en~enwiki, MonsieurET, Dicklyon, Chetvorno, Petr Matas, Harej bot, NaBUru38, Vdonof, Pajz, Dawnseeker2000, Harryzilber, DuncanHill, Nyq, Swpb, Conquerist, Lcabanel, 28bytes, Ringtailed-Fox, Oshwah, The Original Wildbear, Perohanych, Clarince63, Spinningspark, Fanatix, SieBot, BotMultichill, Dravecky, Wiki libs, Jonverve, Scotttroyer, Wnt, WikiHead, Addbot, Fgnievinski, Mneuner, GyroMagician, RTG, Lightbot, Yaman32, Luckas-bot, Yobot, AnomieBOT, Materialscientist, Nasa-verve, Erik9bot, FrescoBot, TobeBot, Jonkerz, Kajervi, RjwilmsiBot, EmausBot, WikitanvirBot, GoingBatty, Evanh2008, ZéroBot, ClueBot NG, Paul Gaskell, Dr. Zombieman, Helpful Pixie Bot, CitationCleanerBot, Jor.langneh, Seanadamcik, Potor111, Mogism, DetroitSeattle, Masonwardle, Maxmichte, HHubi, Wyn.junior, Jason Akimbo, CV9933, KasparBot, Veleggiare, Treinkvist, Qzd and Anonymous: 78

- **Reception report** *Source:* https://en.wikipedia.org/wiki/Reception_report?oldid=700686976 *Contributors:* Bearcat, BarkingFish, Vegaswikian, NickelShoe, SmackBot, Kharker, Neo-Jay, Nitchell, TreasuryTag, Miniapolis, Dravecky, JL-Bot, Kathleen.wright5, Rcooley~enwiki, Dxinginfo, Yobot, Cyberbot II and Anonymous: 3

- **Satellite radio** *Source:* https://en.wikipedia.org/wiki/Satellite_radio?oldid=700673561 *Contributors:* DavidLevinson, Stevertigo, Kku, Dynamism, Mulad, Radiojon, Morwen, Jeffq, Denelson83, SD6-Agent, Bearcat, TMC1221, RedWolf, Rfc1394, Davodd, Pretzelpaws, Bobblewik, Chowbok, Rdsmith4, Bumm13, Bk0, Sfoskett, James Cridland, RickoniX, RossPatterson, Rich Farmbrough, TrbleClef, Vsmith, Loren36, Violetriga, Evice, Kiand, Susvolans, Femto, NetBot, Clawson, BrokenSegue, Cmdrjameson, Brim, Azure Haights, Sparkgap, Pobrien, Jason One, Zellin, Jknam, Edwards, Metron4, Snowolf, Mavros, Ronark, Cburnett, Versageek, Gene Nygaard, Alai, Dismas, Boothy443, Armando, Jeff3000, Rotten, Zpb52, Mandarax, Siqbal, Josh Parris, Rjwilmsi, Bobsky~enwiki, Wikibofh, PinchasC, SMC, RexNL, RobyWayne, Wongm, SteveBaker, Glenn L, WhyBeNormal, Bgwhite, YurikBot, Jcam, SpuriousQ, Bill52270, MilitantRabbit, ZacBowling, Rjensen, Robert Moore, Nick, Jpbowen, Zwobot, Abrooke, Derbeobachter, ScottFish, Csyria, Deville, Jwrivers, GraemeL, Alex Ruddick, John Broughton, SmackBot, Delldot, Brossow, Nil Einne, Betacommand, MPD01605, JorgePeixoto, Roscoryan, Kasakato, Oli Filth, Analogue Kid, Colonies Chris, Radioguy, Frap, JonHarder, Dali, SnappingTurtle, Dreadstar, Dogosaurus, Tv's emory, SpoonBender, Mksword, Mbergman42, 3eguoxn02, Stupid Corn, Es330td, Hu12, Linkspamremover, NaBUru38, ShelfSkewed, Cydebot, Enoch the red, Gogo Dodo, Pv2b, Corpx, Dancter, Ftazero, Bmitchelf, JamesAM, Geothermal, DanielLevitin, JCam, Dbrodbeck, Joe Schmedley, Jimj wpg, Epsmcl, Nathan Mercer, Gambinotoo, Bongwarrior, Nyq, KJRehberg, Granpire Viking Man, Johnbrownsbody, Oroso, PhantomS, Conquerist, MartinBot, R'n'B, Patar knight, Jmccormac, J.delanoy, JaedenStormes, RatSkrew, DH85868993, Jaimeastorga2000, Center4499, VolkovBot, CWii, Mrkmrk, Ai4ijoel, Alinja, Abberley2, TomXP411, Hoffman2121, BrianRecchia, Vchimpanzee, Malcolmxl5, Editore99, Thisis0, Constanzaribas, Hammerwell, ClueBot, SalineBrain, Mild Bill Hiccup, CounterVandalismBot, Skeeball93, Jay Smilkstein, Alexbot, WikiZorro, DumZiBoT, Shoqed, SilvonenBot, Webb87tm, Addbot, Aurust, Satbuff, Keithatncd, Glane23, Lightbot, Nuberger13, Luckas-bot, Yobot, Corentinoger, Mahmudmasri, Materialscientist, The Evil IP address, Nasa-verve, Inputpersona, FrescoBot, Sanel vejzovic, Jlwilder, Mdelfs, MastiBot, كاشف عقیل, ShondellStover, Ginaswim, Dcirovic, Illegitimate Barrister, Xoius, BaskervilleFantastic, ClueBot NG, Bernie44, JordoCo, Goldenshimmer, Cokeebeen, Pietade, Formido576, Grant T B, Paypayis1, HHubi, Bigmike88 and Anonymous: 210

- **Satellite television** *Source:* https://en.wikipedia.org/wiki/Satellite_television?oldid=704337142 *Contributors:* Kpjas, Mav, Koyaanis Qatsi, Mark Ryan, Rjstott, Dachshund, William Avery, Roadrunner, Minesweeper, Pagingmrherman, Mac, CatherineMunro, IMSoP, Dysprosia, LMB, Fvw, Secretlondon, Skybunny, RadicalBender, Robbot, Hankwang, Donreed, Rfc1394, Academic Challenger, Texture, Aleron235, Hadal, HaeB, Radagast, Alan Liefting, Poszwa~enwiki, DocWatson42, Aps~enwiki, Wikilibrarian, Ferkelparade, Fleminra, Rogier, Hansjorn, Yekrats, Jason Quinn, Rchandra, Bobblewik, Lucky 6.9, Thewikipedian, Gadfium, Bumm13, Cab88, MakeRocketGoNow, Chmod007, Grunt, Discospinster, Rhobite, Sladen, Rupertslander, Wikigeo~enwiki, Quiensabe, Bender235, Sum0, Evice, Kaszeta, Kiand, Kloy1334, Lankiveil, Sietse Snel, Roy-Boy, Väsk, Clawson, Chiacomo, Richi, Minghong, Alansohn, Ihatepotsmokinghippies, Patrick Bernier, Gortu, Versageek, Algocu, Ceyockey, Jakes18, Daranz, Bobrayner, Mindmatrix, Brazil4Linux, Stickguy, Drseti, Tabletop, Yueh, Keta, SDC, MarkPos, Kortsleting~enwiki, Mandarax, RichardWeiss, NickF, David Levy, BorgHunter, Joe Decker, Misternuvistor, IRT.BMT.IND, FutureNJGov, Bryan H Bell, Sango123, Heycam, RAMChYLD, Mark83, Gurch, RobyWayne, BMF81, Chobot, Antilived, Bgwhite, Agamemnon2, Borgx, Rapido, MMuzammils, John Quincy Adding Machine, RonH, Hydrargyrum, Ksyrie, CambridgeBayWeather, Rsrikanth05, Pseudomonas, Ali6236, Fizan, Tvtonightokc, Grafen, Welsh, Moppet65535, FivePointPalmExplodingHeart, Nick, Off!, Ma3nocum, Samir, PS2pcGAMER, Daniel C, Lod, Deville, Zzuuzz, PTSE,

Open2universe, KGasso, JRawle, JuJube, GraemeL, Brianh6630, Alex Ruddick, AntL, Junglecat, Thomas Blomberg, DocendoDiscimus, Smack-Bot, Elonka, F, Unyoyega, Darkman007e, Sea diver, Giandrea, Bernard François, Gilliam, Ohnoitsjamie, Chris the speller, Thumperward, Oli Filth, Robocoder, Deli nk, Adpete, Raymie, Antonrojo, A. B., Dethme0w, Tsca.bot, Samazer, Can't sleep, clown will eat me, Egsan Bacon, SheeEttin, Jamse, Nixeagle, JonHarder, UU, Fuhghettaboutit, DinosaursLoveExistence, Teehee123, Vedek Dukat, Matt Whyndham, A5b, Euchiasmus, Mrhazelj, Soptep, Jpogi, Waftycrank, Csari, Ian Dalziel, Loadmaster, Shamrox, Flamerule, AxG, H, Hu12, Iridescent, Joseph Solis in Australia, JoeBot, Mihitha, Linkspamremover, Tawkerbot2, LSX, Holkingers, Didimos, CmdrObot, Macsupport, Matthew Auger, Meek-Mark, Cydebot, Aarongman, Gogo Dodo, Hebrides, ST47, Poloolop, Acs4b, B, Ftazero, Starionwolf, Expediter, Viridae, Duhon~enwiki, Piccolo Modificatore Laborioso, Thijs!bot, Epbr123, Grayshi, Libertyernie2, Mentifisto, AntiVandalBot, Etzkorn1, Tyballer, EyeMD, JAnDbot, Mckee, Harryzilber, MER-C, Chanakyathegreat, DISH, Magioladitis, Dics, Jwateska, AtticusX, Pixel ;-), Alturivijay, KJRehberg, MartDawg, Eldumpo, Boffob, Cpl Syx, Dipper2, LW77, Seba5618, Abebenjoe, R'n'B, Jmccormac, Ikuwara, Pbengani, Trusilver, Terrek, Uncle Dick, Jesant13, Danielk2, AntiSpamBot, Warut, Yughaur, Fountains of Bryn Mawr, Batv0r, Brian Pearson, Sachinairan, KylieTastic, Totsugeki, Chin Man2, Kettlebeller, Dondozie, Stanleyuroy, Funandtrvl, DogFrankBeans, VolkovBot, Mrkmrk, Fences and windows, 911wasaninsidejob, TXiKiBoT, Fxhomie, Saber girl08, B Pete, Haggismn, Jackfork, Hamitr, Lephilippe, SQL, Hedi0058, Falcon8765, Vchimpanzee, Lokioak, Deconstructhis, WereSpielChequers, BillydaFish, Crmadsen, Psychless, Deworrall, Susyr, Tom2789, Nopetro, Mrmazda, Lightmouse, Toddro, Gqegg, MyCustomWebsearch, Mark Eliassen, Triedtool, Celique, Martarius, ClueBot, Snigbrook, Malpass93, The Thing That Should Not Be, Newlloreda, Gaia Octavia Agrippa, Drmies, Blanchardb, StigBot, DragonBot, Jbubfrog, Jusdafax, Dcpc0807, Jotterbot, BOTarate, WikiJedits, Berean Hunter, Johnuniq, Royoakmich, Xzumi, XLinkBot, Dthomsen8, HappyJake, Avoided, WikHead, CohibAA, SilvonenBot, Beach drifter, Navy Blue, Osarius, Addbot, Bibhabasu, Abuelo jack, Wickey-nl, Satbuff, Ronhjones, MrVanBot, Chamal N, Glane23, SpBot, Juanpablosoto, Lightbot, MuZemike, Windward1, Luckas-bot, Timurite, Yobot, Mdavid89, Imagetune, Pablo323, AmeliorationBot, Starbois, AnomieBOT, Exs007, Dogue, Kschroer105, Materialscientist, Nick747, ArthurBot, LilHelpa, Obersachsebot, Nitintomar20, Capricorn42, Vivaelcelta, PSB-satellite, Mahonni, Nasa-verve, Perlyngemark, RibotBOT, Jineshgopinathan, Alikara, FrescoBot, GEBStgo, Sanel vejzovic, MGA73bot, Gerardoruiz, Fraser360, Mutinus, BRUTE, Full-date unlinking bot, Rastoman m, UtubeGodwin, Michaelorder, SeoMac, Miracle Pen, Islandhopper99, Matt2727, Aoidh, Ludojigeho, DARTH SIDIOUS 2, Onel5969, Steve-p-james, TjBot, Salvio giuliano, DASHBot, EmausBot, WikitanvirBot, Somthinggood212, Zollerriia, IESNIPER, Wertov, Tehytjd, ZéroBot, Rickster20010, Sandyuk, H3llBot, AManWithNoPlan, Jaycb1, Warrenj1979, ChuispastonBot, Ch.tsog, Amaraizad, Mjbmrbot, ClueBot NG, Sattracker61, Totally.random.bloke, Manubot, Pejno Simono, DarknessVisitor, JordoCo, Mmmmpa, BG19bot, Hisatdotcom, Asaporito91754, Musi9a, DPL bot, BattyBot, Dklacy1978, EuroCarGT, Satfootprint, Qxukhgiels, Mogism, Amitdix, Ugog Nizdast, Otisenpdl, Stamptrader, HHubi, Odoustar, Demoniccathandler, Railholiday, GeorginaMat, JellydPuppy, Chandankumarpaswa, KasparBot, Συντάκτης Βικιλεξικό and Anonymous: 526

- **Secret broadcast** *Source:* https://en.wikipedia.org/wiki/Secret_broadcast?oldid=693557888 *Contributors:* David Newton, Matt Crypto, Bobblewik, ArnoldReinhold, Jnestorius, Aard, Woohookitty, RHaworth, Tetraminoe, Stefanomione, Edison, Nneonneo, Cliffb, Junglecat, Smack-Bot, Brossow, Imzadi1979, CrypticBacon, Sloman, Bluebot, Otto4711, Catsmoke, Wa3frp, Fyrael, Lightbot, AnomieBOT, Xqbot, Bravo Foxtrot, PhnomPencil, Neøn, BattyBot, TortoiseWrath, 32RB17 and Anonymous: 14

- **Shortwave bands** *Source:* https://en.wikipedia.org/wiki/Shortwave_bands?oldid=685593089 *Contributors:* Malcolm Farmer, SimonP, Randomned, Joe Shupienis, Denelson83, Academic Challenger, Blainster, Davidl9999, Phil1988, Mike Rosoft, Gerry Lynch, Nigelj, Cmdrjameson, Sparkgap, Wtshymanski, Alai, Jakes18, Bluemoose, Eyreland, Haikupoet, Ncc1701zzz, Alan J Shea, Kornbelt, G Clark, Anonym1ty, Hansamurai, Ravenswing, Mysid, Dddstone, Caerwine, Bluebot, Kharker, A. B., Erzahler, Adamantios, Beetstra, Hu12, Tawkerbot2, Ladycathyofwales, Harryzilber, Steve Hosgood, RJASE1, Xnuala, Marknagel, Samslipknot, Thunderbird2, Miniapolis, Finefir2001, Arjayay, Mortense, Quantock, Captain-tucker, Sillyfolkboy, Capricorn42, FrescoBot, DrilBot, F1jmm, GoingBatty, BG19bot, Rvwomersley, Blanco257, Lukas1231 and Anonymous: 35

- **Shortwave broadcasting in the United States** *Source:* https://en.wikipedia.org/wiki/Shortwave_broadcasting_in_the_United_States?oldid= 700492315 *Contributors:* Edward, GABaker, Paul A, Bearcat, Flauto Dolce, Fermion, Closeapple, SlimVirgin, Wtshymanski, Richard Weil, LizardWizard, Andromeda321, Eyreland, SDC, Choess, Astral, Caerwine, SmackBot, TBH, Bluebot, Colonies Chris, A. B., Radiofreemountairy, Syrcatbot, Iridescent, Stereorock, Enwilson, J Milburn, Wws, Dragomiloff, FastLizard4, PKT, Mutiny, Luxomni, Rvmillion, Cgingold, Jeffconn, SmokeySteve, Oromethehuntsman, R'n'B, Peter Chastain, Crazyjudo, Shortride, RingtailedFox, Bolt, Ai4ijoel, Qworty, Callidior, Dwdollar, Dravecky, Illinois2011, Kathleen.wright5, Niceguyedc, NYtalkradio, Pacificfm, Mlaffs, Ds02006, Morriswa, Lightbot, RockinRick, Yobot, Kq4ym, Spike-from-NH, FrescoBot, Dxworld and Anonymous: 57

- **Shortwave listening** *Source:* https://en.wikipedia.org/wiki/Shortwave_listening?oldid=661363479 *Contributors:* GABaker, Altailji, Wiki Wikardo, NightMonkey, Stepp-Wulf, Bneely, Che fox, Sukiari, RussBlau, Mavros, ProhibitOnions, Stephan Leeds, SteinbDJ, Woohookitty, Eyreland, Tetraminoe, Sherubtse, Chobot, Richard-L-James, Alcot, RussBot, Gaius Cornelius, Mysid, Ilmaisin, SIGURD42, That Guy, From That Show!, Veinor, SmackBot, Rutja76, Hmains, Carl.bunderson, Chris the speller, Bluebot, Kharker, A. B., Erzahler, Zvar, TenPoundHammer, Eliyak, T-dot, Kc5fm, Beetstra, Hu12, JoeBot, Haus, MGlosenger, CmdrObot, Jonathan Headland, Woody, LuckyLouie, CosineKitty, Appraiser, Ihbusboy, Magic Speller, SmokeySteve, R'n'B, Mange01, WFinch, Net1360, Canaima, Daveh4h, NW7US, RedBlade7, Finefir2001, Sfan00 IMG, Sv1xv, Alexbot, Rcooley~enwiki, Dxinginfo, DumZiBoT, Robert W. Betts, Twitherspoon, Roxy the dog, Alexius08, Addbot, Willking1979, Quantock, Materialscientist, Porlo, Asfarer, Surv1v4l1st, Dxace1, U of I 1983, Hsnmoom, Trappist the monk, North8000, F1jmm, Steve03Mills, Wexlax20, Nicocorn20, Jkummerwro, Mark garvey, Ego White Tray, Helpful Pixie Bot, Stephenbrown222222, Radiowood, CitationCleanerBot, BattyBot, Editfromwithout, Mchanges! and Anonymous: 49

- **Shortwave relay station** *Source:* https://en.wikipedia.org/wiki/Shortwave_relay_station?oldid=650454854 *Contributors:* Tom harrison, Ivansanchez, Gene Nygaard, Tabletop, Eyreland, Vegaswikian, Wavelength, Colonies Chris, Chetvorno, CmdrObot, TreasuryTag, WRK, Miniapolis, ImageRemovalBot, Avobert, Yobot, ImageTagBot, John of Reading, Midas02, Ivolocy, Khazar2 and Anonymous: 6

- **SINPO code** *Source:* https://en.wikipedia.org/wiki/SINPO_code?oldid=675787966 *Contributors:* Glenn, BarkingFish, Deadworm222, Plaws, Marudubshinki, NatusRoma, YurikBot, Alynna Kasmira, Mysid, Deville, SmackBot, Rutja76, Kharker, Neo-Jay, Erzahler, PetesGuide, Barticus88, Kryptonita, Truthanado, Deconstructhis, Biscuittin, Miniapolis, Dravecky, Krajinaetc, Kwjbot, MystBot, Addbot, Luckas-bot, DarekSz, Xqbot, MastiBot, EmausBot, Сергей-СПб and Anonymous: 8

- **Skip zone** *Source:* https://en.wikipedia.org/wiki/Skip_zone?oldid=681734778 *Contributors:* Eloquence, Timo Honkasalo, Dcljr, Seth Ilys, Ssd, Kevin Rector, Meggar, Merenta, RJFJR, YurikBot, Member, Brandon, SmackBot, Eastlaw, Alaibot, Thijs!bot, Aille, VolkovBot, Cowpip, OlEnglish, Legobot, ZéroBot, Maxwell Verbeek and Anonymous: 17

- **Skywave** *Source:* https://en.wikipedia.org/wiki/Skywave?oldid=694690173 *Contributors:* Tedernst, Michael Hardy, Dcljr, Julesd, Reddi, Radiojon, Motor, Twang, RedWolf, Ssd, Mboverload, Cmdrjameson, Stillnotelf, Vegaswikian, Anonym1ty, Srleffler, 121a0012, Gaius Cornelius, DragonHawk, Brandon, Mkill, Ilmaisin, Benhoyt, KD5TVI, Bluebot, Kharker, Swat671, DinosaursLoveExistence, Springnuts, Chetvorno, Requestion, Jbonnell, Cydebot, Rico402, Harryzilber, Pixel ;-), CodeCat, Kf4yfd, Keith D, VolkovBot, SamMichaels, J3gum, UnitedStatesian, Shuhail, Hertz1888, Binksternet, The Thing That Should Not Be, Piastu, Conical Johnson, SchreiberBike, Addbot, Luckas-bot, Yobot, Materialscientist, Loveless, MSturova, FrescoBot, Steve Quinn, Lukeclimber, Kmarawer, Tom.Reding, Lopifalko, Teravolt, Helvitica Bold, Jeremy112233, Cyberbot II, Khazar2, JakeWi, 32RB17, Kartane, Knife-in-the-drawer and Anonymous: 33

- **SR International – Radio Sweden** *Source:* https://en.wikipedia.org/wiki/SR_International_%E2%80%93_Radio_Sweden?oldid=696619290 *Contributors:* LA2, WhisperToMe, Bearcat, Spamhog, Picapica, Rlongstaff, Rjwilmsi, UkPaolo, SmackBot, Dangherous~enwiki, Chris the speller, Pinots, Jjc104, Harryzilber, The Anomebot2, Cgingold, Flrn, LordAnubisBOT, Ai4ijoel, SteveStrummer, Werldwayd, ElSaxo, Addbot, Next-Genn-Gamer, Yobot, AnomieBOT, Mahmudmasri, Ooopickme, FrescoBot, ProcEnforce, DrilBot, Mean as custard, DASHBot, EmausBot, Radiosweden, Lisner, Radiowood, P6nassin, Ceannlann gorm, JenniferMari-Gardner and Anonymous: 7

- **Stand Up to Cancer** *Source:* https://en.wikipedia.org/wiki/Stand_Up_to_Cancer?oldid=704665438 *Contributors:* Edward, Rl, Enochlau, Hiphats, Calwatch, Discospinster, Closeapple, Zachlipton, Woohookitty, Tim!, MarnetteD, SNIyer12, Sherool, Wavelength, RussBot, Ericorbit, Morphh, Harro, Bratboyz, SmackBot, AnOddName, TimBentley, Kanabekobaton, MJBurrage, Azumanga1, GDVS, Derek R Bullamore, Salamurai, AxG, Telrod, Jetong, CmdrObot, ShelfSkewed, Cyberfray, Reywas92, The Lake Effect, Richhoncho, Bogger, CTF83!, Billymeade, SteveChervitzTrutane, Jedawson2000, RingtailedFox, Sunrainprods, Arcoiris724, Farnishk, Edkollin, Coffee, Hertz1888, 1RodStewartFan, ImageRemovalBot, ClueBot, Justin W Smith, 718 Bot, Mlaffs, XLinkBot, Tubesurfer, Cool-guy357, Hunter Kahn, Addbot, CrisCris84, Ckvasko, SU2C, Andie483, Yobot, LAimposter, AnomieBOT, Paralympic, LilHelpa, Brackenheim, FashionTalk411, FrescoBot, Xyazzyx, Tumacama, Tarametblog, Crazyseiko, GoingBatty, Unreal7, Jw walton, ShrimpMonkey, BornonJune8, Dchicago, ClueBot NG, Splogaton, Codyjane98, Go Phightins!, 20chances, BG19bot, Samdod2427, Kangaroopower, SolomanMcKenzie, Electricburst1996, Jockzain, Fort esc, Frosty, NorthCoastGirl, Skr15081997, Oriole85, JohnGormleyJG, Ellinewilliams231, Digitalsu2c, Bammie73, SonicSupernova81365, WebEditor12, Michael hunter, S3venevan and Anonymous: 66

- **Stephen Williams (Radio Luxembourg)** *Source:* https://en.wikipedia.org/wiki/Stephen_Williams_(Radio_Luxembourg)?oldid=698373844 *Contributors:* Rich Farmbrough, Deacon of Pndapetzim, RussBot, CrazyLegsKC, SmackBot, Fayenatic london, Ghmyrtle, Waacstats, Treasury-Tag, Fragilethreads, Woblosch, Cirt, Mertozoro, Full-date unlinking bot, RjwilmsiBot, Helpful Pixie Bot and Anonymous: 1

- **The Filipino Channel** *Source:* https://en.wikipedia.org/wiki/The_Filipino_Channel?oldid=702671805 *Contributors:* SimonP, Bearcat, Davidcannon, DanielCD, Jonnny, Gronky, Debigboy, Zscout370, Väsk, Of~enwiki, Giraffedata, Chicago god, TheCoffee, Firsfron, Woohookitty, ^demon, Kbdank71, Rjwilmsi, Nanami Kamimura, Vegaswikian, Nihiltres, ApprenticeFan, Sherool, Daduzi, Stalmannen, NawlinWiki, Avraham, Geopgeop, Zzuuzz, Kevin nico, Luckystars, ViperSnake151, SunKing, SmackBot, Reedy, KnowledgeOfSelf, McGeddon, Commander Keane bot, Chris the speller, Glenncando, AWeenieMan, Kanabekobaton, Charleslemark, Iam4Lost, Howard the Duck, Plop123, WayKurat, Ohconfucius, ArglebargleIV, NJA, DI2000, Hu12, Vegassteven, Joseph Solis in Australia, Wjejskenewr, Coffeezombie, Beaker2000, Dapen, Ken Gallager, Cydebot, Jameboy, The 80s chick, John earlm, Nezzadar, E. Ripley, Luna Santin, Dryedmangoez, CobraWiki, Spencer, Mikomouse, Geniac, Baspe, JamesBWatson, Fabrictramp, DarkFalls, GrahamHardy, Neomarianasarias, Loulover, DJCA, Squeakumz, Madapaka, Kwekwc, Pinoysurfer, HkCaGu, MFGV.3, Antonio Lopez, Aspects, Lightmouse, Rabbit 20, Nemo24, ClueBot, Lumaisu, Niceguyedc, SolarWind, Alexbot, Arjayay, Carl Francis, +u3)u!^ 7!3N, Badmachine, DCFan101, Addbot, JnJVideography, Zararo, -iNu-, Niksh98, AnomieBOT, Ciphers, Fetch dickson, Fighter 10, TheTechieGeek63, Tamiera, Cthuang, Supergabbyshoe, Joaquin008, FrescoBot, RafaelPPascual, Chockyboy, Drakedasalla, The GateKeeper07, I dream of horses, Smartkid pinoy, Σ, Athene cheval, Hollyckuhno, Cookieass, Aircorn, John of Reading, WikitanvirBot, Zollerriia, Srgagospears, Eraldtfc, ExtraEdit, Wickednite, Ybes03, Venusraj, Nikbert16, Christian2941, Romar9141, RM0312, Samson5007, Asxsl15, Yududuy, Eric abiog, Epicgenius, Nerhoestebat, Edgardo Valentino D. Olaes, Steven Muhadi, Vladismeer, Samantha-PuckettIndo, ElNiñoMonstruo, Lifesdear, Goodday51, Kyndr352, ABSCBNKapamilya144, Powderwreath and Anonymous: 277

- **The Flattery Show** *Source:* https://en.wikipedia.org/wiki/The_Flattery_Show?oldid=620853083 *Contributors:* Bearcat, Alan Liefting, RussBot, Colonies Chris, John, DI2000, CommonsDelinker, GrahamHardy, SteveStrummer, Deveze, XLinkBot, Kjell Knudde, AvicAWB, Escloupere and Faizan

- **Tropospheric propagation** *Source:* https://en.wikipedia.org/wiki/Tropospheric_propagation?oldid=676154748 *Contributors:* Anders Feder, Docu, Ungvichian, D6, Geof, Rich Farmbrough, Andrewpmk, Woohookitty, DonPMitchell, Tabletop, BD2412, Andlarry, Mrschimpf, DoriSmith, SmackBot, Will Beback, Ken Gallager, Headbomb, Grayshi, Milonica, Mrceleb2007, Bdbd, Webfan29, Dodger67, Arjayay, Mlaffs, XLinkBot, Addbot, Lightbot, OlEnglish, Yobot, AnomieBOT, Nerdluck34, Michael93555, Zvartoshu, Full-date unlinking bot, Jeatwell 67, Defrector, P15able, Neøn and Anonymous: 43

- **Tsunami Aid** *Source:* https://en.wikipedia.org/wiki/Tsunami_Aid?oldid=623305183 *Contributors:* Mulad, Saint-Paddy, Bearcat, Dbenbenn, MakeRocketGoNow, Rhobite, TheFireCheese, Cmdrjameson, Evil Monkey, Pcpcpc, Woohookitty, Ashley Grant, Xepo~enwiki, Tim!, Wahoofive, Oliverkeenan, Wasted Time R, Grafen, SmackBot, Bluebot, Azumanga1, Green lantern40, Dreadstar, JimmB, ViewAskewer, Tascha96, Twinsday, Kenilworth Terrace, SporkBot, BornonJune8, Spicemix and Anonymous: 8

- **Utility station** *Source:* https://en.wikipedia.org/wiki/Utility_station?oldid=681850064 *Contributors:* Ixfd64, Dysprosia, Jakes18, Haikupoet, Jdcooper, Alphachimp, Malcolma, SmackBot, Mairibot, JamesAM, Dravecky, Calor, John of Reading, Neøn, Jberri12 and Anonymous: 10

- **Voice of America Bethany Relay Station** *Source:* https://en.wikipedia.org/wiki/Voice_of_America_Bethany_Relay_Station?oldid= 685312424 *Contributors:* Malcolm Farmer, Mxn, Bearcat, Orangemike, Bobblewik, PedanticallySpeaking, Fahooglewitz1077, Woohookitty, Hillrhpc, Hmains, DIDouglass, Ser Amantio di Nicolao, General Ization, Syrcatbot, WilliamJE, Cydebot, DOSGuy, Jllm06, Nyttend, Mikebanks, KudzuVine, Pubdog, Od Mishehu AWB, Stepshep, Proxy User, Mlaffs, Good Olfactory, Taketa, AnomieBOT, Robinsonbrown, Full-date unlinking bot, SmartOneK, TheClerksWell, NationalRegisterBot, Evinande, Founding Joker and Anonymous: 7

- **Willis Conover** *Source:* https://en.wikipedia.org/wiki/Willis_Conover?oldid=703892330 *Contributors:* AnonMoos, Bearcat, Orangemike, Gamaliel, Rparle, Gyrofrog, Deepanjan nag, Jost Riedel, Gene Nygaard, Woohookitty, Weichbrodt, RussBot, Dissolve, CLW, RustySpear, Pegship, J. Van Meter, SmackBot, Bluebot, Colonies Chris, GRuban, Ohconfucius, MayerG, Eurodog, Besha, Cydebot, PhiLiP, Rothorpe, Waacstats, Bot-Schafter, KudzuVine, VolkovBot, Mercurywoodrose, Dravecky, AuntFlo, RogDel, Good Olfactory, Addbot, DBJohnso, Yobot, HairyPerry, Easydog, RjwilmsiBot, Umansky~enwiki, Proscribe, VIAFbot, KasparBot and Anonymous: 16

- **World Administrative Radio Conference** *Source:* https://en.wikipedia.org/wiki/World_Administrative_Radio_Conference?oldid= 688125065 *Contributors:* Bumm13, Gene Nygaard, Anonym1ty, SmackBot, Radagast83, Harryzilber, Dsergeant, Dravecky, Kathleen.wright5, Arjayay, WikiDreamer Bot, Lotje, KLBot2 and Anonymous: 2

- **World Radio TV Handbook** *Source:* https://en.wikipedia.org/wiki/World_Radio_TV_Handbook?oldid=620207475 *Contributors:* LA2, Arj, Bearcat, PatrikR, Remuel, Ceyockey, Wikiklrsc, Caerwine, Pegship, SmackBot, Rutja76, Mairibot, Bluebot, Ikiroid, Colonies Chris, Emre D., Pit-yacker, Beachyboy, CommonsDelinker, Vchimpanzee, Kathleen.wright5, DragonBot, Addbot, LaaknorBot, Lightbot, Luckas-bot, F1jmm, Helpful Pixie Bot and Anonymous: 9

- **List of international television channels** *Source:* https://en.wikipedia.org/wiki/List_of_international_television_channels?oldid=700069536 *Contributors:* CaribDigita, Firsfron, Woohookitty, Tabletop, BD2412, Ketiltrout, SmackBot, TimBentley, Iam4Lost, Mitchumch, Checco, 16@r, AxG, CmdrObot, Clovis Sangrail, Tikiwont, Dialh, Davehi1, Takuy, Paper Luigi, Download, WikiEditor50, Yobot, Kamatysha, Dubious20, Orenburg1, Cheong Kok Chun, John of Reading, Zollerriia, Dewritech, Erpert, Sandyuk, Jay-Sebastos, Mantchi, ClueBot NG, Turyalazamkhan, Gabriel Yuji, Erlbaeko, TBrandley, BattyBot, ChrisGualtieri, Hmainsbot1, Mo2010, Ramjchandran, ColRad85, Rajesultanpur, Nashmim2000, AnonAnnu, Rp131 and Anonymous: 31

- **List of shortwave radio broadcasters** *Source:* https://en.wikipedia.org/wiki/List_of_shortwave_radio_broadcasters?oldid=702788339 *Contributors:* Yobot, Mahmudmasri, Winner 42, Nikbert16, Bxxiaolin, Timdog13, AnonAnnu and Anonymous: 2

4.2 Images

- **File:2008-07-28_Mast_radiator.jpg** *Source:* https://upload.wikimedia.org/wikipedia/commons/7/72/2008-07-28_Mast_radiator.jpg *License:* GFDL *Contributors:* Own work *Original artist:* Ildar Sagdejev (Specious)

- **File:802_11bg_interference.ogg** *Source:* https://upload.wikimedia.org/wikipedia/commons/2/2e/802_11bg_interference.ogg *License:* CC-BY-SA-3.0 *Contributors:* Own work *Original artist:* flip619

- **File:ATS_909_worldband_receiver.jpg** *Source:* https://upload.wikimedia.org/wikipedia/commons/6/65/ATS_909_worldband_receiver.jpg *License:* CC-BY-SA-3.0 *Contributors:* Photographed by uploader. *Original artist:* Oona Räisänen (w:User:Mysid)

- **File:Alfred_Norton_Goldsmith_&_Guglielmo_Marconi_1922.jpg** *Source:* https://upload.wikimedia.org/wikipedia/commons/f/f8/ Alfred_Norton_Goldsmith_%26_Guglielmo_Marconi_1922.jpg *License:* Public domain *Contributors:* This image is available from the United States Library of Congress's Prints and Photographs division under the digital ID cph.3b03232.
This tag does not indicate the copyright status of the attached work. A normal copyright tag is still required. See Commons:Licensing for more information. *Original artist:* Unattributed

- **File:All_you_need_is_love.jpg** *Source:* https://upload.wikimedia.org/wikipedia/en/3/36/All_you_need_is_love.jpg *License:* Fair use *Contributors:*
Apple.
Original artist: ?

- **File:Amateur_Radio_International_Agreements.png** *Source:* https://upload.wikimedia.org/wikipedia/en/e/ea/Amateur_Radio_ International_Agreements.png *License:* CC-BY-3.0 *Contributors:*
Derived from WIKIPEDIA sample blank map (http://en.wikipedia.org/wiki/File:BlankMap-World6-Equirectangular.svg) *Original artist:*
Vincent Chapman

- **File:Amateurfunkstation.jpg** *Source:* https://upload.wikimedia.org/wikipedia/commons/2/22/Amateurfunkstation.jpg *License:* CC-BY-SA-3.0 *Contributors:* Emil Neuerer, DJ4PI *Original artist:* Emil Neuerer, DJ4PI

- **File:Ambox_PR.svg** *Source:* https://upload.wikimedia.org/wikipedia/commons/b/b3/Ambox_PR.svg *License:* Public domain *Contributors:* self-made in Adobe Illustrator and Inkscape *Original artist:* penubag

- **File:Ambox_current_red.svg** *Source:* https://upload.wikimedia.org/wikipedia/commons/9/98/Ambox_current_red.svg *License:* CC0 *Contributors:* self-made, inspired by Gnome globe current event.svg, using Information icon3.svg and Earth clip art.svg *Original artist:* Vipersnake151, penubag, Tkgd2007 (clock)

- **File:Ambox_important.svg** *Source:* https://upload.wikimedia.org/wikipedia/commons/b/b4/Ambox_important.svg *License:* Public domain *Contributors:* Own work, based off of Image:Ambox scales.svg *Original artist:* Dsmurat (talk · contribs)

- **File:Amfm3-en-de.gif** *Source:* https://upload.wikimedia.org/wikipedia/commons/a/a4/Amfm3-en-de.gif *License:* CC BY-SA 2.5 *Contributors:* Own work *Original artist:* Berserkerus

- **File:Amsterdam_RAI_EC.jpg** *Source:* https://upload.wikimedia.org/wikipedia/commons/3/35/Amsterdam_RAI_EC.jpg *License:* GFDL *Contributors:* Own work *Original artist:* author: *Adam Kliczek, http://zatrzymujeczas.pl (CC-BY-SA-3.0)*

- **File:Analog_TV_EMI.jpg** *Source:* https://upload.wikimedia.org/wikipedia/commons/6/6e/Analog_TV_EMI.jpg *License:* Public domain *Contributors:* Own work *Original artist:* Shaddack

- **File:Antenna_Setup.jpg** *Source:* https://upload.wikimedia.org/wikipedia/en/6/6c/Antenna_Setup.jpg *License:* PD *Contributors:*
Own work
Original artist:
Dipper2 (talk) (Uploads)

- **File:Assoc_Intl_Broadcasting_logo.jpg** *Source:* https://upload.wikimedia.org/wikipedia/en/4/4b/Assoc_Intl_Broadcasting_logo.jpg *License:* PD *Contributors:* ? *Original artist:* ?

- **File:Euronews_HQ_©euronews_photo_Stéphane_Audras_2015.10.15_(169).JPG** *Source:* https://upload.wikimedia.org/wikipedia/commons/b/b7/Euronews_HQ_%C2%A9euronews_photo_St%C3%A9phane_Audras_2015.10.15_%28169%29.JPG *License:* CC BY-SA 4.0 *Contributors:* Own work *Original artist:* Lydie22

- **File:Euronews_grey_logo.svg** *Source:* https://upload.wikimedia.org/wikipedia/en/9/9d/Euronews_grey_logo.svg *License:* Fair use *Contributors:*

  http://www.comhem.se/publicpages/companyinfo/comhem_nytt/comhem_nytt_host08.pdf *Original artist:* ?

- **File:Euronews_logo_globe.png** *Source:* https://upload.wikimedia.org/wikipedia/en/5/53/Euronews_logo_globe.png *License:* Fair use *Contributors:*
  File was found on their official site.
  Original artist: ?

- **File:Euronews_logo_text.svg** *Source:* https://upload.wikimedia.org/wikipedia/en/9/93/Euronews_logo_text.svg *License:* Fair use *Contributors:*
  PDF file found on official site [1] *Original artist:* ?

- **File:European_Broadcasting_Area.png** *Source:* https://upload.wikimedia.org/wikipedia/commons/0/03/European_Broadcasting_Area.png *License:* CC0 *Contributors:* Own work *Original artist:* Vanjagenije

- **File:Ferritantenne_2.jpg** *Source:* https://upload.wikimedia.org/wikipedia/commons/c/ca/Ferritantenne_2.jpg *License:* CC-BY-SA-3.0 *Contributors:* Transferred from da.wikipedia *Original artist:* Glenn at da.wikipedia

- **File:Flag_of_Abkhazia.svg** *Source:* https://upload.wikimedia.org/wikipedia/commons/2/27/Flag_of_Abkhazia.svg *License:* Public domain *Contributors:* Own work , see URL http://www.abkhaziagov.org/ru/state/sovereignty/flag_b.jpg *Original artist:* Drawn by User:Achim1999

- **File:Flag_of_Albania.svg** *Source:* https://upload.wikimedia.org/wikipedia/commons/3/36/Flag_of_Albania.svg *License:* Public domain *Contributors:* ? *Original artist:* ?

- **File:Flag_of_Algeria.svg** *Source:* https://upload.wikimedia.org/wikipedia/commons/7/77/Flag_of_Algeria.svg *License:* Public domain *Contributors:* SVG implementation of the 63-145 Algerian law "*on Characteristics of the Algerian national emblem*" ("Caractéristiques du Drapeau Algérien", in English). *Original artist:* This graphic was originaly drawn by User:SKopp.

- **File:Flag_of_Andorra.svg** *Source:* https://upload.wikimedia.org/wikipedia/commons/1/19/Flag_of_Andorra.svg *License:* Public domain *Contributors:* Llibre de normes gràfiques per a la reproducció i aplicació dels signes d'Estat per als quals el Govern és autoritat competent (Aprovat pel Govern en la sessió del dia 5 de maig de 1999) *Original artist:* HansenBCN

- **File:Flag_of_Argentina.svg** *Source:* https://upload.wikimedia.org/wikipedia/commons/1/1a/Flag_of_Argentina.svg *License:* Public domain *Contributors:* Here, based on: http://manuelbelgrano.gov.ar/bandera/creacion-de-la-bandera-nacional/ *Original artist:* Government of Argentina

- **File:Flag_of_Armenia.svg** *Source:* https://upload.wikimedia.org/wikipedia/commons/2/2f/Flag_of_Armenia.svg *License:* Public domain *Contributors:* Own work *Original artist:* SKopp

- **File:Flag_of_Australia.svg** *Source:* https://upload.wikimedia.org/wikipedia/en/b/b9/Flag_of_Australia.svg *License:* Public domain *Contributors:* ? *Original artist:* ?

- **File:Flag_of_Austria.svg** *Source:* https://upload.wikimedia.org/wikipedia/commons/4/41/Flag_of_Austria.svg *License:* Public domain *Contributors:* Own work, http://www.bmlv.gv.at/abzeichen/dekorationen.shtml *Original artist:* User:SKopp

- **File:Flag_of_Azerbaijan.svg** *Source:* https://upload.wikimedia.org/wikipedia/commons/d/dd/Flag_of_Azerbaijan.svg *License:* Public domain *Contributors:* http://www.elibrary.az/docs/remz/pdf/remz_bayraq.pdf and http://www.meclis.gov.az/?/az/topcontent/21 *Original artist:* SKopp and others

- **File:Flag_of_Bangladesh.svg** *Source:* https://upload.wikimedia.org/wikipedia/commons/f/f9/Flag_of_Bangladesh.svg *License:* Public domain *Contributors:* http://www.dcaa.com.bd/Modules/CountryProfile/BangladeshFlag.aspx *Original artist:* User:SKopp

- **File:Flag_of_Belarus.svg** *Source:* https://upload.wikimedia.org/wikipedia/commons/8/85/Flag_of_Belarus.svg *License:* Public domain *Contributors:* http://www.tnpa.by/ViewFileText.php?UrlRid=52178&UrlOnd=%D1%D2%C1%20911-2008 *Original artist:* Zscout370

- **File:Flag_of_Belgium_(civil).svg** *Source:* https://upload.wikimedia.org/wikipedia/commons/9/92/Flag_of_Belgium_%28civil%29.svg *License:* Public domain *Contributors:* ? *Original artist:* ?

- **File:Flag_of_Bosnia_and_Herzegovina.svg** *Source:* https://upload.wikimedia.org/wikipedia/commons/b/bf/Flag_of_Bosnia_and_Herzegovina.svg *License:* Public domain *Contributors:* Own work *Original artist:* Kseferovic

- **File:Flag_of_Brazil.svg** *Source:* https://upload.wikimedia.org/wikipedia/en/0/05/Flag_of_Brazil.svg *License:* PD *Contributors:* ? *Original artist:* ?

- **File:Flag_of_Bulgaria.svg** *Source:* https://upload.wikimedia.org/wikipedia/commons/9/9a/Flag_of_Bulgaria.svg *License:* Public domain *Contributors:* The flag of Bulgaria. The colors are specified at http://www.government.bg/cgi-bin/e-cms/vis/vis.pl?s=001&p=0034&n=000005&g= as: *Original artist:* SKopp

- **File:Flag_of_Canada.svg** *Source:* https://upload.wikimedia.org/wikipedia/en/c/cf/Flag_of_Canada.svg *License:* PD *Contributors:* ? *Original artist:* ?

- **File:Flag_of_Croatia.svg** *Source:* https://upload.wikimedia.org/wikipedia/commons/1/1b/Flag_of_Croatia.svg *License:* Public domain *Contributors:* http://www.sabor.hr/Default.aspx?sec=4317 *Original artist:* Nightstallion, Elephantus, Neoneo13, Denelson83, Rainman, R-41, Minestrone, Lupo, Zscout370,
  <a href='//commons.wikimedia.org/wiki/User:MaGa' title='User:MaGa'>Ma</a><a href='//commons.wikimedia.org/wiki/File:Croatian_squares_Ljubicic.png' class='image'><img alt='Croatian squares Ljubicic.png' src='https://upload.wikimedia.org/wikipedia/commons/thumb/7/7f/Croatian_squares_Ljubicic.png/15px-Croatian_squares_Ljubicic.png' width='15' height='15' srcset='https:

- **File:Flag_of_Indonesia.svg** *Source:* https://upload.wikimedia.org/wikipedia/commons/9/9f/Flag_of_Indonesia.svg *License:* Public domain *Contributors:* Law: s:id:Undang-Undang Republik Indonesia Nomor 24 Tahun 2009 (http://badanbahasa.kemdiknas.go.id/lamanbahasa/sites/default/files/UU_2009_24.pdf) *Original artist:* Drawn by User:SKopp, rewritten by User:Gabbe

- **File:Flag_of_Iran.svg** *Source:* https://upload.wikimedia.org/wikipedia/commons/c/ca/Flag_of_Iran.svg *License:* Public domain *Contributors:* URL http://www.isiri.org/portal/files/std/1.htm and an English translation / interpretation at URL http://flagspot.net/flags/ir'.html *Original artist:* Various

- **File:Flag_of_Iraq.svg** *Source:* https://upload.wikimedia.org/wikipedia/commons/f/f6/Flag_of_Iraq.svg *License:* Public domain *Contributors:*

- This image is based on the CIA Factbook, and the website of Office of the President of Iraq, vectorized by User:Militaryace *Original artist:* Unknown, published by Iraqi governemt, vectorized by User:Militaryace based on the work of User:Hoshie

- **File:Flag_of_Ireland.svg** *Source:* https://upload.wikimedia.org/wikipedia/commons/4/45/Flag_of_Ireland.svg *License:* Public domain *Contributors:* Drawn by User:SKopp *Original artist:* ?

- **File:Flag_of_Israel.svg** *Source:* https://upload.wikimedia.org/wikipedia/commons/d/d4/Flag_of_Israel.svg *License:* Public domain *Contributors:* http://www.mfa.gov.il/MFA/History/Modern%20History/Israel%20at%2050/The%20Flag%20and%20the%20Emblem *Original artist:* "The Provisional Council of State Proclamation of the Flag of the State of Israel" of 25 Tishrei 5709 (28 October 1948) provides the official specification for the design of the Israeli flag.

- **File:Flag_of_Italy.svg** *Source:* https://upload.wikimedia.org/wikipedia/en/0/03/Flag_of_Italy.svg *License:* PD *Contributors:* ? *Original artist:* ?

- **File:Flag_of_Japan.svg** *Source:* https://upload.wikimedia.org/wikipedia/en/9/9e/Flag_of_Japan.svg *License:* PD *Contributors:* ? *Original artist:* ?

- **File:Flag_of_Jersey.svg** *Source:* https://upload.wikimedia.org/wikipedia/commons/1/1c/Flag_of_Jersey.svg *License:* Public domain *Contributors:* ? *Original artist:* ?

- **File:Flag_of_Jordan.svg** *Source:* https://upload.wikimedia.org/wikipedia/commons/c/c0/Flag_of_Jordan.svg *License:* Public domain *Contributors:* ? *Original artist:* ?

- **File:Flag_of_Kosovo.svg** *Source:* https://upload.wikimedia.org/wikipedia/commons/1/1f/Flag_of_Kosovo.svg *License:* CC-BY-SA-3.0 *Contributors:* Originally from Image:Flag of Kosovo.png. *Original artist:* Cradel (current version), earlier version by Ningyou

- **File:Flag_of_Kurdistan.svg** *Source:* https://upload.wikimedia.org/wikipedia/commons/3/35/Flag_of_Kurdistan.svg *License:* Public domain *Contributors:* Own work *Original artist:* <a href='//validator.w3.org/' data-x-rel='nofollow'><img alt='W3C' src='https://upload.wikimedia.org/wikipedia/commons/thumb/5/51/Invalid_SVG_1.1_%28pink%29.svg/88px-Invalid_SVG_1.1_%28pink%29.svg.png' width='88' height='30' style='vertical-align: top' srcset='https://upload.wikimedia.org/wikipedia/commons/thumb/5/51/Invalid_SVG_1.1_%28pink%29.svg/132px-Invalid_SVG_1.1_%28pink%29.svg.png 1.5x, https://upload.wikimedia.org/wikipedia/commons/thumb/5/51/Invalid_SVG_1.1_%28pink%29.svg/176px-Invalid_SVG_1.1_%28pink%29.svg.png 2x' data-file-width='91' data-file-height='31' /></a>iThe source code of the previous SVG was **invalid** due to **12** errors.

- **File:Flag_of_Latvia.svg** *Source:* https://upload.wikimedia.org/wikipedia/commons/8/84/Flag_of_Latvia.svg *License:* Public domain *Contributors:* Own work *Original artist:* SKopp

- **File:Flag_of_Lebanon.svg** *Source:* https://upload.wikimedia.org/wikipedia/commons/5/59/Flag_of_Lebanon.svg *License:* Public domain *Contributors:* ? *Original artist:* Traced based on the CIA World Factbook with some modification done to the colours based on information at Vexilla mundi.

- **File:Flag_of_Libya.svg** *Source:* https://upload.wikimedia.org/wikipedia/commons/0/05/Flag_of_Libya.svg *License:* Public domain *Contributors:* File:Flag of Libya (1951).svg *Original artist:* The source code of this SVG is <a data-x-rel='nofollow' class='external text' href='//validator.w3.org/check?uri=https%3A%2F%2Fcommons.wikimedia.org%2Fwiki%2FSpecial%3AFilepath%2FFlag_of_Libya.svg,<span>,&,</span>,ss=1'>valid</a>.

- **File:Flag_of_Liechtenstein.svg** *Source:* https://upload.wikimedia.org/wikipedia/commons/4/47/Flag_of_Liechtenstein.svg *License:* Public domain *Contributors:* ? *Original artist:* ?

- **File:Flag_of_Lithuania.svg** *Source:* https://upload.wikimedia.org/wikipedia/commons/1/11/Flag_of_Lithuania.svg *License:* Public domain *Contributors:* Own work *Original artist:* SuffKopp

- **File:Flag_of_Luxembourg.svg** *Source:* https://upload.wikimedia.org/wikipedia/commons/d/da/Flag_of_Luxembourg.svg *License:* Public domain *Contributors:* Own work http://www.legilux.public.lu/leg/a/archives/1972/0051/a051.pdf#page=2, colors from http://www.legilux.public.lu/leg/a/archives/1993/0731609/0731609.pdf *Original artist:* Drawn by User:SKopp

- **File:Flag_of_Macedonia.svg** *Source:* https://upload.wikimedia.org/wikipedia/commons/f/f8/Flag_of_Macedonia.svg *License:* Public domain *Contributors:* Own work *Original artist:* User:SKopp, rewritten by User:Gabbe

- **File:Flag_of_Madeira.svg** *Source:* https://upload.wikimedia.org/wikipedia/commons/a/a4/Flag_of_Madeira.svg *License:* CC BY-SA 1.0 *Contributors:* ? *Original artist:* ?

- **File:Flag_of_Malta.svg** *Source:* https://upload.wikimedia.org/wikipedia/commons/7/73/Flag_of_Malta.svg *License:* CC0 *Contributors:* ? *Original artist:* ?

- **File:Flag_of_Mexico_(1934-1968).svg** *Source:* https://upload.wikimedia.org/wikipedia/commons/8/8f/Flag_of_Mexico_%281934-1968%29.svg *License:* Public domain *Contributors:* This vector image was created with Inkscape. *Original artist:* TownDown

- **File:Flag_of_Moldova.svg** *Source:* https://upload.wikimedia.org/wikipedia/commons/2/27/Flag_of_Moldova.svg *License:* Public domain *Contributors:* vector coat of arms image traced by User:Nameneko from Image:Moldova gerb large.png. Construction sheet can be found at http://flagspot.net/flags/md.html#const *Original artist:* Nameneko and others

- **File:Flag_of_Monaco.svg** *Source:* https://upload.wikimedia.org/wikipedia/commons/e/ea/Flag_of_Monaco.svg *License:* Public domain *Contributors:* ? *Original artist:* ?

- **File:Flag_of_Mongolia.svg** *Source:* https://upload.wikimedia.org/wikipedia/commons/4/4c/Flag_of_Mongolia.svg *License:* Public domain *Contributors:* Current version is SVG implementation of the Mongolian flag as described by Mongolian National Standard **MNS 6262:2011** (Mongolian State Flag. General requirements [1]
Original artist: User:Zscout370

- **File:Flag_of_Montenegro.svg** *Source:* https://upload.wikimedia.org/wikipedia/commons/6/64/Flag_of_Montenegro.svg *License:* Public domain *Contributors:* Own work *Original artist:* B1mbo, Froztbyte

- **File:Flag_of_Morocco.svg** *Source:* https://upload.wikimedia.org/wikipedia/commons/2/2c/Flag_of_Morocco.svg *License:* Public domain *Contributors:* Flag of the Kingdom of Morocco

  Moroccan royal decree (17 November 1915)
  Original artist: Denelson83, Zscout370

- **File:Flag_of_Nagorno-Karabakh.svg** *Source:* https://upload.wikimedia.org/wikipedia/commons/8/8d/Flag_of_Nagorno-Karabakh.svg *License:* Public domain *Contributors:* ? *Original artist:* ?

- **File:Flag_of_New_Zealand.svg** *Source:* https://upload.wikimedia.org/wikipedia/commons/3/3e/Flag_of_New_Zealand.svg *License:* Public domain *Contributors:* http://www.mch.govt.nz/files/NZ%20Flag%20-%20proportions.JPG *Original artist:* Zscout370, Hugh Jass and many others

- **File:Flag_of_Nigeria.svg** *Source:* https://upload.wikimedia.org/wikipedia/commons/7/79/Flag_of_Nigeria.svg *License:* Public domain *Contributors:* ? *Original artist:* ?

- **File:Flag_of_North_Korea.svg** *Source:* https://upload.wikimedia.org/wikipedia/commons/5/51/Flag_of_North_Korea.svg *License:* Public domain *Contributors:* Template:🇰🇵 🇰🇵🇰🇵🇰🇵 🇰🇵🇰 🇰🇵🇰🇵🇰 *Original artist:* Zscout370

- **File:Flag_of_Norway.svg** *Source:* https://upload.wikimedia.org/wikipedia/commons/d/d9/Flag_of_Norway.svg *License:* Public domain *Contributors:* Own work *Original artist:* Dbenbenn

- **File:Flag_of_Pakistan.svg** *Source:* https://upload.wikimedia.org/wikipedia/commons/3/32/Flag_of_Pakistan.svg *License:* Public domain *Contributors:* The drawing and the colors were based from flagspot.net. *Original artist:* User:Zscout370

- **File:Flag_of_Palestine.svg** *Source:* https://upload.wikimedia.org/wikipedia/commons/0/00/Flag_of_Palestine.svg *License:* Public domain *Contributors:* Own work. Based on Law No. 5 for the year 2006 amending some provisions of Law No. 22 for the year 2005 on the Sanctity of the Palestinian Flag *Original artist:* Orionist, previous versions by Makaristos, Mysid, etc.

- **File:Flag_of_Poland.svg** *Source:* https://upload.wikimedia.org/wikipedia/en/1/12/Flag_of_Poland.svg *License:* Public domain *Contributors:* ? *Original artist:* ?

- **File:Flag_of_Portugal.svg** *Source:* https://upload.wikimedia.org/wikipedia/commons/5/5c/Flag_of_Portugal.svg *License:* Public domain *Contributors:* http://jorgesampaio.arquivo.presidencia.pt/pt/republica/simbolos/bandeiras/index.html#imgs *Original artist:* Columbano Bordalo Pinheiro (1910; generic design); Vítor Luís Rodrigues; António Martins-Tuválkin (2004; this specific vector set: see sources)

- **File:Flag_of_Romania.svg** *Source:* https://upload.wikimedia.org/wikipedia/commons/7/73/Flag_of_Romania.svg *License:* Public domain *Contributors:* Own work *Original artist:* AdiJapan

- **File:Flag_of_Russia.svg** *Source:* https://upload.wikimedia.org/wikipedia/en/f/f3/Flag_of_Russia.svg *License:* PD *Contributors:* ? *Original artist:* ?

- **File:Flag_of_SFR_Yugoslavia.svg** *Source:* https://upload.wikimedia.org/wikipedia/commons/7/71/Flag_of_SFR_Yugoslavia.svg *License:* Public domain *Contributors:* Own work *Original artist:* Flag designed by Đorđe Andrejević-Kun[3]

- **File:Flag_of_San_Marino.svg** *Source:* https://upload.wikimedia.org/wikipedia/commons/b/b1/Flag_of_San_Marino.svg *License:* Public domain *Contributors:* Own work: [/Users/bicio/Desktop/Cailungo logo 40°.jpg] *Original artist:* Zscout370

- **File:Flag_of_Saudi_Arabia.svg** *Source:* https://upload.wikimedia.org/wikipedia/commons/0/0d/Flag_of_Saudi_Arabia.svg *License:* CC0 *Contributors:* the actual flag *Original artist:* Unknown<a href='//www.wikidata.org/wiki/Q4233718' title='wikidata:Q4233718'><img alt='wikidata:Q4233718' src='https://upload.wikimedia.org/wikipedia/commons/thumb/f/ff/Wikidata-logo.svg/20px-Wikidata-logo.svg.png' width='20' height='11' srcset='https://upload.wikimedia.org/wikipedia/commons/thumb/f/ff/Wikidata-logo.svg/30px-Wikidata-logo.svg.png 1.5x, https://upload.wikimedia.org/wikipedia/commons/thumb/f/ff/Wikidata-logo.svg/40px-Wikidata-logo.svg.png 2x' data-file-width='1050' data-file-height='590' /></a>

- **File:Flag_of_Serbia.svg** *Source:* https://upload.wikimedia.org/wikipedia/commons/f/ff/Flag_of_Serbia.svg *License:* Public domain *Contributors:* From http://www.parlament.gov.rs/content/cir/o_skupstini/simboli/simboli.asp. *Original artist:* sodipodi.com

- **File:Flag_of_Serbia_and_Montenegro.svg** *Source:* https://upload.wikimedia.org/wikipedia/commons/9/90/Flag_of_Serbia_and_Montenegro.svg *License:* Public domain *Contributors:* Transferred from en.wikipedia to Commons. *Original artist:* The original uploader was Milan B. at English Wikipedia

- **File:Flag_of_Slovakia.svg** *Source:* https://upload.wikimedia.org/wikipedia/commons/e/e6/Flag_of_Slovakia.svg *License:* Public domain *Contributors:* Own work; here, colors *Original artist:* SKopp

- **File:Flag_of_Slovenia.svg** *Source:* https://upload.wikimedia.org/wikipedia/commons/f/f0/Flag_of_Slovenia.svg *License:* Public domain *Contributors:* Own work construction sheet from http://flagspot.net/flags/si%27.html#coa *Original artist:* User:Achim1999

- **File:Flag_of_South_Africa.svg** *Source:* https://upload.wikimedia.org/wikipedia/commons/a/af/Flag_of_South_Africa.svg *License:* Public domain *Contributors:* Per specifications in the Constitution of South Africa, Schedule 1 - National flag *Original artist:* Flag design by Frederick Brownell, image by Wikimedia Commons users

- **File:Flag_of_South_Korea.svg** *Source:* https://upload.wikimedia.org/wikipedia/commons/0/09/Flag_of_South_Korea.svg *License:* Public domain *Contributors:* Ordinance Act of the Law concerning the National Flag of the Republic of Korea, Construction and color guidelines (Russian/English) ← This site is not exist now.(2012.06.05) *Original artist:* Various

- **File:Flag_of_South_Ossetia.svg** *Source:* https://upload.wikimedia.org/wikipedia/commons/1/12/Flag_of_South_Ossetia.svg *License:* Public domain *Contributors:* The law on State flag of South Ossetia *Original artist:* Various

- **File:Flag_of_Spain.svg** *Source:* https://upload.wikimedia.org/wikipedia/en/9/9a/Flag_of_Spain.svg *License:* PD *Contributors:* ? *Original artist:* ?

- **File:Flag_of_Spain_(1945_-_1977).svg** *Source:* https://upload.wikimedia.org/wikipedia/commons/a/ae/Flag_of_Spain_%281945_-_1977%29.svg *License:* GFDL *Contributors:* Own work *Original artist:* SanchoPanzaXXI

- **File:Flag_of_Sri_Lanka.svg** *Source:* https://upload.wikimedia.org/wikipedia/commons/1/11/Flag_of_Sri_Lanka.svg *License:* Public domain *Contributors:* SLS 693 - National flag of Sri Lanka *Original artist:* Zscout370

- **File:Flag_of_Sweden.svg** *Source:* https://upload.wikimedia.org/wikipedia/en/4/4c/Flag_of_Sweden.svg *License:* PD *Contributors:* ? *Original artist:* ?

- **File:Flag_of_Switzerland.svg** *Source:* https://upload.wikimedia.org/wikipedia/commons/f/f3/Flag_of_Switzerland.svg *License:* Public domain *Contributors:* PDF Colors Construction sheet *Original artist:* User:Marc Mongenet

Credits:

- **File:Flag_of_Syria.svg** *Source:* https://upload.wikimedia.org/wikipedia/commons/5/53/Flag_of_Syria.svg *License:* Public domain *Contributors:* see below *Original artist:* see below

- **File:Flag_of_Thailand.svg** *Source:* https://upload.wikimedia.org/wikipedia/commons/a/a9/Flag_of_Thailand.svg *License:* Public domain *Contributors:* Own work *Original artist:* Zscout370

- **File:Flag_of_Transnistria.svg** *Source:* https://upload.wikimedia.org/wikipedia/commons/5/58/Flag_of_Transnistria.svg *License:* Public domain *Contributors:*

- Drawing based from http://www.president-pmr.org/english/index_e.htm *Original artist:* Zscout370 (SVG)

- **File:Flag_of_Tunisia.svg** *Source:* https://upload.wikimedia.org/wikipedia/commons/c/ce/Flag_of_Tunisia.svg *License:* Public domain *Contributors:* http://www.w3.org/ *Original artist:* entraîneur: BEN KHALIFA WISSAM

- **File:Flag_of_Turkey.svg** *Source:* https://upload.wikimedia.org/wikipedia/commons/b/b4/Flag_of_Turkey.svg *License:* Public domain *Contributors:* Turkish Flag Law (Türk Bayrağı Kanunu), Law nr. 2893 of 22 September 1983. Text (in Turkish) at the website of the Turkish Historical Society (Türk Tarih Kurumu) *Original artist:* David Benbennick (original author)

- **File:Flag_of_Ukraine.svg** *Source:* https://upload.wikimedia.org/wikipedia/commons/4/49/Flag_of_Ukraine.svg *License:* Public domain *Contributors:* ДСТУ 4512:2006 — Державний прапор України. Загальні технічні умови *Original artist:* Government of Ukraine

- **File:Flag_of_Venezuela.svg** *Source:* https://upload.wikimedia.org/wikipedia/commons/0/06/Flag_of_Venezuela.svg *License:* Public domain *Contributors:* official websites *Original artist:* Zscout370

- **File:Flag_of_Vietnam.svg** *Source:* https://upload.wikimedia.org/wikipedia/commons/2/21/Flag_of_Vietnam.svg *License:* Public domain *Contributors:* http://vbqppl.moj.gov.vn/law/vi/1951_to_1960/1955/195511/195511300001 http://vbqppl.moj.gov.vn/vbpq/Lists/Vn%20bn%20php%20lut/View_Detail.aspx?ItemID=820 *Original artist:* Lưu Ly vẽ lại theo nguồn trên

- **File:Flag_of_the_Azores.svg** *Source:* https://upload.wikimedia.org/wikipedia/commons/6/6c/Flag_of_the_Azores.svg *License:* Public domain *Contributors:* Own work. Original rendition derived from the legal description of the flag. *Original artist:* Tonyjeff

- **File:Flag_of_the_Canary_Islands.svg** *Source:* https://upload.wikimedia.org/wikipedia/commons/b/b0/Flag_of_the_Canary_Islands.svg *License:* Public domain *Contributors:* ? *Original artist:* ?

- **File:Flag_of_the_Czech_Republic.svg** *Source:* https://upload.wikimedia.org/wikipedia/commons/c/cb/Flag_of_the_Czech_Republic.svg *License:* Public domain *Contributors:*

  - -xfi-'s file
  - -xfi-'s code
  - Zirland's codes of colors

Original artist:
(of code): SVG version by cs:-xfi-.

- **File:Flag_of_the_Faroe_Islands.svg** *Source:* https://upload.wikimedia.org/wikipedia/commons/3/3c/Flag_of_the_Faroe_Islands.svg *License:* Public domain *Contributors:* ? *Original artist:* ?

- **File:Flag_of_the_Isle_of_Man.svg** *Source:* https://upload.wikimedia.org/wikipedia/commons/b/bc/Flag_of_the_Isle_of_Man.svg *License:* CC0 *Contributors:* Sodipodi flag collection, OpenClipart *Original artist:* Edited by Reisio, Alkari, e.a.

- **File:Flag_of_the_Netherlands.svg** *Source:* https://upload.wikimedia.org/wikipedia/commons/2/20/Flag_of_the_Netherlands.svg *License:* Public domain *Contributors:* Own work *Original artist:* Zscout370

- **File:Flag_of_the_People'{}s_Republic_of_China.svg** *Source:* https://upload.wikimedia.org/wikipedia/commons/f/fa/Flag_of_the_People%27s_Republic_of_China.svg *License:* Public domain *Contributors:* Own work, http://www.protocol.gov.hk/flags/eng/n_flag/design.html *Original artist:* Drawn by User:SKopp, redrawn by User:Denelson83 and User:Zscout370

- **File:Flag_of_the_Philippines.svg** *Source:* https://upload.wikimedia.org/wikipedia/commons/9/99/Flag_of_the_Philippines.svg *License:* Public domain *Contributors:* The design was taken from [1] and the colors were also taken from a Government website *Original artist:* User:Achim1999

- **File:Flag_of_the_Republic_of_China.svg** *Source:* https://upload.wikimedia.org/wikipedia/commons/7/72/Flag_of_the_Republic_of_China.svg *License:* Public domain *Contributors:* [1] *Original artist:* User:SKopp

- **File:Flag_of_the_Soviet_Union.svg** *Source:* https://upload.wikimedia.org/wikipedia/commons/a/a9/Flag_of_the_Soviet_Union.svg *License:* Public domain *Contributors:* http://pravo.levonevsky.org/ *Original artist:* CCCP

- **File:Flag_of_the_Turkish_Republic_of_Northern_Cyprus.svg** *Source:* https://upload.wikimedia.org/wikipedia/commons/1/1e/Flag_of_the_Turkish_Republic_of_Northern_Cyprus.svg *License:* CC BY-SA 2.5 *Contributors:* ? *Original artist:* ?

- **File:Flag_of_the_United_Kingdom.svg** *Source:* https://upload.wikimedia.org/wikipedia/en/a/ae/Flag_of_the_United_Kingdom.svg *License:* PD *Contributors:* ? *Original artist:* ?

- **File:Flag_of_the_United_Nations.svg** *Source:* https://upload.wikimedia.org/wikipedia/commons/2/2f/Flag_of_the_United_Nations.svg *License:* Public domain *Contributors:* Flag of the United Nations from the Open Clip Art website. Modifications by Denelson83, Zscout370 and Madden. Official construction sheet here.
United Nations (1962) *The United Nations flag code and regulations, as amended November 11, 1952,* New York OCLC: 7548838. *Original artist:* Wilfried Huss / Anonymous

- **File:Flag_of_the_United_States.svg** *Source:* https://upload.wikimedia.org/wikipedia/en/a/a4/Flag_of_the_United_States.svg *License:* PD *Contributors:* ? *Original artist:* ?

- **File:Flag_of_the_Vatican_City.svg** *Source:* https://upload.wikimedia.org/wikipedia/commons/0/00/Flag_of_the_Vatican_City.svg *License:* CC0 *Contributors:* http://files.mojeeuro.meu.zoznam.sk/200000288-390ab3a04d/2_Commemorative_coin_Vatican_city_2010.jpg labelbasis *Original artist:* Unknown<a href='//www.wikidata.org/wiki/Q4233718' title='wikidata:Q4233718'><img alt='wikidata:Q4233718' src='https://upload.wikimedia.org/wikipedia/commons/thumb/f/ff/Wikidata-logo.svg/20px-Wikidata-logo.svg.png' width='20' height='11' srcset='https://upload.wikimedia.org/wikipedia/commons/thumb/f/ff/Wikidata-logo.svg/30px-Wikidata-logo.svg.png 1.5x, https://upload.wikimedia.org/wikipedia/commons/thumb/f/ff/Wikidata-logo.svg/40px-Wikidata-logo.svg.png 2x' data-file-width='1050' data-file-height='590' /></a>

- **File:Flowerpowerportfolio.jpg** *Source:* https://upload.wikimedia.org/wikipedia/commons/7/75/Flowerpowerportfolio.jpg *License:* Public domain *Contributors:* Own work *Original artist:* Rightleftright

- **File:Folder_Hexagonal_Icon.svg** *Source:* https://upload.wikimedia.org/wikipedia/en/4/48/Folder_Hexagonal_Icon.svg *License:* Cc-by-sa-3.0 *Contributors:* ? *Original artist:* ?

- **File:Frq_Band_Comparison.png** *Source:* https://upload.wikimedia.org/wikipedia/commons/c/c1/Frq_Band_Comparison.png *License:* CC BY-SA 4.0 *Contributors:* Own work *Original artist:* Treinkvist

- **File:Gascony_Show_Logo.jpg** *Source:* https://upload.wikimedia.org/wikipedia/en/b/bc/Gascony_Show_Logo.jpg *License:* CC-BY-SA-3.0 *Contributors:*

Own work

Original artist:

Deveze

- **File:Gnome-mime-sound-openclipart.svg** *Source:* https://upload.wikimedia.org/wikipedia/commons/8/87/Gnome-mime-sound-openclipart.svg *License:* Public domain *Contributors:* Own work. Based on File:Gnome-mime-audio-openclipart.svg, which is public domain. *Original artist:* User:Eubulides

- **File:Great_Seal_of_the_United_States_(obverse).svg** *Source:* https://upload.wikimedia.org/wikipedia/commons/5/5c/Great_Seal_of_the_United_States_%28obverse%29.svg *License:* Public domain *Contributors:* Extracted from PDF version of *Our Flag,* available here (direct PDF URL here.) *Original artist:* U.S. Government

- **File:Greenwich_Time_Signal1970.jpg** *Source:* https://upload.wikimedia.org/wikipedia/commons/7/7b/Greenwich_Time_Signal1970.jpg *License:* GFDL *Contributors:* Photo by User:geni *Original artist:* Geni dec 2008

- **File:Greenwich_Time_Signal_pips.png** *Source:* https://upload.wikimedia.org/wikipedia/commons/d/d0/Greenwich_Time_Signal_pips.png *License:* Public domain *Contributors:* ? *Original artist:* ?

- **File:Grundig_Satellit_Professional_400.JPG** *Source:* https://upload.wikimedia.org/wikipedia/commons/8/87/Grundig_Satellit_Professional_400.JPG *License:* CC BY-SA 3.0 *Contributors:* Own work *Original artist:* User:Mattes

- **File:Gts_(bbc)_pips.ogg** *Source:* https://upload.wikimedia.org/wikipedia/commons/0/0f/Gts_%28bbc%29_pips.ogg *License:* Public domain *Contributors:* en:Image:Gts (bbc) pips.ogg *Original artist:* en:User:AlexJ

- **File:Guglielmo,_Marchese_Marconi._Colour_lithograph_by_Sir_L._War_Wellcome_V0003849.jpg** *Source:* https://upload.wikimedia.org/wikipedia/commons/8/8c/Guglielmo%2C_Marchese_Marconi._Colour_lithograph_by_Sir_L._War_Wellcome_V0003849.jpg *License:* CC BY 2.0 *Contributors:* http://wellcomeimages.org/indexplus/image/V0003849.html *Original artist:* Leslie Ward

- **File:Guglielmo_Marconi_1901_wireless_signal.jpg** *Source:* https://upload.wikimedia.org/wikipedia/commons/7/76/Guglielmo_Marconi_1901_wireless_signal.jpg *License:* Public domain *Contributors:* This image comes from the Google-hosted **LIFE Photo Archive** where it is available under the filename 4a204d82f07524bd. *Original artist:* Published on LIFE

- **File:Guglielmo_Marconi_Memorial.JPG** *Source:* https://upload.wikimedia.org/wikipedia/commons/9/9d/Guglielmo_Marconi_Memorial.JPG *License:* CC BY-SA 3.0 *Contributors:* Transferred from en.wikipedia *Original artist:* <a href='//en.wikipedia.org/wiki/User:AgnosticPreachersKid' class='extiw' title='en:User:AgnosticPreachersKid'>**APK**</a> <a href='//en.wikipedia.org/wiki/User_talk:AgnosticPreachersKid' class='extiw' title='en:User talk:AgnosticPreachersKid'>**i kissed a girl and i didn't like it**</a>

- **File:Guglielmo_Marconi_Signature.svg** *Source:* https://upload.wikimedia.org/wikipedia/commons/f/f4/Guglielmo_Marconi_Signature.svg *License:* Public domain *Contributors:* Own work by uploader traced by hand *Original artist:* Connormah, G. Marconi

- **File:HD_Television.svg** *Source:* https://upload.wikimedia.org/wikipedia/commons/c/c7/HD_Television.svg *License:* LGPL *Contributors:*

- File:Television.svg *Original artist:* Wylve (<a href='//commons.wikimedia.org/wiki/User_talk:Wylve' title='User talk:Wylve'>talk</a>)

- **File:HR_ALLISS_system.jpg** *Source:* https://upload.wikimedia.org/wikipedia/en/9/9b/HR_ALLISS_system.jpg *License:* Cc-by-sa-3.0 *Contributors:* ? *Original artist:* ?

- **File:Handshake_(Workshop_Cologne_'06).jpeg** *Source:* https://upload.wikimedia.org/wikipedia/commons/a/a3/Handshake_%28Workshop_Cologne_%2706%29.jpeg *License:* CC-BY-SA-3.0 *Contributors:* Own work *Original artist:* Tobias Wolter

- **File:Hughes_Direcway_LNB.jpg** *Source:* https://upload.wikimedia.org/wikipedia/commons/b/be/Hughes_Direcway_LNB.jpg *License:* Public domain *Contributors:* Own work *Original artist:* Chetvorno

- **File:ICOM_IC-P7_dscn2510a.jpg** *Source:* https://upload.wikimedia.org/wikipedia/commons/1/18/ICOM_IC-P7_dscn2510a.jpg *License:* Public domain *Contributors:* No machine-readable source provided. Own work assumed (based on copyright claims). *Original artist:* No machine-readable author provided. Bernd in Japan~commonswiki assumed (based on copyright claims).

- **File:INTELSAT_I_(Early_Bird).jpg** *Source:* https://upload.wikimedia.org/wikipedia/commons/a/ab/INTELSAT_I_%28Early_Bird%29.jpg *License:* Public domain *Contributors:* Great Images in NASA Description *Original artist:* NASA

- **File:ISS-24_Doug_Wheelock_uses_ham_radio_system_1.jpg** *Source:* https://upload.wikimedia.org/wikipedia/commons/f/fd/ISS-24_Doug_Wheelock_uses_ham_radio_system_1.jpg *License:* Public domain *Contributors:* http://spaceflight.nasa.gov/gallery/images/station/crew-24/html/iss024e013398.html *Original artist:* NASA

- **File:Ibcred.jpg** *Source:* https://upload.wikimedia.org/wikipedia/en/d/d0/Ibcred.jpg *License:* Fair use *Contributors:*
The logo may be obtained from International Broadcasting Convention.
Original artist: ?

- **File:International_amateur_radio_symbol.svg** *Source:* https://upload.wikimedia.org/wikipedia/commons/2/2c/International_amateur_radio_symbol.svg *License:* Public domain *Contributors:* Own work *Original artist:* Denelson83

- **File:John_Slattery_Gascony_Show.jpg** *Source:* https://upload.wikimedia.org/wikipedia/en/c/cd/John_Slattery_Gascony_Show.jpg *License:* CC-BY-SA-3.0 *Contributors:*
Own work
Original artist:
Deveze

- **File:Johnrobles.jpg** *Source:* https://upload.wikimedia.org/wikipedia/en/6/6d/Johnrobles.jpg *License:* PD *Contributors:* ? *Original artist:* ?

- **File:Kyiv_province_location_map.svg** *Source:* https://upload.wikimedia.org/wikipedia/commons/0/0f/Kyiv_province_location_map.svg *License:* Public domain *Contributors:* Own work *Original artist:* Artemco

- **File:Logo_Agência_Estado.png** *Source:* https://upload.wikimedia.org/wikipedia/commons/a/a3/Logo_Ag%C3%AAncia_Estado.png *License:* CC BY-SA 3.0 *Contributors:* Own work *Original artist:* maxhhl

- **File:Logotipo_Oficial_da_Agência_Brasil.jpg** *Source:* https://upload.wikimedia.org/wikipedia/commons/3/39/Logotipo_Oficial_da_Ag%C3%AAncia_Brasil.jpg *License:* CC BY-SA 3.0 *Contributors:* http://plid.blogspot.com/2010/11/agencia-brasil-da-destaque-ao-plid-em.html *Original artist:* EBC - Empresa Brasil de Comunicação

- **File:Map_of_Ohio_highlighting_Butler_County.svg** *Source:* https://upload.wikimedia.org/wikipedia/commons/2/21/Map_of_Ohio_highlighting_Butler_County.svg *License:* Public domain *Contributors:* The maps use data from nationalatlas.gov, specifically countyp020.tar.gz on the Raw Data Download page. The maps also use state outline data from statesp020.tar.gz. The Florida maps use hydrogm020.tar.gz to display Lake Okeechobee. *Original artist:* David Benbennick

- **File:Marconi'{}s_first_radio_transmitter.jpg** *Source:* https://upload.wikimedia.org/wikipedia/commons/3/36/Marconi%27s_first_radio_transmitter.jpg *License:* Public domain *Contributors:* Retrieved January 28, 2016 from <a data-x-rel='nofollow' class='external text' href='http://www.americanradiohistory.com/Archive-Radio-Broadcast/Radio-Broadcast-1926-11.pdf'>Guglielmo Marconi, "Looking back over thirty years of radio", *Radio Broadcast* magazine, Doubleday, Page, and Co., New York, Vol. 10, No. 1, November 1926, p. 31</a> on http://www.americanradiohistory.com *Original artist:* Guglielmo Marconi

- **File:Marconi_Rock_Salvan,_Switzerland.jpg** *Source:* https://upload.wikimedia.org/wikipedia/commons/6/6f/Marconi_Rock_Salvan%2C_Switzerland.jpg *License:* CC BY-SA 3.0 *Contributors:* Own work *Original artist:* YellowFratello

- **File:Marconi_at_newfoundland.jpg** *Source:* https://upload.wikimedia.org/wikipedia/commons/2/27/Marconi_at_newfoundland.jpg *License:* Public domain *Contributors:* ? *Original artist:* ?

- **File:Marconi_in_London.jpg** *Source:* https://upload.wikimedia.org/wikipedia/commons/c/c9/Marconi_in_London.jpg *License:* CC BY-SA 3.0 *Contributors:* Own work *Original artist:* Yamen

- **File:Marconi_portrait.jpg** *Source:* https://upload.wikimedia.org/wikipedia/commons/d/d5/Marconi_portrait.jpg *License:* Public domain *Contributors:* Strand Magazine 1910 *Original artist:* Sir Benjamin Stone (1838 – 2 July 1914)

- **File:Merge-arrow.svg** *Source:* https://upload.wikimedia.org/wikipedia/commons/a/aa/Merge-arrow.svg *License:* Public domain *Contributors:* ? *Original artist:* ?

- **File:Montreal-tower-top.thumb2-crop.jpg** *Source:* https://upload.wikimedia.org/wikipedia/commons/0/0a/Montreal-tower-top.thumb2-crop.jpg *License:* Public domain *Contributors:*

- Montreal-tower-top.thumb2.jpg *Original artist:* Montreal-tower-top.thumb2.jpg: Original uploader was Aarchiba at en.wikipedia

- **File:Mw0rkbshack.jpg** *Source:* https://upload.wikimedia.org/wikipedia/commons/1/16/Mw0rkbshack.jpg *License:* CC BY-SA 3.0 *Contributors:* Own work *Original artist:* Mw0rkb

- **File:Newspaper_nicu_buculei_01.svg** *Source:* https://upload.wikimedia.org/wikipedia/commons/0/0d/Newspaper_nicu_buculei_01.svg *License:* CC0 *Contributors:* OpenClipart *Original artist:* nicu buculei

- **File:No_Skip_Example.gif** *Source:* https://upload.wikimedia.org/wikipedia/en/e/ed/No_Skip_Example.gif *License:* PD *Contributors:* self-made

  Original artist:

  C. Oler

- **File:Nobel_Prize.png** *Source:* https://upload.wikimedia.org/wikipedia/en/e/ed/Nobel_Prize.png *License:* ? *Contributors:* Derivative of File:NobelPrize.JPG *Original artist:*

  Photograph: JonathunderMedal: Erik Lindberg (1873-1966)

- **File:Nuvola_apps_ksim.png** *Source:* https://upload.wikimedia.org/wikipedia/commons/8/8d/Nuvola_apps_ksim.png *License:* LGPL *Contributors:* http://icon-king.com *Original artist:* David Vignoni / ICON KING

- **File:Olympic_Rings.svg** *Source:* https://upload.wikimedia.org/wikipedia/en/b/b1/Olympic_Rings.svg *License:* PD-US *Contributors:* ? *Original artist:* ?

- **File:Olympic_Swimming_Marathon_(7769099394).jpg** *Source:* https://upload.wikimedia.org/wikipedia/commons/0/0f/Olympic_Swimming_Marathon_%287769099394%29.jpg *License:* CC BY 2.0 *Contributors:* Olympic Swimming Marathon *Original artist:* illang

- **File:P_vip.svg** *Source:* https://upload.wikimedia.org/wikipedia/en/6/69/P_vip.svg *License:* PD *Contributors:* ? *Original artist:* ?

- **File:People_icon.svg** *Source:* https://upload.wikimedia.org/wikipedia/commons/3/37/People_icon.svg *License:* CC0 *Contributors:* OpenClipart *Original artist:* OpenClipart

- **File:Phone_icon_rotated.svg** *Source:* https://upload.wikimedia.org/wikipedia/commons/d/df/Phone_icon_rotated.svg *License:* Public domain *Contributors:* Originally uploaded on en.wikipedia *Original artist:* Originally uploaded by Beao (Transferred by varnent)

- **File:Portal-puzzle.svg** *Source:* https://upload.wikimedia.org/wikipedia/en/f/fd/Portal-puzzle.svg *License:* Public domain *Contributors:* ? *Original artist:* ?

- **File:Post_Office_Engineers.jpg** *Source:* https://upload.wikimedia.org/wikipedia/commons/e/e2/Post_Office_Engineers.jpg *License:* CC BY 3.0 *Contributors:* Cardiff Council Flat Holm Project *Original artist:* Cardiff Council Flat Holm Project

- **File:QSL-VOA-Thessaloniki-1972.jpg** *Source:* https://upload.wikimedia.org/wikipedia/commons/1/1c/QSL-VOA-Thessaloniki-1972.jpg *License:* Public domain *Contributors:* Own collection *Original artist:* VoA

- **File:Question_book-new.svg** *Source:* https://upload.wikimedia.org/wikipedia/en/9/99/Question_book-new.svg *License:* Cc-by-sa-3.0 *Contributors:*

  Created from scratch in Adobe Illustrator. Based on Image:Question book.png created by User:Equazcion *Original artist:* Tkgd2007

- **File:RADIOBUDAPESTPENNANTLATE80s.JPG** *Source:* https://upload.wikimedia.org/wikipedia/en/f/fa/RADIOBUDAPESTPENNANTLATE80s.JPG *License:* PD *Contributors:*

  I (Mickeylove73 (talk)) created this work entirely by myself. *Original artist:* mickeylove73

- **File:RCI-BC-ANZ431-dbu.png** *Source:* https://upload.wikimedia.org/wikipedia/en/f/fe/RCI-BC-ANZ431-dbu.png *License:* PD *Contributors:* ? *Original artist:* ?

- **File:RCI-sask-LAT-dbu.png** *Source:* https://upload.wikimedia.org/wikipedia/en/4/42/RCI-sask-LAT-dbu.png *License:* CC-BY-2.5 *Contributors:* ? *Original artist:* ?

- **File:Radio_Moscow_logo.png** *Source:* https://upload.wikimedia.org/wikipedia/commons/7/7f/Radio_Moscow_logo.png *License:* Public domain *Contributors:* ? *Original artist:* ?

- **File:Radio_icon.png** *Source:* https://upload.wikimedia.org/wikipedia/commons/1/1d/Radio_icon.png *License:* Public domain *Contributors:* ? *Original artist:* ?

- **File:Red_pog.svg** *Source:* https://upload.wikimedia.org/wikipedia/en/0/0c/Red_pog.svg *License:* Public domain *Contributors:* ? *Original artist:* ?

- **File:Satellite_dish_1_C-Band.jpg** *Source:* https://upload.wikimedia.org/wikipedia/commons/7/7b/Satellite_dish_1_C-Band.jpg *License:* CC-BY-SA-3.0 *Contributors:* Picture taken by en:User:Bogdangiusca *Original artist:* en:User:Bogdangiusca, german translation Andreas -horn- Hornig

- **File:Set-top_Box,_2013.png** *Source:* https://upload.wikimedia.org/wikipedia/commons/5/56/Set-top_Box%2C_2013.png *License:* CC BY-SA 3.0 *Contributors:* http://inview.tv/ *Original artist:* Inview Technology

- **File:Skip_Zone_Example.gif** *Source:* https://upload.wikimedia.org/wikipedia/en/e/eb/Skip_Zone_Example.gif *License:* PD *Contributors:* self-made

  Original artist:

  C. Oler

- **File:Sky_minidish.JPG** *Source:* https://upload.wikimedia.org/wikipedia/commons/2/2d/Sky_minidish.JPG *License:* CC-BY-SA-3.0 *Contributors:* ? *Original artist:* ?

4.3　Content license

www.ingramcontent.com/pod-product-compliance
Lightning Source LLC
Chambersburg PA
CBHW080654190526
45169CB00006B/2113